AF598665

FLUID MECHANICS

FLUID MECHANICS

Edited by

Frank Kreith

CRC Press
Boca Raton London New York Washington, D.C.

Library of Congress Cataloging-in-Publication Data

Fluid mechanics / editor-in-chief, Frank Kreith.
p. cm.
Includes bibliographical references and index.
ISBN 0-8493-0055-X
1. Fluid mechanics. I. Kreith, Frank.
TA357.F577 1999
620.1′06—dc21 99-050067
CIP

This book contains information obtained from authentic and highly regarded sources. Reprinted material is quoted with permission, and sources are indicated. A wide variety of references are listed. Reasonable efforts have been made to publish reliable data and information, but the author and the publisher cannot assume responsibility for the validity of all materials or for the consequences of their use.

Direct all inquiries to CRC Press LLC, 2000 N.W. Corporate Blvd., Boca Raton, Florida 33431.

The material in this book was taken from, The CRC Handbook of Mechanical Engineering edited by Frank Kreith.

International Standard Book Number 0-8493-0055-X
Library of Congress Card Number 99-050067
Printed in the United States of America 1 2 3 4 5 6 7 8 9 0
Printed on acid-free paper

Preface

The purpose of this book is to provide, in a single volume, a ready reference for the practicing engineer in industry, government, and academia, on the traditional area of fluid mechanics. Many figures and illustrations accompany the readable text, and the index and table of contents are very detailed, making this an especiallly accessible and convenient resource. The book offers numerous examples that clarify problem-solving processes and are applicable to engineering practices. The ease of use and descriptive text enable readers to rely heavily on this one resource for all of their fluid mechanics needs, and provide the necessary background and current information needed to keep pace with modern technical engineering needs.

Frank Kreith
Editor-in-Chief

Contributors

Stanley A. Berger
University of California
Berkeley, California

Bharat Bhushan
Ohio State University
Columbus, Ohio

Robert F. Boehm
University of Nevada
Las Vegas, Nevada

E. Richard Booser
Consulting Engineer
Scotia, New York

Massimo Capobianchi
SUNY Stony Brook
Stony Brook, New York

John C. Chen
Lehigh University
Bethlehem, Pennsylvania

Stuart W. Churchill
University of Pennsylvania
Philadelphia, Pennsylvania

Thomas F. Irvine, Jr.
SUNY Stony Brook
Stony Brook, New York

Francis E. Kennedy
Dartmouth College
Hanover, New Hampshire
and JFK and Associates
Boulder, Colorado

Frank Kreith
University of Colorado
Boulder, Colorado

Ajay Kumar
NASA Langley Research Center
Hampton, Virginia

Alan T. McDonald
Purdue University
West Lafayette, Indiana

Paul Norton
National Renewable Energy Laboratory
Golden, Colorado

Rolf D. Reitz
University of Wisconsin
Madison, Wisconsin

Sherif A. Sherif
University of Florida
Gainesville, Florida

J. Paul Tullis
Utah State University
Logan, Utah

Frank M. White
University of Rhode Island
Kingston, Rhode Island

Donald F. Wilcock
Consulting Engineer, Tribolock
Slingerlands, New York

Contents

CHAPTER 1 Fluid Statics

Stanley A. Berger **1**

Equilibrium of a Fluid Element • Hydrostatic Pressure • Manometry • Hydrostatic Forces on Submerged Objects • Hydrostatic Forces in Layered Fluids • Buoyancy • Stability of Submerged and Floating Bodies • Pressure Variation in Rigid-Body Motion of a Fluid

CHAPTER 2 Equations of Motion and Potential Flow

Stanley A. Berger **11**

Integral Relations for a Control Volume • Reynolds Transport Theorem • Conservation of Mass • Conservation of Momentum • Conservation of Energy • Differential Relations for Fluid Motion • Mass Conservation–Continuity Equation • Momentum Conservation • Analysis of Rate of Deformation • Relationship Between Forces and Rate of Deformation • The Navier–Stokes Equations • Energy Conservation — The Mechanical and Thermal Energy Equations • Boundary Conditions • Vorticity in Incompressible Flow • Stream Function • Inviscid Irrotational Flow: Potential Flow

CHAPTER 3 Similitude: Dimensional Analysis and Data Correlation

Stuart W. Churchill 29

Dimensional Analysis • Correlation of Experimental Data and Theoretical Values

CHAPTER 4 Hydraulics of Pipe Systems

J. Paul Tullis **47**

Basic Computations • Pipe Design • Valve Selection • Pump Selection • Other Considerations

CHAPTER 5 Open Channel Flow

Frank M. White **65**

Definition • Uniform Flow • Critical Flow • Hydraulic Jump • Weirs • Gradually Varied Flow

CHAPTER 6 External Incompressible Flow

Alan T. McDonald 75

Introduction and Scope • Boundary Layers • Drag • Lift • Boundary Layer Control • Computation vs. Experiment

CHAPTER 7 Compressible Flow

Ajay Kumar 87

Introduction • One-Dimensional Flow • Normal Shock Wave • One-Dimensional Flow with Heat Addition • Quasi-One-Dimensional Flow • Two-Dimensional Supersonic Flow

CHAPTER 8 Multiphase Flow

John C. Chen 105

Introduction • Fundamentals • Gas–Liquid Two-Phase Flow • Gas–Solid, Liquid–Solid Two-Phase Flows

CHAPTER 9 Non-Newtonian Flow

Thomas F. Irvine Jr. and Massimo Capobianchi 123

Introduction • Classification of Non-Newtonian Fluids • Apparent Viscosity • Constitutive Equations • Rheological Property Measurements • Fully Developed Laminar Pressure Drops for Time-Independent Non-Newtonian Fluids • Fully Developed Turbulent Flow Pressure Drops • Viscoelastic Fluids

CHAPTER 10 Tribology, Lubrication, and Bearing Design

Francis E. Kennedy, E. Richard Booser, and Donald F. Wilcock 137

Introduction • Sliding Friction and its Consequences • Lubricant Properties • Fluid Film Bearings • Dry and Semilubricated Bearings • Rolling Element Bearings • Lubricant Supply Methods

CHAPTER 11 Pumps and Fans

Robert F. Boehm 181

Introduction • Pumps • Fans

CHAPTER 12 Liquid Atomization and Spraying

Rolf D. Reitz 189

Spray Characterization • Atomizer Design Considerations • Atomizer Types

CHAPTER 13 Flow Measurement

Alan T. McDonald and Sherif A. Sherif **199**

Direct Methods • Restriction Flow Meters for Flow in Ducts • Linear Flow Meters • Traversing Methods • Viscosity Measurements

CHAPTER 14 Micro/Nanotribology

Bharat Bhushan **211**

Introduction • Experimental Techniques • Surface Roughness, Adhesion, and Friction • Scratching, Wear, and Indentation • Boundary Lubrication

APPENDICES

APPENDIX A **A-1**

APPENDIX B **B-35**

APPENDIX C **C-38**

APPENDIX D **D-74**

APPENDIX E **E-75**

INDEX **I-1**

1

Fluid Statics

Stanley A. Berger

University of California, Berkeley

Fluid Statics 1
Equilibrium of a Fluid Element • Hydrostatic Pressure • Manometry • Hydrostatic Forces on Submerged Objects • Hydrostatic Forces in Layered Fluids • Buoyancy • Stability of Submerged and Floating Bodies • Pressure Variation in Rigid-Body Motion of a Fluid

1 Fluid Statics

Equilibrium of a Fluid Element

If the sum of the external forces acting on a fluid element is zero, the fluid will be either at rest or moving as a solid body — in either case, we say the fluid element is in equilibrium. In this section we consider fluids in such an equilibrium state. For fluids in equilibrium the only internal stresses acting will be normal forces, since the shear stresses depend on velocity gradients, and all such gradients, by the definition of equilibrium, are zero. If one then carries out a balance between the normal surface stresses and the body forces, assumed proportional to volume or mass, such as gravity, acting on an elementary prismatic fluid volume, the resulting equilibrium equations, after shrinking the volume to zero, show that the normal stresses at a point are the same in all directions, and since they are known to be negative, this common value is denoted by $-p$, p being the pressure.

Hydrostatic Pressure

If we carry out an equilibrium of forces on an elementary volume element $dxdydz$, the forces being pressures acting on the faces of the element and gravity acting in the $-z$ direction, we obtain

$$\frac{\partial p}{\partial x} = \frac{\partial p}{\partial y} = 0, \quad \text{and} \quad \frac{\partial p}{\partial z} = -\rho g = -\gamma \tag{1.1}$$

0-8493-0055-X/00/$0.00+$.50

The first two of these imply that the pressure is the same in all directions at the same vertical height in a gravitational field. The third, where γ is the specific weight, shows that the pressure increases with depth in a gravitational field, the variation depending on $\rho(z)$. For homogeneous fluids, for which $\rho =$ constant, this last equation can be integrated immediately, yielding

$$p_2 - p_1 = -\rho g(z_2 - z_1) = -\rho g(h_2 - h_1) \tag{1.2}$$

or

$$p_2 + \rho g h_2 = p_1 + \rho g h_1 = \text{constant} \tag{1.3}$$

where h denotes the elevation. These are the equations for the hydrostatic pressure distribution.

When applied to problems where a liquid, such as the ocean, lies below the atmosphere, with a constant pressure p_{atm}, h is usually measured from the ocean/atmosphere interface and p at any distance h below this interface differs from p_{atm} by an amount

$$p - p_{atm} = \rho g h \tag{1.4}$$

Pressures may be given either as *absolute pressure,* pressure measured relative to absolute vacuum, or *gauge pressure,* pressure measured relative to atmospheric pressure.

Manometry

The hydrostatic pressure variation may be employed to measure pressure differences in terms of heights of liquid columns — such devices are called manometers and are commonly used in wind tunnels and a host of other applications and devices. Consider, for example the U-tube manometer shown in Figure 1.1 filled with liquid of specific weight γ, the left leg open to the atmosphere and the right to the region whose pressure p is to be determined. In terms of the quantities shown in the figure, in the left leg

$$p_0 - \rho g h_2 = p_{atm} \tag{1.5a}$$

and in the right leg

$$p_0 - \rho g h_1 = p \tag{1.5b}$$

the difference being

$$p - p_{atm} = -\rho g(h_1 - h_2) = -\rho g d = -\gamma d \tag{1.6}$$

and determining p in terms of the height difference $d = h_1 - h_2$ between the levels of the fluid in the two legs of the manometer.

Hydrostatic Forces on Submerged Objects

We now consider the force acting on a submerged object due to the hydrostatic pressure. This is given by

$$\boldsymbol{F} = \iint p\, d\boldsymbol{A} = \iint p \cdot \boldsymbol{n}\, dA = \iint \rho g h\, d\boldsymbol{A} + p_0 \iint d\boldsymbol{A} \tag{1.7}$$

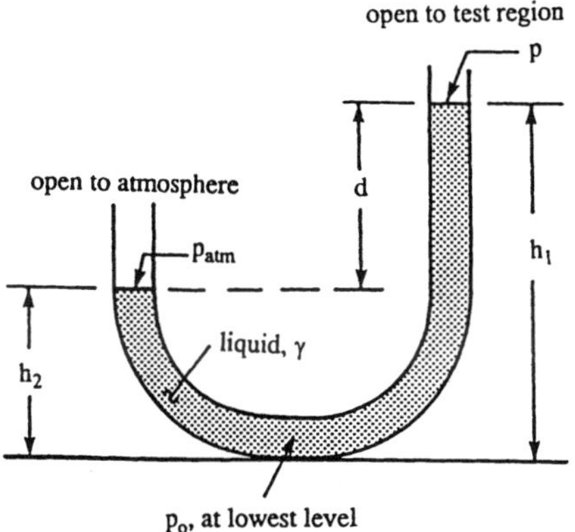

FIGURE 1.1 U-tube manometer.

where h is the variable vertical depth of the element $\boldsymbol{dA}$ and p_0 is the pressure at the surface. In turn we consider plane and nonplanar surfaces.

Forces on Plane Surfaces

Consider the planar surface A at an angle θ to a free surface shown in Figure 1.2. The force on one side of the planar surface, from Equation (1.7), is

$$\boldsymbol{F} = \rho g \boldsymbol{n} \iint_A h\, dA + p_0 A \boldsymbol{n} \tag{1.8}$$

but $h = y \sin\theta$, so

$$\iint_A h\, dA = \sin\theta \iint_A y\, dA = y_c A \sin\theta = h_c A \tag{1.9}$$

where the subscript c indicates the distance measured to the centroid of the area A. Thus, the total force (on one side) is

$$\boldsymbol{F} = \gamma h_c \boldsymbol{A} + p_0 \boldsymbol{A} \tag{1.10}$$

Thus, the magnitude of the force is independent of the angle θ, and is equal to the pressure at the centroid, $\gamma h_c + p_0$, times the area. If we use gauge pressure, the term $p_0 A$ in Equation (1.10) can be dropped.

Since p is not evenly distributed over A, but varies with depth, $\boldsymbol{F}$ does not act through the centroid. The point action of $\boldsymbol{F}$, called the *center of pressure*, can be determined by considering moments in Figure 1.2. The moment of the hydrostatic force acting on the elementary area dA about the axis perpendicular to the page passing through the point 0 on the free surface is

$$y\, dF = y(\gamma y \sin\theta\, dA) = \gamma y^2 \sin\theta\, dA \tag{1.11}$$

so if y_{cp} denotes the distance to the center of pressure,

$$y_{cp} F = \gamma \sin\theta \iint y^2\, dA = \gamma \sin\theta\, I_x \tag{1.12}$$

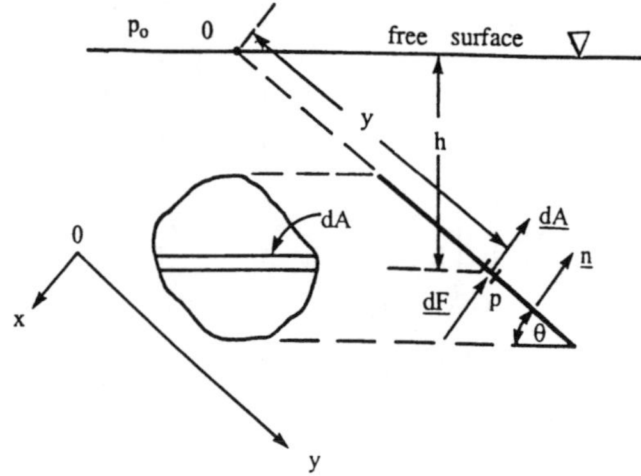

FIGURE 1.2 Hydrostatic force on a plane surface.

where I_x is the moment of inertia of the plane area with respect to the axis formed by the intersection of the plane containing the planar surface and the free surface (say $0x$). Dividing by $F = \gamma h_c A = \gamma y_c \sin\theta A$ gives

$$y_{cp} = \frac{I_x}{y_c A} \tag{1.13}$$

By using the parallel axis theorem $I_x = I_{xc} + Ay_c^2$, where I_{xc} is the moment of inertia with respect to an axis parallel to $0x$ passing through the centroid, Equation (1.13) becomes

$$y_{cp} = y_c + \frac{I_{xc}}{y_c A} \tag{1.14}$$

which shows that, in general, the center of pressure lies below the centroid.

Similarly, we find x_{cp} by taking moments about the y axis, specifically

$$x_{cp} F = \gamma \sin\theta \iint xy \, dA = \gamma \sin\theta \, I_{xy} \tag{1.15}$$

or

$$x_{cp} = \frac{I_{xy}}{y_c A} \tag{1.16}$$

where I_{xy} is the product of inertia with respect to the x and y axes. Again, the parallel axis theorem $I_{xy} = I_{xyc} + Ax_c y_c$, where the subscript c denotes the value at the centroid, allows Equation (1.16) to be written

$$x_{cp} = x_c + \frac{I_{xyc}}{y_c A} \tag{1.17}$$

This completes the determination of the center of pressure (x_{cp}, y_{cp}). Note that if the submerged area is symmetrical with respect to an axis passing through the centroid and parallel to either the x or y axes that $I_{xyc} = 0$ and $x_{cp} = x_c$; also that as y_c increases, $y_{cp} \to y_c$.

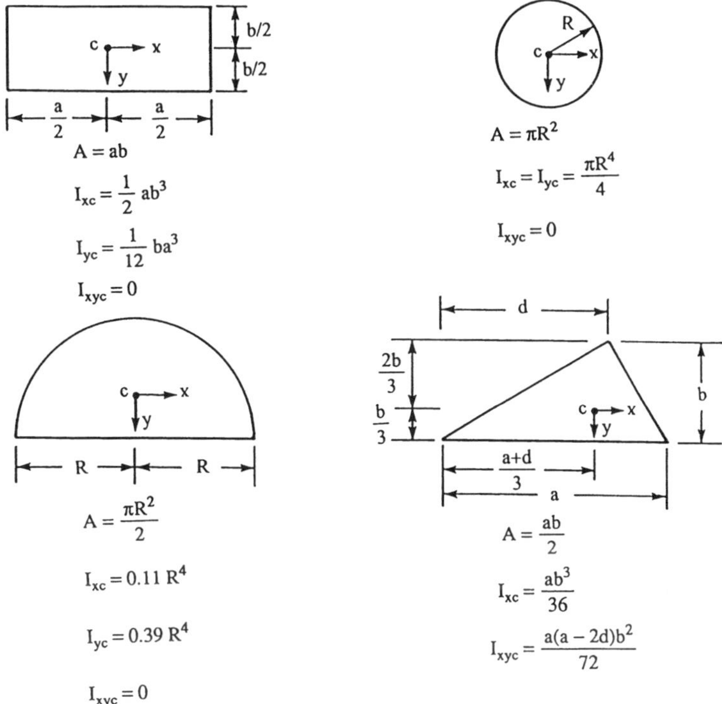

FIGURE 1.3 Centroidal moments of inertia and coordinates for some common areas.

Centroidal moments of inertia and centroidal coordinates for some common areas are shown in Figure 1.3.

Forces on Curved Surfaces

On a curved surface the forces on individual elements of area differ in direction so a simple summation of them is not generally possible, and the most convenient approach to calculating the pressure force on the surface is by separating it into its horizontal and vertical components.

A free-body diagram of the forces acting on the volume of fluid lying above a curved surface together with the conditions of static equilibrium of such a column leads to the results that:

1. The horizontal components of force on a curved submerged surface are equal to the forces exerted on the planar areas formed by the projections of the curved surface onto vertical planes normal to these components, the lines of action of these forces calculated as described earlier for planar surfaces; and
2. The vertical component of force on a curved submerged surface is equal in magnitude to the weight of the entire column of fluid lying above the curved surface, and acts through the center of mass of this volume of fluid.

Since the three components of force, two horizontal and one vertical, calculated as above, need not meet at a single point, there is, in general, no single resultant force. They can, however, be combined into a single force at any arbitrary point of application together with a moment about that point.

Hydrostatic Forces in Layered Fluids

All of the above results which employ the linear hydrostatic variation of pressure are valid only for homogeneous fluids. If the fluid is heterogeneous, consisting of individual layers each of constant density, then the pressure varies linearly with a different slope in each layer and the preceding analyses must be remedied by computing and summing the separate contributions to the forces and moments.

Buoyancy

The same principles used above to compute hydrostatic forces can be used to calculate the net pressure force acting on completely submerged or floating bodies. These laws of buoyancy, the principles of Archimedes, are that:

1. A completely submerged body experiences a vertical upward force equal to the weight of the displaced fluid; and
2. A floating or partially submerged body displaces its own weight in the fluid in which it floats (i.e., the vertical upward force is equal to the body weight).

The line of action of the buoyancy force in both (1) and (2) passes through the centroid of the displaced volume of fluid; this point is called the *center of buoyancy*. (This point need not correspond to the center of mass of the body, which could have nonuniform density. In the above it has been assumed that the displaced fluid has a constant γ. If this is not the case, such as in a layered fluid, the magnitude of the buoyant force is still equal to the weight of the displaced fluid, but the line of action of this force passes through the center of gravity of the displaced volume, not the centroid.)

If a body has a weight exactly equal to that of the volume of fluid it displaces, it is said to be *neutrally buoyant* and will remain at rest at any point where it is immersed in a (homogeneous) fluid.

Stability of Submerged and Floating Bodies

Submerged Body

A body is said to be in stable equilibrium if, when given a slight displacement from the equilibrium position, the forces thereby created tend to restore it back to its original position. The forces acting on a submerged body are the buoyancy force, F_B, acting through the center of buoyancy, denoted by CB, and the weight of the body, W, acting through the center of gravity denoted by CG (see Figure 1.4). We see from Figure 1.4 that if the CB lies above the CG a rotation from the equilibrium position creates a restoring couple which will rotate the body back to its original position — thus, this is a *stable* equilibrium situation. The reader will readily verify that when the CB lies below the CG, the couple that results from a rotation from the vertical increases the displacement from the equilibrium position — thus, this is an *unstable* equilibrium situation.

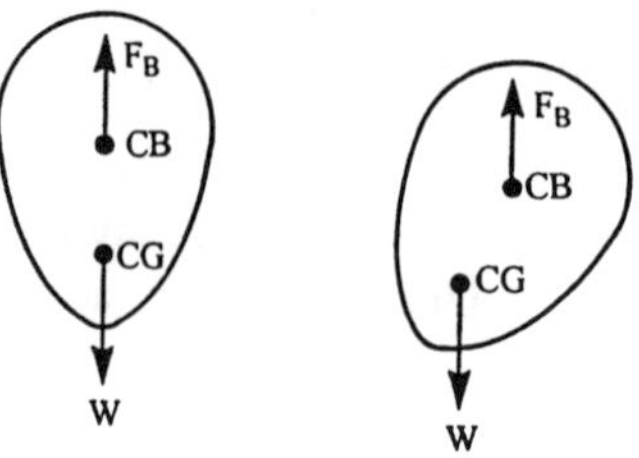

FIGURE 1.4 Stability for a submerged body.

Partially Submerged Body

The stability problem is more complicated for floating bodies because as the body rotates the location of the center of buoyancy may change. To determine stability in these problems requires that we determine the location of the *metacenter*. This is done for a symmetric body by tilting the body through a small angle $\Delta\theta$ from its equilibrium position and calculating the new location of the center of buoyancy CB′; the point of intersection of a vertical line drawn upward from CB′ with the line of symmetry of the floating body is the metacenter, denoted by M in Figure 1.5, and it is independent of $\Delta\theta$ for small angles. If M lies above the CG of the body, we see from Figure 1.5 that rotation of the body leads to a restoring couple, whereas M lying below the CG leads to a couple which will increase the displacement. Thus,

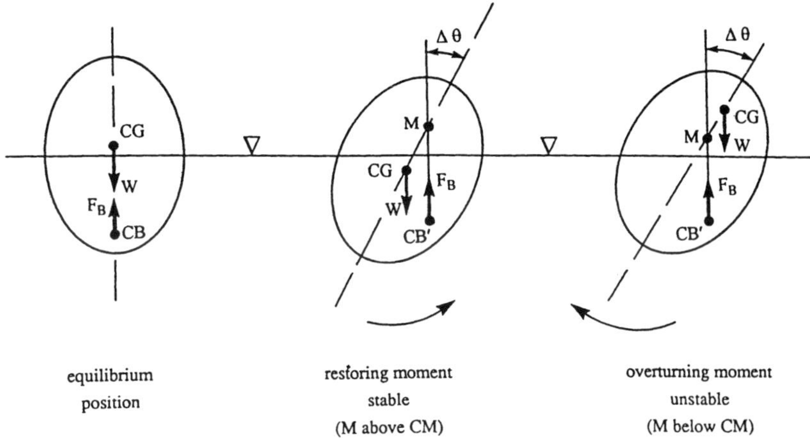

FIGURE 1.5 Stability for a partially submerged body.

the stability of the equilibrium depends on whether M lies above or below the CG. The directed distance from CG to M is called the *metacentric height*, so equivalently the equilibrium is stable if this vector is positive and unstable if it is negative; stability increases as the metacentric height increases. For geometrically complex bodies, such as ships, the computation of the metacenter can be quite complicated.

Pressure Variation in Rigid-Body Motion of a Fluid

In rigid-body motion of a fluid all the particles translate and rotate as a whole, there is no relative motion between particles, and hence no viscous stresses since these are proportional to velocity gradients. The equation of motion is then a balance among pressure, gravity, and the fluid acceleration, specifically.

$$\nabla p = \rho(\boldsymbol{g} - \boldsymbol{a}) \tag{1.18}$$

where $\boldsymbol{a}$ is the uniform acceleration of the body. Equation (1.18) shows that the lines of constant pressure, including a free surface if any, are perpendicular to the direction $\boldsymbol{g} - \boldsymbol{a}$. Two important applications of this are to a fluid in uniform linear translation and rigid-body rotation. While such problems are not, strictly speaking, fluid statics problems, their analysis and the resulting pressure variation results are similar to those for static fluids.

Uniform Linear Acceleration

For a fluid partially filling a large container moving to the right with constant acceleration $\boldsymbol{a} = (a_x, a_y)$ the geometry of Figure 1.6 shows that the magnitude of the pressure gradient in the direction $\boldsymbol{n}$ normal to the accelerating free surface, in the direction $\boldsymbol{g} - \boldsymbol{a}$, is

$$\frac{dp}{dn} = \rho\left[a_x^2 + \left(g + a_y\right)^2\right]^{1/2} \tag{1.19}$$

and the free surface is oriented at an angle to the horizontal

$$\theta = \tan^{-1}\left(\frac{a_x}{g + a_y}\right) \tag{1.20}$$

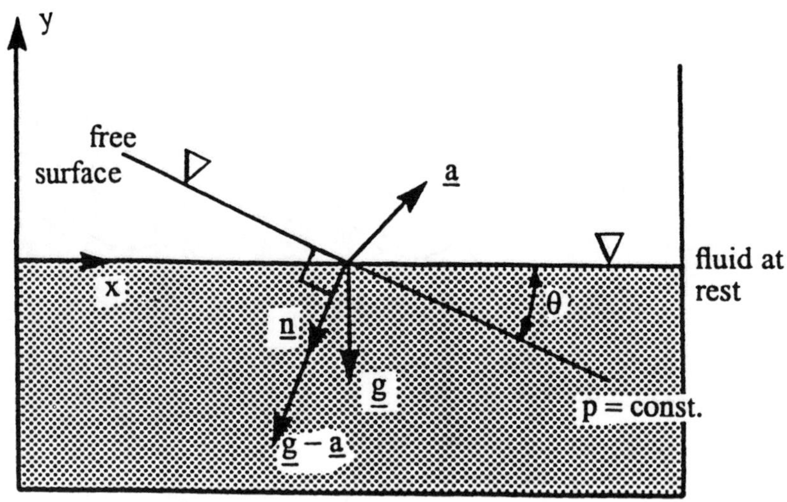

FIGURE 1.6 A fluid with a free surface in uniform linear acceleration.

Rigid-Body Rotation

Consider the fluid-filled circular cylinder rotating uniformly with angular velocity $\Omega = \Omega e_r$ (Figure 1.7). The only acceleration is the centripetal acceleration $\Omega \times \Omega \times r) = - r\Omega^2 e_r$, so Equation 1.18 becomes

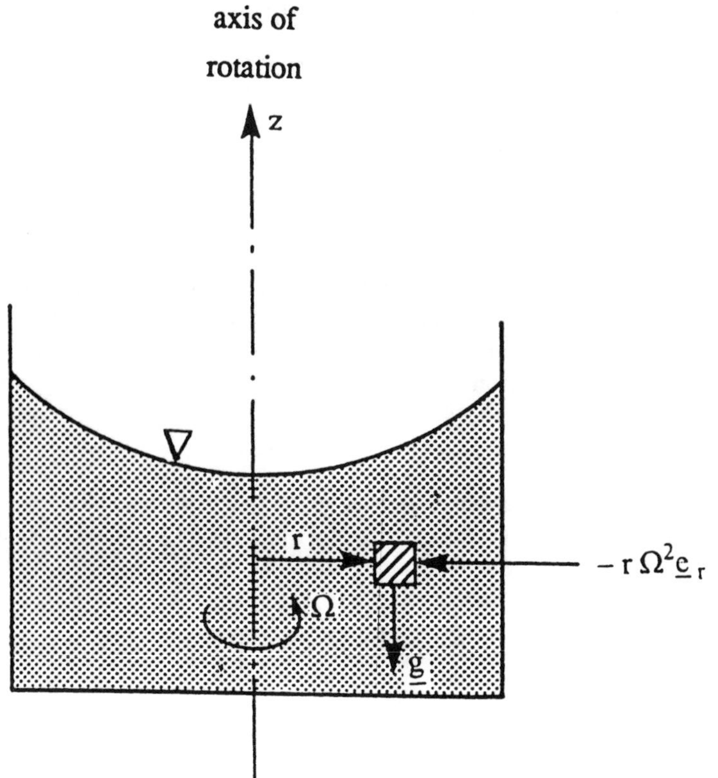

FIGURE 1.7 A fluid with a free surface in rigid-body rotation.

$$\nabla p = \frac{\partial p}{\partial r}\boldsymbol{e}_r + \frac{\partial p}{\partial z}\boldsymbol{e}_z = \rho(\boldsymbol{g} - \boldsymbol{a}) = \rho\left(r\Omega^2\boldsymbol{e}_r - g\boldsymbol{e}_z\right) \tag{1.21}$$

or

$$\frac{\partial p}{\partial r} = \rho r\Omega^2, \qquad \frac{\partial p}{\partial z} = -\rho g = -\gamma \tag{1.22}$$

Integration of these equations leads to

$$p = p_o - \gamma z + \frac{1}{2}\rho r^2\Omega^2 \tag{1.23}$$

where p_o is some reference pressure. This result shows that at any fixed r the pressure varies hydrostatically in the vertical direction, while the constant pressure surfaces, including the free surface, are paraboloids of revolution.

Further Information

The reader may find more detail and additional information on the topics in this section in any one of the many excellent introductory texts on fluid mechanics, such as

White, F.M. 1994. *Fluid Mechanics,* 3rd ed., McGraw-Hill, New York.

Munson, B.R., Young, D.F., and Okiishi, T.H. 1994. *Fundamentals of Fluid Mechanics,* 2nd ed., John Wiley & Sons, New York.

2

Equations of Motion and Potential Flow

Stanley A. Berger
University of California

2 Equations of Motion and Potential Flow 11
Integral Relations for a Control Volume • Reynolds Transport Theorem • Conservation of Mass • Conservation of Momentum • Conservation of Energy • Differential Relations for Fluid Motion • Mass Conservation–Continuity Equation • Momentum Conservation • Analysis of Rate of Deformation • Relationship between Forces and Rate of Deformation • The Navier–Stokes Equations • Energy Conservation — The Mechanical and Thermal Energy Equations • Boundary Conditions • Vorticity in Incompressible Flow • Stream Function • Inviscid Irrotational Flow: Potential Flow

2 Equations of Motion and Potential Flow

Integral Relations for a Control Volume

Like most physical conservation laws those governing motion of a fluid apply to material particles or systems of such particles. This so-called Lagrangian viewpoint is generally not as useful in practical fluid flows as an analysis through fixed (or deformable) control volumes — the Eulerian viewpoint. The relationship between these two viewpoints can be deduced from the Reynolds transport theorem, from which we also most readily derive the governing integral and differential equations of motion.

Reynolds Transport Theorem

The *extensive* quantity B, a scalar, vector, or tensor, is defined as any property of the fluid (e.g., momentum, energy) and b as the corresponding value per unit mass (the *intensive* value). The Reynolds transport theorem for a moving and arbitrarily deforming control volume CV, with boundary CS (see Figure 2.1), states that

$$\frac{d}{dt}\left(B_{\text{system}}\right) = \frac{d}{dt}\left(\iiint_{\text{CV}} \rho b \, d\upsilon\right) + \iint_{\text{CS}} \rho b\left(\boldsymbol{V}_r \cdot \boldsymbol{n}\right) dA \tag{2.1}$$

0-8493-0055-X/00/$0.00+$.50

where B_{system} is the total quantity of B in the system (any mass of fixed identity), $\boldsymbol{n}$ is the outward normal to the CS, $\boldsymbol{V}_r = \boldsymbol{V}(\boldsymbol{r}, t) - \boldsymbol{V}_{\text{CS}}(\boldsymbol{r}, t)$, the velocity of the fluid particle, $\boldsymbol{V}(\boldsymbol{r}, t)$, relative to that of the CS, $\boldsymbol{V}_{\text{CS}}(\boldsymbol{r}, t)$, and d/dt on the left-hand side is the derivative following the fluid particles, i.e., the fluid mass comprising the system. The theorem states that the time rate of change of the total B in the system is equal to the rate of change within the CV plus the net flux of B through the CS. To distinguish between the d/dt which appears on the two sides of Equation (2.1) but which have different interpretations, the derivative on the left-hand side, following the system, is denoted by D/Dt and is called the material derivative. This notation is used in what follows. For any function $f(x, y, z, t)$,

$$\frac{Df}{Dt} = \frac{\partial f}{\partial t} + \boldsymbol{V} \cdot \nabla f$$

For a CV fixed with respect to the reference frame, Equation (2.1) reduces to

$$\frac{D}{Dt}\left(B_{\text{system}}\right) = \frac{d}{dt} \underset{\substack{\text{CV} \\ \text{(fixed)}}}{\iiint} (\rho b)\, d\upsilon + \underset{\text{CS}}{\iint} \rho b (\boldsymbol{V} \cdot \boldsymbol{n})\, dA \tag{2.2}$$

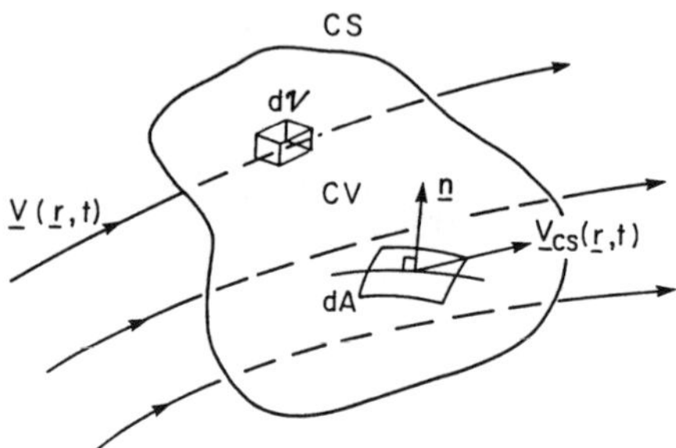

FIGURE .2.1 Control volume.

(The time derivative operator in the first term on the right-hand side may be moved inside the integral, in which case it is then to be interpreted as the partial derivative ∂/∂t.)

Conservation of Mass

If we apply Equation (2.2) for a fixed control volume, with B_{system} the total mass in the system, then since conservation of mass requires that $DB_{\text{system}}/Dt = 0$ there follows, since $b = B_{\text{system}}/m = 1$,

$$\underset{\substack{\text{CV} \\ \text{(fixed)}}}{\iiint} \frac{\partial \rho}{\partial t}\, d\upsilon + \underset{\text{CS}}{\iint} \rho (\boldsymbol{V} \cdot \boldsymbol{n})\, dA = 0 \tag{2.3}$$

This is the integral form of the conservation of mass law for a fixed control volume. For a steady flow, Equation (2.3) reduces to

$$\underset{\text{CS}}{\iint} \rho (\boldsymbol{V} \cdot \boldsymbol{n})\, dA = 0 \tag{2.4}$$

whether compressible or incompressible. For an incompressible flow, ρ = constant, so

$$\iint_{CS} (\boldsymbol{V}\cdot\boldsymbol{n})\, dA = 0 \tag{2.5}$$

whether the flow is steady or unsteady.

Conservation of Momentum

The conservation of (linear) momentum states that

$$F_{\text{total}} \equiv \sum(\text{external forces acting on the fluid system}) = \frac{D\boldsymbol{M}}{Dt} \equiv \frac{D}{Dt}\left(\iiint_{\text{system}} \rho\boldsymbol{V}\, d\upsilon\right) \tag{2.6}$$

where $\boldsymbol{M}$ is the total system momentum. For an arbitrarily moving, deformable control volume it then follows from Equation 2.1) with b set to $\boldsymbol{V}$,

$$\boldsymbol{F}_{\text{total}} = \frac{d}{dt}\left(\iiint_{CV} \rho\boldsymbol{V}\, d\upsilon\right) + \iint_{CS} \rho\boldsymbol{V}(\boldsymbol{V}_r\cdot\boldsymbol{n})\, dA \tag{2.7}$$

This expression is only valid in an inertial coordinate frame. To write the equivalent expression for a noninertial frame we must use the relationship between the acceleration $\boldsymbol{a}_I$ in an inertial frame and the acceleration $\boldsymbol{a}_R$ in a noninertial frame,

$$\boldsymbol{a}_I = \boldsymbol{a}_R + \frac{d^2\boldsymbol{R}}{dt^2} + 2\Omega\times\boldsymbol{V} + \Omega\times(\Omega\times\boldsymbol{r}) + \frac{d\Omega}{dt}\times\boldsymbol{r} \tag{2.8}$$

where $\boldsymbol{R}$ is the position vector of the origin of the noninertial frame with respect to that of the inertial frame, Ω is the angular velocity of the noninertial frame, and $\boldsymbol{r}$ and $\boldsymbol{V}$ the position and velocity vectors in the noninertial frame. The third term on the right-hand side of Equation (2.8) is the Coriolis acceleration, and the fourth term is the centrifugal acceleration. For a noninertial frame Equation (2.7) is then

$$\begin{aligned}\boldsymbol{F}_{\text{total}} - \iiint_{\text{system}}\left[\frac{d^2\boldsymbol{R}}{dt^2} + 2\Omega\times\boldsymbol{V} + \Omega\times(\Omega\times\boldsymbol{r}) + \frac{d\Omega}{dt}\times\boldsymbol{r}\right]\rho\, d\upsilon &= \frac{D}{Dt}\left(\iiint_{\text{system}} \rho\boldsymbol{V}\, d\upsilon\right)\\ &= \frac{d}{dt}\left(\iiint_{CV} \rho\boldsymbol{V}\, d\upsilon\right) + \iint_{CS} \rho\boldsymbol{V}\cdot(\boldsymbol{V}_r\cdot\boldsymbol{n})\, dA\end{aligned} \tag{2.9}$$

where the frame acceleration terms of Equation (2.8) have been brought to the left-hand side because to an observer in the noninertial frame they act as "apparent" body forces.

For a fixed control volume in an inertial frame for steady flow it follows from the above that

$$\boldsymbol{F}_{\text{total}} = \iint_{CS} \rho\boldsymbol{V}(\boldsymbol{V}\cdot\boldsymbol{n})\, dA \tag{2.10}$$

This expression is the basis of many control volume analyses for fluid flow problems.

The cross product of $\boldsymbol{r}$, the position vector with respect to a convenient origin, with the momentum Equation (2.6) written for an elementary particle of mass dm, noting that $(d\boldsymbol{r}/dt) \times \boldsymbol{V} = 0$, leads to the integral moment of momentum equation

$$\sum \boldsymbol{M} - \boldsymbol{M}_I = \frac{D}{Dt} \iiint\limits_{\text{system}} \rho(\boldsymbol{r} \times \boldsymbol{V})\, d\upsilon \tag{2.11}$$

where $\Sigma \boldsymbol{M}$ is the sum of the moments of all the external forces acting on the system about the origin of $\boldsymbol{r}$, and $\boldsymbol{M}_I$ is the moment of the apparent body forces (see Equation (2.9)). The right-hand side can be written for a control volume using the appropriate form of the Reynolds transport theorem.

Conservation of Energy

The conservation of energy law follows from the first law of thermodynamics for a moving system

$$\dot{Q} - \dot{W} = \frac{D}{Dt}\left(\iiint\limits_{\text{system}} \rho e\, d\upsilon \right) \tag{2.12}$$

where $\dot{Q}$ is the rate at which heat is added to the system, $\dot{W}$ the rate at which the system works on its surroundings, and e is the total energy per unit mass. For a particle of mass dm the contributions to the specific energy e are the internal energy u, the kinetic energy $V^2/2$, and the potential energy, which in the case of gravity, the only body force we shall consider, is gz, where z is the vertical displacement opposite to the direction of gravity. (We assume no energy transfer owing to chemical reaction as well as no magnetic or electric fields.) For a fixed control volume it then follows from Equation (2.2) [with $b = e = u + (V^2/2) + gz$] that

$$\dot{Q} - \dot{W} = \frac{d}{dt}\left(\iiint\limits_{\text{CV}} \rho\left(u + \frac{1}{2}V^2 + gz \right) d\upsilon \right) + \iint\limits_{\text{CS}} \rho\left(u + \frac{1}{2}V^2 + gz \right)(\boldsymbol{V} \cdot \boldsymbol{n})\, dA \tag{2.13}$$

Problem

An incompressible fluid flows through a pump at a volumetric flow rate $\hat{Q}$. The (head) loss between sections 1 and 2 (see Figure 2.2) is equal to $\beta \rho V_1^2/2$ (V is the average velocity at the section). Calculate the power that must be delivered by the pump to the fluid to produce a given increase in pressure, $\Delta p = p_2 - p_1$.

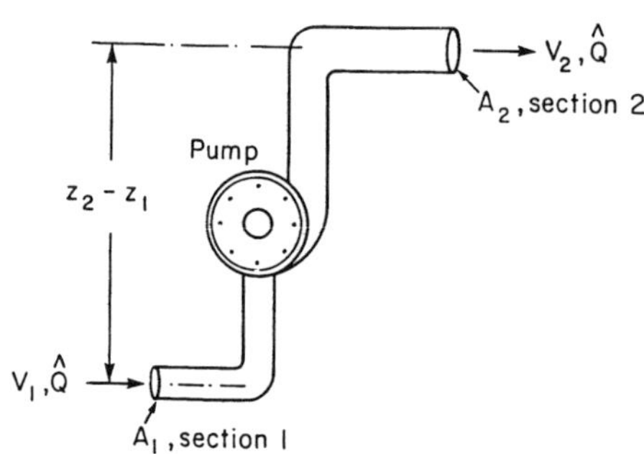

FIGURE 2.2 Pump producing pressure increase.

Solution: The principal equation needed is the energy Equation (2.13). The term $\dot{W}$, the rate at which the system does work on its surroundings, for such problems has the form

$$\dot{W} = -\dot{W}_{\text{shaft}} + \iint_{\text{CS}} p\boldsymbol{V}\cdot\boldsymbol{n}\, dA \tag{P.2.1}$$

where $\dot{W}_{\text{shaft}}$ represents the work done on the fluid by a moving shaft, such as by turbines, propellers, fans, etc., and the second term on the right side represents the rate of working by the normal stress, the pressure, at the boundary. For a steady flow in a control volume coincident with the physical system boundaries and bounded at its ends by sections 1 and 2, Equation (2.13) reduces to ($u \equiv 0$),

$$\dot{Q} + \dot{W}_{\text{shaft}} - \iint_{\text{CS}} p\boldsymbol{V}\cdot\boldsymbol{n}\, dA = \iint_{\text{CS}} \left(\frac{1}{2}\rho V^2 + \gamma z\right)(\boldsymbol{V}\cdot\boldsymbol{n})\, dA \tag{P.2.2}$$

Using average quantities at sections 1 and 2, and the continuity Equation (2.5), which reduces in this case to

$$V_1A_1 = V_2A_2 = \hat{Q} \tag{P.2.3}$$

We can write Equation (P.2.2) as

$$\dot{Q} + \dot{W}_{\text{shaft}} - (p_2 - p_1)\hat{Q} = \left[\frac{1}{2}\rho\left(V_2^2 - V_1^2\right) + \gamma(z_2 - z_1)\right]\hat{Q} \tag{P.2.4}$$

$\dot{Q}$, the rate at which heat is added to the system, is here equal to $-\beta\rho V_1^2/2$, the head loss between sections 1 and 2. Equation (P.2.4) then can be rewritten

$$\dot{W}_{\text{shaft}} = \beta\rho\frac{V_1^2}{2} + (\Delta p)\hat{Q} + \frac{1}{2}\rho\left(V_2^2 - V_1^2\right)\hat{Q} + \gamma(z_2 - z_1)\hat{Q}$$

or, in terms of the given quantities,

$$\dot{W}_{\text{shaft}} = \frac{\beta\rho\hat{Q}^2}{A_1^2} + (\Delta p)\hat{Q} + \frac{1}{2}\rho\frac{\hat{Q}^3}{A_2^2}\left(1 - \frac{A_2^2}{A_1^2}\right) + \gamma(z_2 - z_1)\hat{Q} \tag{P.2.5}$$

Thus, for example, if the fluid is water ($\rho \approx 1000$ kg/m^3, $\gamma = 9.8$ kN/m^3), $\hat{Q} = 0.5$ m^3/sec, the heat loss is $0.2\rho V_1^2/2$, and $\Delta p = p_2 - p_1 = 2 \times 10^5$N/m^2 = 200 kPa, $A_1 = 0.1$ m^2 = $A_2/2$, $(z_2 - z_1) = 2$ m, we find, using Equation (P.2.5)

$$\begin{aligned}\dot{W}_{\text{shaft}} &= \frac{0.2(1000)(0.5)^2}{(0.1)^2} + \left(2\times 10^5\right)(0.5) + \frac{1}{2}(1000)\frac{(0.5)^3}{(0.2)^2}(1-4) + \left(9.8\times 10^3\right)(2)(0.5) \\ &= 5{,}000 + 10{,}000 - 4{,}688 + 9{,}800 = 20{,}112 \text{ Nm/sec} \\ &= 20{,}112 \text{ W} = \frac{20{,}112}{745.7}\text{hp} = 27 \text{ hp}\end{aligned}$$

Differential Relations for Fluid Motion

In the previous section the conservation laws were derived in integral form. These forms are useful in calculating, generally using a control volume analysis, gross features of a flow. Such analyses usually require some *a priori* knowledge or assumptions about the flow. In any case, an approach based on integral conservation laws cannot be used to determine the point-by-point variation of the dependent variables, such as velocity, pressure, temperature, etc. To do this requires the use of the differential forms of the conservation laws, which are presented below.

Mass Conservation–Continuity Equation

Applying Gauss's theorem (the divergence theorem) to Equation (2.3) we obtain

$$\iiint\limits_{\substack{\text{CV}\\(\text{fixed})}} \left[\frac{\partial \rho}{\partial t} + \nabla\cdot(\rho \boldsymbol{V})\right] d\upsilon = 0 \tag{2.14}$$

which, because the control volume is arbitrary, immediately yields

$$\frac{\partial \rho}{\partial t} + \nabla\cdot(\rho \boldsymbol{V}) = 0 \tag{2.15}$$

This can also be written as

$$\frac{D\rho}{Dt} + \rho\nabla\cdot \boldsymbol{V} = 0 \tag{2.16}$$

using the fact that

$$\frac{D\rho}{Dt} = \frac{\partial \rho}{\partial t} + \boldsymbol{V}\cdot\nabla\rho \tag{2.17}$$

Special cases:

1. Steady flow $[(\partial/\partial t)\;(\;) \equiv 0]$

$$\nabla\cdot(\rho \boldsymbol{V}) = 0 \tag{2.18}$$

2. Incompressible flow $(D\rho/Dt \equiv 0)$

$$\nabla\cdot \boldsymbol{V} = 0 \tag{2.19}$$

Momentum Conservation

We note first, as a consequence of mass conservation for a system, that the right-hand side of Equation (2.6) can be written as

$$\frac{D}{Dt}\left(\iiint\limits_{\text{system}} \rho \boldsymbol{V}\, d\upsilon\right) \equiv \iiint\limits_{\text{system}} \rho\frac{D\boldsymbol{V}}{Dt}\, d\upsilon \tag{2.20}$$

The total force acting on the system which appears on the left-hand side of Equation (2.6) is the sum of body forces $\boldsymbol{F}_b$ and surface forces $\boldsymbol{F}_s$. The body forces are often given as forces per unit mass (e.g., gravity), and so can be written

$$\boldsymbol{F}_b = \iiint\limits_{\text{system}} \rho \boldsymbol{f}\, d\upsilon \tag{2.21}$$

The surface forces are represented in terms of the second-order stress tensor* $\underline{\underline{\sigma}} = \{\sigma_{ij}\}$, where σ_{ij} is defined as the force per unit area in the i direction on a planar element whose normal lies in the j direction. From elementary angular momentum considerations for an infinitesimal volume it can be shown that σ_{ij} is a symmetric tensor, and therefore has only six independent components. The total surface force exerted on the system by its surroundings is then

$$\boldsymbol{F}_s = \iint\limits_{\substack{\text{system}\\ \text{surface}}} \underline{\underline{\sigma}} \cdot \boldsymbol{n}\, dA, \text{ with } i-\text{component } F_{s_i} = \iint \sigma_{ij} n_j\, dA \tag{2.22}$$

The integral momentum conservation law Equation (2.6) can then be written

$$\iiint\limits_{\text{system}} \rho \frac{D\boldsymbol{V}}{Dt}\, d\upsilon = \iiint\limits_{\text{system}} \rho \boldsymbol{f}\, d\upsilon + \iint\limits_{\substack{\text{system}\\ \text{surface}}} \underline{\underline{\sigma}} \cdot \boldsymbol{n}\, dA \tag{2.23}$$

The application of the divergence theorem to the last term on the right-side of Equation (2.23) leads to

$$\iiint\limits_{\text{system}} \rho \frac{D\boldsymbol{V}}{Dt}\, d\upsilon = \iiint\limits_{\text{system}} \rho \boldsymbol{f}\, d\upsilon + \iiint\limits_{\text{system}} \nabla \cdot \underline{\underline{\sigma}}\, d\upsilon \tag{2.24}$$

where $\nabla \cdot \underline{\underline{\sigma}} \equiv \{\partial\sigma_{ij}/x_j\}$. Since Equation (2.24) holds for any material volume, it follows that

$$\rho \frac{D\boldsymbol{V}}{Dt} = \rho \boldsymbol{f} + \nabla \cdot \underline{\underline{\sigma}} \tag{2.25}$$

(With the decomposition of $\boldsymbol{F}_{\text{total}}$ above, Equation (2.10) can be written

$$\iiint\limits_{\text{CV}} \rho \boldsymbol{f}\, d\upsilon + \iint\limits_{\text{CS}} \underline{\underline{\sigma}} \cdot \boldsymbol{n}\, dA = \iint\limits_{\text{CS}} \rho \boldsymbol{V}(\boldsymbol{V} \cdot \boldsymbol{n})\, dA \tag{2.26}$$

If ρ is uniform and $\boldsymbol{f}$ is a conservative body force, i.e., $\boldsymbol{f} = -\nabla\Psi$, where Ψ is the force potential, then Equation (2.26), after application of the divergence theorem to the body force term, can be written

$$\iint\limits_{\text{CS}} \left(-\rho\Psi\boldsymbol{n} + \underline{\underline{\sigma}} \cdot \boldsymbol{n}\right) dA = \iint\limits_{\text{CS}} \rho \boldsymbol{V}(\boldsymbol{V} \cdot \boldsymbol{n})\, dA \tag{2.27}$$

* We shall assume the reader is familiar with elementary Cartesian tensor analysis and the associated subscript notation and conventions. The reader for whom this is not true should skip the details and concentrate on the final principal results and equations given at the ends of the next few subsections.

It is in this form, involving only integrals over the surface of the control volume, that the integral form of the momentum equation is used in control volume analyses, particularly in the case when the body force term is absent.)

Analysis of Rate of Deformation

The principal aim of the following two subsections is to derive a relationship between the stress and the rate of strain to be used in the momentum Equation (2.25). The reader less familiar with tensor notation may skip these sections, apart from noting some of the terms and quantities defined therein, and proceed directly to Equations (2.38) or (2.39).

The relative motion of two neighboring points P and Q, separated by a distance η, can be written (using $\boldsymbol{u}$ for the local velocity)

$$\boldsymbol{u}(Q) = \boldsymbol{u}(P) + (\nabla \boldsymbol{u})\eta$$

or, equivalently, writing $\nabla \boldsymbol{u}$ as the sum of antisymmetric and symmetric tensors,

$$\boldsymbol{u}(Q) = \boldsymbol{u}(P) + \frac{1}{2}\left((\nabla \boldsymbol{u}) - (\nabla \boldsymbol{u})^*\right)\eta + \frac{1}{2}\left((\nabla \boldsymbol{u}) + (\nabla \boldsymbol{u})^*\right)\eta \tag{2.28}$$

where $\nabla \boldsymbol{u} = \{\partial u_i / \partial x_j\}$, and the superscript * denotes transpose, so $(\nabla \boldsymbol{u})^* = \{\partial u_j / \partial x_i\}$. The second term on the right-hand side of Equation (2.28) can be rewritten in terms of the *vorticity*, $\nabla \times \boldsymbol{u}$, so Equation (2.28) becomes

$$\boldsymbol{u}(Q) = \boldsymbol{u}(P) + \frac{1}{2}(\nabla \times \boldsymbol{u}) \times \eta + \frac{1}{2}\left((\nabla \boldsymbol{u}) + (\nabla \boldsymbol{u})^*\right)\eta \tag{2.29}$$

which shows that the local rate of deformation consists of a rigid-body translation, a rigid-body rotation with angular velocity $^1/_2\ (\nabla \times \boldsymbol{u})$, and a velocity or rate of deformation. The coefficient of η in the last term in Equation (2.29) is defined as the rate-of-strain tensor and is denoted by $\underline{\underline{e}}$, in subscript form

$$e_{ij} = \frac{1}{2}\left(\frac{\partial u_i}{\partial x_j} + \frac{\partial u_j}{\partial x_i}\right) \tag{2.30}$$

From $\underline{\underline{e}}$ we can define a rate-of-strain central quadric, along the principal axes of which the deforming motion consists of a pure extension or compression.

Relationship Between Forces and Rate of Deformation

We are now in a position to determine the required relationship between the stress tensor $\underline{\underline{\sigma}}$ and the rate of deformation. Assuming that in a static fluid the stress reduces to a (negative) hydrostatic or thermodynamic pressure, equal in all directions, we can write

$$\underline{\underline{\sigma}} = -p\underline{\underline{I}} + \underline{\underline{\tau}} \quad \text{or} \quad \underline{\underline{\sigma}}_{ij} = -p\delta_{ij} + \tau_{ij} \tag{2.31}$$

where $\underline{\underline{\tau}}$ is the viscous part of the total stress and is called the deviatoric stress tensor, $\underline{\underline{I}}$ is the identity tensor, and δ_{ij} is the corresponding Kronecker delta ($\delta_{ij} = 0$ if $i \neq j$; $\delta_{ij} = 1$ if $i = j$). We make further assumptions that (1) the fluid exhibits no preferred directions; (2) the stress is independent of any previous history of distortion; and (3) that the stress depends only on the local thermodynamic state and the kinematic state of the immediate neighborhood. Precisely, we assume that $\underline{\tau}$ is linearly proportional to the first spatial derivatives of $\boldsymbol{u}$, the coefficient of proportionality depending only on the local thermodynamic state. These assumptions and the relations below which follow from them are appropriate for a Newtonian fluid. Most common fluids, such as air and water under most conditions, are Newtonian, but there are many other fluids, including many which arise in industrial applications, which exhibit so-called non-Newtonian properties. The study of such non-Newtonian fluids, such as viscoelastic fluids, is the subject of the field of rheology.

With the Newtonian fluid assumptions above, and the symmetry of $\underline{\tau}$ which follows from the symmetry of $\underline{\underline{\sigma}}$, one can show that the viscous part $\underline{\tau}$ of the total stress can be written as

$$\underline{\tau} = \lambda(\nabla\cdot\boldsymbol{u})\underline{I} + 2\mu\underline{e} \tag{2.32}$$

so the total stress for a Newtonian fluid is

$$\underline{\underline{\sigma}} = -p\underline{I} + \lambda(\nabla\cdot\boldsymbol{u})\underline{I} + 2\mu\underline{e} \tag{2.33}$$

or, in subscript notation

$$\sigma_{ij} = -p\delta_{ij} + \lambda\left(\frac{\partial u_k}{\partial x_k}\right)\delta_{ij} + \mu\left(\frac{\partial u_i}{\partial x_j} + \frac{\partial u_j}{\partial x_i}\right) \tag{2.34}$$

(the Einstein summation convention is assumed here, namely, that a repeated subscript, such as in the second term on the right-hand side above, is summed over; note also that $\nabla\cdot\boldsymbol{u} = \partial u_k/\partial x_k = e_{kk}$.) The coefficient λ is called the "second viscosity" and μ the "absolute viscosity," or more commonly the "dynamic viscosity," or simply the "viscosity." For a Newtonian fluid λ and μ depend only on local thermodynamic state, primarily on the temperature.

We note, from Equation (2.34), that whereas in a fluid at rest the pressure is an isotropic normal stress, this is not the case for a moving fluid, since in general $\sigma_{11} \neq \sigma_{22} \neq \sigma_{33}$. To have an analogous quantity to p for a moving fluid we define the pressure in a moving fluid as the negative mean normal stress, denoted, say, by $\bar{p}$

$$\bar{p} = -\frac{1}{3}\sigma_{ii} \tag{2.35}$$

(σ_{ii} is the trace of $\underline{\underline{\sigma}}$ and an invariant of $\underline{\underline{\sigma}}$, independent of the orientation of the axes). From Equation (2.34)

$$\bar{p} = -\frac{1}{3}\sigma_{ii} = p - \left(\lambda + \frac{2}{3}\mu\right)\nabla\cdot\boldsymbol{u} \tag{2.36}$$

For an incompressible fluid $\nabla\cdot\boldsymbol{u} = 0$ and hence $\bar{p} \equiv p$. The quantity $(\lambda + {}^2/_3\mu)$ is called the bulk viscosity. If one assumes that the deviatoric stress tensor τ_{ij} makes no contribution to the mean normal stress, it follows that $\lambda + {}^2/_3\mu = 0$, so again $\bar{p} = p$. This condition, $\lambda = -{}^2/_3\mu$, is called the Stokes assumption or hypothesis. If neither the incompressibility nor the Stokes assumptions are made, the difference between $\bar{p}$ and p is usually still negligibly small because $(\lambda + {}^2/_3\mu)\ \nabla\cdot\boldsymbol{u} << p$ in most fluid flow problems. If the Stokes hypothesis is made, as is often the case in fluid mechanics, Equation (2.34) becomes

$$\sigma_{ij} = -p\delta_{ij} + 2\mu\left(e_{ij} - \frac{1}{3}e_{kk}\delta_{ij}\right) \tag{2.37}$$

The Navier–Stokes Equations

Substitution of Equation (2.33) into (2.25), since $\nabla \cdot (\phi \underline{I}) = \nabla\phi$, for any scalar function ϕ, yields (replacing $\boldsymbol{u}$ in Equation (2.33) by $\boldsymbol{V}$)

$$\rho\frac{D\boldsymbol{V}}{Dt} = \rho\boldsymbol{f} - \nabla p + \nabla(\lambda\nabla\cdot\boldsymbol{V}) + \nabla\cdot\left(2\mu\underline{\underline{e}}\right) \tag{2.38}$$

These equations are the Navier–Stokes equations (although the name is as often given to the full set of governing conservation equations). With the Stokes assumption ($\lambda = -{}^2/_3\mu$), Equation (2.38) becomes

$$\rho\frac{D\boldsymbol{V}}{Dt} = \rho\boldsymbol{f} - \nabla p + \nabla\cdot\left[2\mu\left(\underline{e} - \frac{1}{3}e_{kk}\underline{I}\right)\right] \tag{2.39}$$

If the Eulerian frame is not an inertial frame, then one must use the transformation to an inertial frame either using Equation (2.8) or the "apparent" body force formulation, Equation (2.9).

Energy Conservation — The Mechanical and Thermal Energy Equations

In deriving the differential form of the energy equation we begin by assuming that heat enters or leaves the material or control volume by heat conduction across the boundaries, the heat flux per unit area being $\boldsymbol{q}$. It then follows that

$$\dot{Q} = -\iint \boldsymbol{q}\cdot\boldsymbol{n}\, dA = -\iiint \nabla\cdot\boldsymbol{q}\, d\upsilon \tag{2.40}$$

The work-rate term $\dot{W}$ can be decomposed into the rate of work done against body forces, given by

$$-\iiint \rho\boldsymbol{f}\cdot\boldsymbol{V}\, d\upsilon \tag{2.41}$$

and the rate of work done against surface stresses, given by

$$-\iint_{\substack{\text{system}\\ \text{surface}}} \boldsymbol{V}\cdot\left(\underline{\sigma}\boldsymbol{n}\right) dA \tag{2.42}$$

Substitution of these expressions for $\dot{Q}$ and $\dot{W}$ into Equation (2.12), use of the divergence theorem, and conservation of mass lead to

$$\rho\frac{D}{Dt}\left(u + \frac{1}{2}V^2\right) = -\nabla\cdot\boldsymbol{q} + \rho\boldsymbol{f}\cdot\boldsymbol{V} + \nabla\cdot\left(\boldsymbol{V}\underline{\sigma}\right) \tag{2.43}$$

(note that a potential energy term is no longer included in e, the total specific energy, as it is accounted for by the body force rate-of-working term $\rho\boldsymbol{f}\cdot\boldsymbol{V}$).

Equation (2.43) is the total energy equation showing how the energy changes as a result of working by the body and surface forces and heat transfer. It is often useful to have a purely thermal energy equation. This is obtained by subtracting from Equation (2.43) the dot product of $\boldsymbol{V}$ with the momentum Equation (2.25), after expanding the last term in Equation (2.43), resulting in

$$\rho\frac{Du}{Dt} = \frac{\partial V_i}{\partial x_j}\sigma_{ij} - \nabla\cdot \boldsymbol{q} \tag{2.44}$$

With $\sigma_{ij} = -p\delta_{ij} + \tau_{ij}$, and the use of the continuity equation in the form of Equation (2.16), the first term on the right-hand side of Equation (2.44) may be written

$$\frac{\partial V_i}{\partial x_j}\sigma_{ij} = -\rho\frac{D\left(\frac{p}{\rho}\right)}{Dt} + \frac{Dp}{Dt} + \Phi \tag{2.45}$$

where Φ is the rate of dissipation of mechanical energy per unit mass due to viscosity, and is given by

$$\Phi \equiv \frac{\partial V_i}{\partial x_j}\tau_{ij} = 2\mu\left(e_{ij}e_{ij} - \frac{1}{3}e_{kk}^2\right) = 2\mu\left(e_{ij} - \frac{1}{3}e_{kk}\delta_{ij}\right)^2 \tag{2.46}$$

With the introduction of Equation (2.45), Equation (2.44) becomes

$$\rho\frac{De}{Dt} = -p\nabla\cdot \boldsymbol{V} + \Phi - \nabla\cdot \boldsymbol{q} \tag{2.47}$$

or

$$\rho\frac{Dh}{Dt} = \frac{Dp}{Dt} + \Phi - \nabla\cdot \boldsymbol{q} \tag{2.48}$$

where $h = e + (p/\rho)$ is the specific enthalpy. Unlike the other terms on the right-hand side of Equation (2.47), which can be negative or positive, Φ is always nonnegative and represents the increase in internal energy (or enthalpy) owing to irreversible degradation of mechanical energy. Finally, from elementary thermodynamic considerations

$$\frac{Dh}{Dt} = T\frac{DS}{Dt} + \frac{1}{\rho}\frac{Dp}{Dt}$$

where S is the entropy, so Equation (2.48) can be written

$$\rho T\frac{DS}{Dt} = \Phi - \nabla\cdot \boldsymbol{q} \tag{2.49}$$

If the heat conduction is assumed to obey the Fourier heat conduction law, so $\boldsymbol{q} = -k\nabla T$, where k is the thermal conductivity, then in all of the above equations

$$-\nabla\cdot \boldsymbol{q} = \nabla\cdot(k\nabla T) = k\nabla^2 T \tag{2.50}$$

the last of these equalities holding only if k = constant.

In the event the thermodynamic quantities vary little, the coefficients of the constitutive relations for $\underline{\sigma}$ and $\boldsymbol{q}$ may be taken to be constant and the above equations simplified accordingly.

We note also that if the flow is incompressible, then the mass conservation, or continuity, equation simplifies to

$$\nabla \cdot \boldsymbol{V} = 0 \tag{2.51}$$

and the momentum Equation (2.38) to

$$\rho \frac{D\boldsymbol{V}}{Dt} = \rho \boldsymbol{f} - \nabla p + \mu \nabla^2 \boldsymbol{V} \tag{2.52}$$

where ∇^2 is the Laplacian operator. The small temperature changes, compatible with the incompressibility assumption, are then determined, for a perfect gas with constant k and specific heats, by the energy equation rewritten for the temperature, in the form

$$\rho c_v \frac{DT}{Dt} = k\nabla^2 T + \Phi \tag{2.53}$$

Boundary Conditions

The appropriate boundary conditions to be applied at the boundary of a fluid in contact with another medium depends on the nature of this other medium — solid, liquid, or gas. We discuss a few of the more important cases here in turn:

1. *At a solid surface:* $\boldsymbol{V}$ and T are continuous. Contained in this boundary condition is the "no-slip" condition, namely, that the tangential velocity of the fluid in contact with the boundary of the solid is equal to that of the boundary. For an inviscid fluid the no-slip condition does not apply, and only the normal component of velocity is continuous. If the wall is permeable, the tangential velocity is continuous and the normal velocity is arbitrary; the temperature boundary condition for this case depends on the nature of the injection or suction at the wall.
2. *At a liquid/gas interface:* For such cases the appropriate boundary conditions depend on what can be assumed about the gas the liquid is in contact with. In the classical liquid free-surface problem, the gas, generally atmospheric air, can be ignored and the necessary boundary conditions are that (a) the normal velocity in the liquid at the interface is equal to the normal velocity of the interface and (b) the pressure in the liquid at the interface exceeds the atmospheric pressure by an amount equal to

$$\Delta p = p_{\text{liquid}} - p_{\text{atm}} = \sigma \left(\frac{1}{R_1} + \frac{1}{R_2} \right) \tag{2.54}$$

   where R_1 and R_2 are the radii of curvature of the intercepts of the interface by two orthogonal planes containing the vertical axis. If the gas is a vapor which undergoes nonnegligible interaction and exchanges with the liquid in contact with it, the boundary conditions are more complex. Then, in addition to the above conditions on normal velocity and pressure, the shear stress (momentum flux) and heat flux must be continuous as well.

For interfaces in general the boundary conditions are derived from continuity conditions for each "transportable" quantity, namely continuity of the appropriate intensity across the interface and continuity of the normal component of the flux vector. Fluid momentum and heat are two such transportable quantities, the associated intensities are velocity and temperature, and the associated flux vectors are stress and heat flux. (The reader should be aware of circumstances where these simple criteria do not apply, for example, the velocity slip and temperature jump for a rarefied gas in contact with a solid surface.)

Vorticity in Incompressible Flow

With μ = constant, ρ = constant, and $\boldsymbol{f} = -\boldsymbol{g} = -g\boldsymbol{k}$ the momentum equation reduces to the form (see Equation (2.52))

$$\rho\frac{D\boldsymbol{V}}{Dt} = -\nabla p - \rho g\boldsymbol{k} + \mu\nabla^2\boldsymbol{V} \tag{2.55}$$

With the vector identities

$$(\boldsymbol{V}\cdot\nabla)\boldsymbol{V} = \nabla\left(\frac{V^2}{2}\right) - \boldsymbol{V}\times(\nabla\times\boldsymbol{V}) \tag{2.56}$$

and

$$\nabla^2\boldsymbol{V} = \nabla(\nabla\cdot\boldsymbol{V}) - \nabla\times(\nabla\times\boldsymbol{V}) \tag{2.57}$$

and defining the *vorticity*

$$\boldsymbol{\zeta} \equiv \nabla\times\boldsymbol{V} \tag{2.58}$$

Equation (2.55) can be written, noting that for incompressible flow $\nabla\cdot\boldsymbol{V} = 0$,

$$\rho\frac{\partial\boldsymbol{V}}{\partial t} + \nabla\left(p + \frac{1}{2}\rho V^2 + \rho g z\right) = \rho\boldsymbol{V}\times\boldsymbol{\zeta} - \mu\nabla\times\boldsymbol{\zeta} \tag{2.59}$$

The flow is said to be *irrotational* if

$$\boldsymbol{\zeta} \equiv \nabla\times\boldsymbol{V} = 0 \tag{2.60}$$

from which it follows that a *velocity potential* Φ can be defined

$$\boldsymbol{V} = \nabla\Phi \tag{2.61}$$

Setting $\zeta = 0$ in Equation (2.59), using Equation (2.61), and then integrating with respect to all the spatial variables, leads to

$$\rho\frac{\partial\Phi}{\partial t} + \left(p + \frac{1}{2}\rho V^2 + \rho g z\right) = F(t) \tag{2.62}$$

(the arbitrary function $F(t)$ introduced by the integration can either be absorbed in Φ, or is determined by the boundary conditions). Equation (2.62) is the unsteady *Bernoulli equation* for irrotational, incompressible flow. (Irrotational flows are always potential flows, even if the flow is compressible. Because the viscous term in Equation (2.59) vanishes identically for $\zeta = 0$, it would appear that the above Bernoulli equation is valid even for viscous flow. Potential solutions of hydrodynamics are in fact exact solutions of the full Navier–Stokes equations. Such solutions, however, are not valid near solid boundaries or bodies because the no-slip condition generates vorticity and causes nonzero ζ; the potential flow solution is invalid in all those parts of the flow field that have been "contaminated" by the spread of the vorticity by convection and diffusion. See below.)

The curl of Equation (2.59), noting that the curl of any gradient is zero, leads to

$$\rho\frac{\partial\zeta}{\partial t}=\rho\nabla\times(V\times\zeta)-\mu\nabla\times\nabla\times\zeta \tag{2.63}$$

but

$$\begin{aligned}\nabla^2\zeta&=\nabla(\nabla\cdot\zeta)-\nabla\times\nabla\times\zeta\\&=-\nabla\times\nabla\times\zeta\end{aligned} \tag{2.64}$$

since div curl ( ) ≡ 0, and therefore also

$$\nabla\times(V\times\zeta)\equiv\zeta(\nabla V)+V\nabla\cdot\zeta-V\nabla\zeta-\zeta\nabla\cdot V \tag{2.65}$$

$$=\zeta(\nabla V)-V\nabla\zeta \tag{2.66}$$

Equation (2.63) can then be written

$$\frac{D\zeta}{Dt}=(\zeta\cdot\nabla)V+\nu\nabla^2\zeta \tag{2.67}$$

where $\nu=\mu/\rho$ is the kinematic viscosity. Equation (2.67) is the vorticity equation for incompressible flow. The first term on the right, an inviscid term, increases the vorticity by vortex stretching. In inviscid, two-dimensional flow both terms on the right-hand side of Equation (2.67) vanish, and the equation reduces to $D\zeta/Dt=0$, from which it follows that the vorticity of a fluid particle remains constant as it moves. This is Helmholtz's theorem. As a consequence it also follows that if $\zeta=0$ initially, $\zeta\equiv0$ always; i.e., initially irrotational flows remain irrotational (for inviscid flow). Similarly, it can be proved that $D\Gamma/Dt=0$; i.e., the circulation around a material closed circuit remains constant, which is Kelvin's theorem.

If $\nu\neq0$, Equation (2.67) shows that the vorticity generated, say, at solid boundaries, diffuses and stretches as it is convected.

We also note that for steady flow the Bernoulli equation reduces to

$$p+\frac{1}{2}\rho V^2+\rho gz=\text{constant} \tag{2.68}$$

valid for steady, irrotational, incompressible flow.

Stream Function

For two-dimensional flows the continuity equation, e.g., for plane, incompressible flows ($V=(u,\nu)$)

$$\frac{\partial u}{\partial x}+\frac{\partial\nu}{\partial y}=0 \tag{2.69}$$

can be identically satisfied by introducing a stream function ψ, defined by

$$u=\frac{\partial\psi}{\partial y},\qquad \nu=-\frac{\partial\psi}{\partial x} \tag{2.70}$$

Physically ψ is a measure of the flow between streamlines. (Stream functions can be similarly defined to satisfy identically the continuity equations for incompressible cylindrical and spherical axisymmetric flows; and for these flows, as well as the above planar flow, also when they are compressible, but only then if they are steady.) Continuing with the planar case, we note that in such flows there is only a single nonzero component of vorticity, given by

$$\boldsymbol{\zeta} = (0, 0, \zeta_z) = \left(0, 0, \frac{\partial v}{\partial x} - \frac{\partial u}{\partial y}\right) \tag{2.71}$$

With Equation (2.70)

$$\zeta_z = -\frac{\partial^2 \psi}{\partial x^2} - \frac{\partial^2 \psi}{\partial y^2} = -\nabla^2 \psi \tag{2.72}$$

For this two-dimensional flow Equation (2.67) reduces to

$$\frac{\partial \zeta_z}{\partial t} + u\frac{\partial \zeta_z}{\partial x} + v\frac{\partial \zeta_z}{\partial y} = \nu\left(\frac{\partial^2 \zeta_z}{\partial x^2} + \frac{\partial^2 \zeta_z}{\partial y^2}\right) \tag{2.73}$$

Substitution of Equation (2.72) into Equation (2.73) yields an equation for the stream function

$$\frac{\partial(\nabla^2 \psi)}{\partial t} + \frac{\partial \psi}{\partial y}\frac{\partial(\nabla^2 \psi)}{\partial x} - \frac{\partial \psi}{\partial x}\frac{\partial}{\partial y}(\nabla^2 \psi) = \nu \nabla^4 \psi \tag{2.74}$$

where $\nabla^4 = \nabla^2(\nabla^2)$. For uniform flow past a solid body, for example, this equation for Ψ would be solved subject to the boundary conditions:

$$\begin{aligned} &\frac{\partial \psi}{\partial x} = 0, \quad \frac{\partial \psi}{\partial y} = V_\infty \quad \text{at infinity} \\ &\frac{\partial \psi}{\partial x} = 0, \quad \frac{\partial \psi}{\partial y} = 0 \quad \text{at the body (no-slip)} \end{aligned} \tag{2.75}$$

For the special case of irrotational flow it follows immediately from Equations (2.70) and (2.71) with $\zeta_z = 0$, that ψ satisfies the Laplace equation

$$\nabla^2 \psi = \frac{\partial^2 \psi}{\partial x^2} + \frac{\partial^2 \psi}{\partial y^2} = 0 \tag{2.76}$$

Inviscid Irrotational Flow: Potential Flow

For irrotational flows we have already noted that a velocity potential Φ can be defined such that $\boldsymbol{V} = \nabla\Phi$. If the flow is also incompressible, so $\nabla \cdot \boldsymbol{V} = 0$, it then follows that

$$\nabla \cdot (\nabla \Phi) = \nabla^2 \Phi = 0 \tag{2.77}$$

so Φ satisfies Laplace's equation. (Note that unlike the stream function ψ, which can only be defined for two-dimensional flows, the above considerations for Φ apply to flow in two and three dimensions. On the other hand, the existence of ψ does not require the flow to be irrotational, whereas the existence of Φ does.)

Since Equation (2.77) with appropriate conditions on $\boldsymbol{V}$ at boundaries of the flow completely determines the velocity field, and the momentum equation has played no role in this determination, we see that inviscid irrotational flow — *potential theory* — is a purely kinematic theory. The momentum equation only enters after Φ is known in order to calculate the pressure field consistent with the velocity field $\boldsymbol{V} = \nabla\Phi$.

For both two- and three-dimensional flows the determination of Φ makes use of the powerful techniques of potential theory, well developed in the mathematical literature. For two-dimensional planar flows the techniques of complex variable theory are available, since Φ may be considered as either the real or imaginary part of an analytic function (the same being true for ψ, since for such two-dimensional flows Φ and ψ are conjugate variables.)

Because the Laplace equation, obeyed by both Φ and ψ, is linear, complex flows may be built up from the superposition of simple flows; this property of inviscid irrotational flows underlies nearly all solution techniques in this area of fluid mechanics.

Problem

A two-dimensional inviscid irrotational flow has the velocity potential

$$\Phi = x^2 - y^2 \tag{P.2.6}$$

What two-dimensional potential flow does this represent?

Solution. It follows from Equations (2.61) and (2.70) that for two-dimensional flows, in general

$$u = \frac{\partial \Phi}{\partial x} = \frac{\partial \psi}{\partial y}, \quad v = \frac{\partial \Phi}{\partial y} = -\frac{\partial \psi}{\partial x} \tag{P.2.7}$$

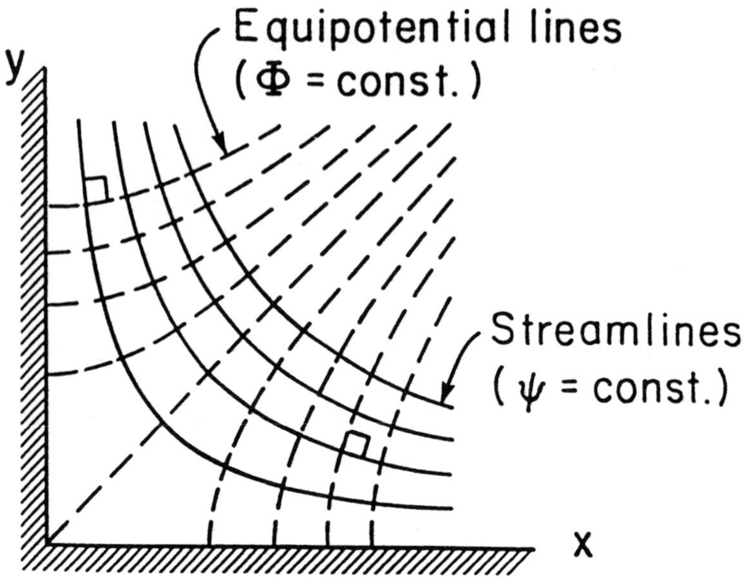

FIGURE 2.3 Potential flow in a 90° corner.

It follows from using Equation (P.2.6) that

$$u = \frac{\partial \psi}{\partial y} = 2x, \quad v = -\frac{\partial \psi}{\partial x} = -2y \tag{P.2.8}$$

Integration of Equation (P.2.8) yields

$$\psi = 2xy \tag{P.2.9}$$

The streamlines, ψ = constant, and equipotential lines, Φ = constant, both families of hyperbolas and each family the orthogonal trajectory of the other, are shown in Figure 2.3. Because the x and y axes are streamlines, Equations (P.2.6) and (P.2.9) represent the inviscid irrotational flow in a right-angle corner. By symmetry, they also represent the planar flow in the upper half-plane directed toward a stagnation point at $x = y = 0$ (see Figure 2.4). In polar coordinates (r, θ), with corresponding velocity components (u_r, u_θ), this flow is represented by

$$\Phi = r^2 \cos 2\theta, \quad \psi = r^2 \sin 2\theta \tag{P.2.10}$$

with

$$\begin{aligned} u_r &= \frac{\partial \Phi}{\partial r} = \frac{1}{r}\frac{\partial \psi}{\partial \theta} = 2r\cos 2\theta \\ u_\theta &= \frac{1}{r}\frac{\partial \Phi}{\partial \theta} = -\frac{\partial \psi}{\partial r} = -2r\sin 2\theta \end{aligned} \tag{P.2.11}$$

For two-dimensional planar potential flows we may also use complex variables, writing the complex potential $f(z) = \Phi + i\psi$ as a function of the complex variable $z = x + iy$, where the complex velocity is given by $f'(z) = w(z) = u - iv$. For the flow above

$$f(z) = z^2 \tag{P.2.12}$$

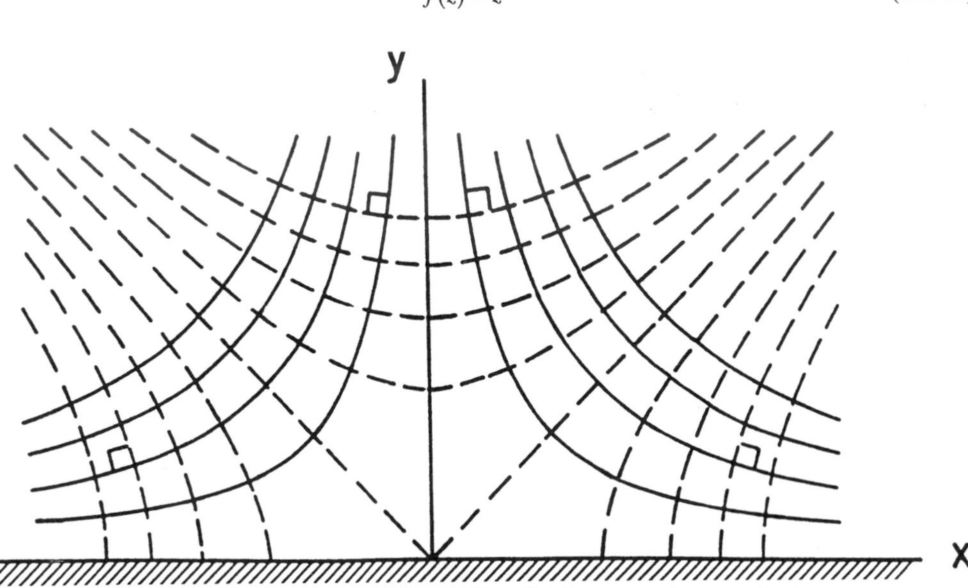

FIGURE 2.4 Potential flow impinging against a flat (180°) wall (plane stagnation-point flow).

Expressions such as Equation (P.2.12), where the right-hand side is an analytic function of z, may also be regarded as a conformal mapping, which makes available as an aid in solving two-dimensional potential problems all the tools of this branch of mathematics.

Further Information

More detail and additional information on the topics in this section may be found in more advanced books on fluid dynamics, such as

Batchelor, G.K. 1967. *An Introduction to Fluid Dynamics,* Cambridge University Press, Cambridge, England.

Warsi, Z.U.A. 1993. *Fluid Dynamics. Theoretical and Computational Approaches,* CRC Press, Boca Raton, FL.

Sherman, F.S. 1990. *Viscous Flow,* McGraw-Hill, New York.

Panton, R.L. 1984. *Incompressible Flow,* John Wiley & Sons, New York.

3

Similitude: Dimensional Analysis and Data Correlation

Stuart W. Churchill

University of Pennsylvania

3 Similitude: Dimensional Analysis and Data Correlation 29
Dimensional Analysis • Correlation of Experimental Data and Theoretical Values

3 Similitude: Dimensional Analysis and Data Correlation

Dimensional Analysis

Similitude refers to the formulation of a description for physical behavior that is general and independent of the individual dimensions, physical properties, forces, etc. In this subsection the treatment of similitude is restricted to *dimensional analysis*; for a more general treatment see Zlokarnik (1991). The full power and utility of dimensional analysis is often underestimated and underutilized by engineers. This technique may be applied to a complete mathematical model or to a simple listing of the variables that define the behavior. Only the latter application is described here. For a description of the application of dimensional analysis to a mathematical model see Hellums and Churchill (1964).

General Principles

Dimensional analysis is based on the principle that all additive or equated terms of a complete relationship between the variables must have the same net dimensions. The analysis starts with the preparation of a list of the individual dimensional variables (dependent, independent, and parametric) that are presumed to define the behavior of interest. The performance of dimensional analysis in this context is reasonably simple and straightforward; the principal difficulty and uncertainty arise from the identification of the variables to be included or excluded. If one or more important variables are inadvertently omitted, the reduced description achieved by dimensional analysis will be incomplete and inadequate as a guide for the correlation of a full range of experimental data or theoretical values. The familiar band of plotted

0-8493-0055-X/00/$0.00+$.50

values in many graphical correlations is more often a consequence of the omission of one or more variables than of inaccurate measurements. If, on the other hand, one or more irrelevant or unimportant variables are included in the listing, the consequently reduced description achieved by dimensional analysis will result in one or more unessential dimensionless groupings. Such excessive dimensionless groupings are generally less troublesome than missing ones because the redundancy will ordinarily be revealed by the process of correlation. Excessive groups may, however, suggest unnecessary experimental work or computations, or result in misleading correlations. For example, real experimental scatter may inadvertently and incorrectly be correlated in all or in part with the variance of the excessive grouping.

In consideration of the inherent uncertainty in selecting the appropriate variables for dimensional analysis, it is recommended that this process be interpreted as a *speculative* and subject to correction of the basis of experimental data or other information. Speculation may also be utilized as a formal technique to identify the effect of eliminating a variable or of combining two or more. The latter aspect of speculation, which may be applied either to the original listing of dimensional variables or to the resulting set of dimensionless groups, is often of great utility in identifying possible limiting behavior or dimensionless groups of marginal significance. The systematic speculative elimination of all but the most certain variables, one at a time, followed by regrouping, is recommended as a general practice. The additional effort as compared with the original dimensional analysis is minimal, but the possible return is very high. A general discussion of this process may be found in Churchill (1981).

The minimum number of independent dimensionless groups i required to describe the fundamental and parametric behavior is (Buckingham, 1914)

$$i = n - m \tag{3.1}$$

where n is the number of variables and m is the number of fundamental dimensions such as mass M, length L, time θ, and temperature T that are introduced by the variables. The inclusion of redundant dimensions such as force F and energy E that may be expressed in terms of mass, length, time, and temperature is at the expense of added complexity and is to be avoided. (Of course, mass could be replaced by force or temperature by energy as alternative fundamental dimensions.) In some rare cases i is actually greater than $n - m$. Then

$$i = n - k \tag{3.2}$$

where k is the maximum number of the chosen variables that cannot be combined to form a dimensionless group. Determination of the minimum number of dimensionless groups is helpful if the groups are to be chosen by inspection, but is unessential if the algebraic procedure described below is utilized to determine the groups themselves since the number is then obvious from the final result.

The *particular* minimal set of dimensionless groups is arbitrary in the sense that two or more of the groups may be multiplied together to any positive, negative, or fractional power as long as the number of independent groups is unchanged. For example, if the result of a dimensional analysis is

$$\phi\left\{XY^{1/2}, Z/Y^2, Z\right\} = 0 \tag{3.3}$$

where X, Y, and Z are independent dimensionless groups, an equally valid expression is

$$\phi\{X, Y, Z\} = 0 \tag{3.4}$$

Dimensional analysis itself does not provide any insight as to the best choice of equivalent dimensionless groupings, such as between those of Equations (3.3) and (3.4). However, isolation of each of the variables that are presumed to be the most important in a separate group may be convenient in terms of interpretation and correlation. Another possible criterion in choosing between alternative groupings may be the relative invariance of a particular one. The functional relationship provided by Equation (3.3) may equally well be expressed as

$$X = \phi\{Y, Z\} \tag{3.5}$$

where X is implied to be the dependent grouping and Y and Z to be independent or parametric groupings.

Three primary methods of determining a minimal set of dimensionless variables are (1) by inspection; (2) by combination of the residual variables, one at a time, with the set of chosen variables that cannot be combined to obtain a dimensionless group; and (3) by an algebraic procedure. These methods are illustrated in the examples that follow.

Example 3.1: Fully Developed Flow of Water Through a Smooth Round Pipe

Choice of Variables. The shear stress τ_w on the wall of the pipe may be postulated to be a function of the density ρ and the dynamic viscosity μ of the water, the inside diameter D of the pipe, and the space-mean of the time-mean velocity u_m. The limitation to fully developed flow is equivalent to a postulate of independence from distance x in the direction of flow, and the specification of a smooth pipe is equivalent to the postulate of independence from the roughness e of the wall. The choice of τ_w rather than the pressure drop per unit length $-dP/dx$ avoids the need to include the acceleration due to gravity g and the elevation z as variables. The choice of u_m rather than the volumetric rate of flow V, the mass rate of flow w, or the mass rate of flow per unit area G is arbitrary but has some important consequences as noted below. The postulated dependence may be expressed functionally as $\phi\{\tau_w, \rho, \mu, D, u_m\} = 0$ or $\tau_w = \phi\{\rho, \mu, D, u_m\}$.

Tabulation. Next prepare a tabular listing of the variables and their dimensions:

	τ_w	ρ	μ	D	u_m
M	1	1	1	0	0
L	–1	–3	–1	1	1
θ	–2	0	–1	0	–1
T	0	0	0	0	0

Minimal Number of Groups. The number of postulated variables is 5. Since the temperature does not occur as a dimension for any of the variables, the number of fundamental dimensions is 3. From Equation (3.1), the minimal number of dimensionless groups is $5 - 3 = 2$. From inspection of the above tabulation, a dimensionless group cannot be formed from as many as three variables such as D, μ, and ρ. Hence, Equation (3.2) also indicates that $i = 5 - 3 = 2$.

Method of Inspection. By inspection of the tabulation or by trial and error it is evident that only two independent dimensionless groups may be formed. One such set is

$$\phi\left\{\frac{\tau_w}{\rho u_m^2}, \frac{D u_m \rho}{\mu}\right\} = 0$$

Method of Combination. The residual variables τ_w and μ may be combined in turn with the noncombining variables ρ, D, and u_m to obtain two groups such as those above.

Algebraic Method. The algebraic method makes formal use of the postulate that the functional relationship between the variables may in general be represented by a power series. In this example such a power series may be expressed as

$$\tau_w = \sum_{i=1}^{N} A_i \rho^{a_i} \mu^{b_i} D^{c_i} u_m^{d_i}$$

where the coefficients A_i are dimensionless. Each additive term on the right-hand side of this expression must have the same net dimensions as τ_w. Hence, for the purposes of dimensional analysis, only the first term need be considered and the indexes may be dropped. The resulting highly restricted expression is $\tau_w = A\rho^a\mu^b D^c u_m^d$. Substituting the dimensions for the variables gives

$$\frac{M}{L\theta^2} = A\left(\frac{M}{L^3}\right)^a \left(\frac{M}{L\theta}\right)^b L^c \left(\frac{L}{\theta}\right)^d$$

Equating the sum of the exponents of M, L, and θ on the right-hand side of the above expression with those of the left-hand side produces the following three simultaneous linear algebraic equations: $1 = a + b$; $-1 = -3a - b + c + d$; and $-2 = -b - d$, which may be solved for a, c, and d in terms of b to obtain $a = 1 - b$, $c = -b$, and $d = 2 - b$. Substitution then gives $\tau_w = A\rho^{1-b}\mu^b D^{-b} u_m^{2-b}$ which may be regrouped as

$$\frac{\tau_w}{\rho u_m^2} = A\left(\frac{\mu}{D u_m \rho}\right)^b$$

Since this expression is only the first term of a power series, it should *not* be interpreted to imply that $\tau_w / \rho u_m^2$ is necessarily proportional to some power at $\mu/Du_m\rho$ but instead only the equivalent of the expression derived by the method of inspection. The inference of a power dependence between the dimensionless groups is the most common and serious error in the use of the algebraic method of dimensional analysis.

Speculative Reductions. Eliminating ρ as a variable on speculative grounds to

$$\phi\left\{\frac{\tau_w D}{\mu u_m}\right\} = 0$$

or its exact equivalent:

$$\frac{\tau_w D}{\mu u_m} = A$$

The latter expression with $A = 8$ is actually the exact solution for the laminar regime ($Du_m\rho/\mu < 1800$). A relationship that does not include ρ may alternatively be derived directly from the solution by the method of inspection as follows. First, ρ is eliminated from one group, say $\tau_w / \rho u_m^2$, by multiplying it with $Du_m\rho/\mu$ to obtain

$$\phi\left\{\frac{\tau_w D}{\mu u_m}, \frac{D u_m \rho}{\mu}\right\} = 0$$

The remaining group containing ρ is now simply dropped. Had the original expression been composed of three independent groups each containing ρ, that variable would have to be eliminated from two of them before dropping the third one.

The relationships that are obtained by the speculative elimination of μ, D, and u_m, one at a time, do not appear to have any range of physical validity. Furthermore, if w or G had been chosen as the independent variable rather than u_m, the limited relationship for the laminar regime would not have been obtained by the elimination of ρ.

Alternative Forms. The solution may also be expressed in an infinity of other forms such as

$$\phi\left\{\frac{\tau_w D^2 \rho}{\mu^2}, \frac{D u_m \rho}{\mu}\right\} = 0$$

If τ_w is considered to be the principal dependent variable and u_m the principal independent variable, this latter form is preferable in that these two quantities do not then appear in the same grouping. On the other hand, if D is considered to be the principal independent variable, the original formulation is preferable. The variance of $\tau_w / \rho u_m^2$ is less than that of $\tau_w D/\mu u_m$ and $\tau_w D^2\rho/\mu^2$ in the turbulent regime while that of $\tau_w D/\mu u_m$ is zero in the laminar regime. Such considerations may be important in devising convenient graphical correlations.

Alternative Notations. The several solutions above are more commonly expressed as

$$\phi\left\{\frac{f}{2}, \mathrm{Re}\right\} = 0$$

$$\phi\left\{\frac{f\,\mathrm{Re}}{2}, \mathrm{Re}\right\} = 0$$

or

$$\phi\left\{\frac{f\,\mathrm{Re}^2}{2}, \mathrm{Re}\right\} = 0$$

where $f = 2\,\tau_w / \rho u_m^2$ is the *Fanning friction factor* and $Re = D u_m \rho/\mu$ is the *Reynolds number.*

The more detailed forms, however, are to be preferred for purposes of interpretation or correlation because of the explicit appearance of the individual, physically measurable variables.

Addition of a Variable. The above results may readily be extended to incorporate the roughness e of the pipe as a variable. If two variables have the same dimensions, they will always appear as a dimensionless group in the form of a ratio, in this case e appears most simply as e/D. Thus, the solution becomes

$$\phi\left\{\frac{\tau_w}{\rho u_m^2}, \frac{D u_m \rho}{\mu}, \frac{e}{D}\right\} = 0$$

Surprisingly, as contrasted with the solution for a smooth pipe, the speculative elimination of μ and hence of the group $D u_m \rho/\mu$ now results in a valid asymptote for $D u_m \rho/\mu \rightarrow \infty$ and all finite values of e/D, namely,

$$\phi\left\{\frac{\tau_w}{\rho u_m^2}, \frac{e}{D}\right\} = 0$$

Example 3.2: Fully Developed Forced Convection in Fully Developed Flow in a Round Tube

It may be postulated for this process that $h = \phi\{D, u_m, \rho, \mu, k, c_p\}$, where here h is the local heat transfer coefficient, and c_p and k are the specific heat capacity and thermal conductivity, respectively, of the fluid. The corresponding tabulation is

	h	D	u_m	ρ	μ	k	c_p
M	1	0	0	1	1	1	0
L	0	1	1	−3	−1	1	2
θ	−3	0	−1	0	−1	−3	−2
T	−1	0	0	0	0	−1	−1

The number of variables is 7 and the number of independent dimensions is 4, as is the number of variables such as D, u_m, ρ, and k that cannot be combined to obtain a dimensionless group. Hence, the minimal number of dimensionless groups is 7 – 4 = 3. The following acceptable set of dimensionless groups may be derived by any of the procedures illustrated in Example 1:

$$\frac{hD}{k} = \phi\left\{\frac{Du_m\rho}{\mu}, \frac{c_p\mu}{k}\right\}$$

Speculative elimination of μ results in

$$\frac{hD}{k} = \phi\left\{\frac{Du_m\rho c_p}{k}\right\}$$

which has often erroneously been inferred to be a valid asymptote for $c_p\mu/k \to 0$. Speculative elimination of D, u_m, ρ, k, and c_p individually also does not appear to result in expressions with any physical validity. However, eliminating c_p and ρ or u_m gives a valid result for the laminar regime, namely,

$$\frac{hD}{k} = A$$

The general solutions for flow and convection in a smooth pipe may be combined to obtain

$$\frac{hD}{k} = \phi\left\{\frac{\tau_w D^2\rho}{\mu^2}, \frac{c_p\mu}{k}\right\}$$

which would have been obtained directly had u_m been replaced by τ_w in the original tabulation. This latter expression proves to be superior in terms of speculative reductions. Eliminating D results in

$$\frac{h\mu}{k(\tau_w\rho)^{1/2}} = \phi\left\{\frac{c_p\mu}{k}\right\}$$

which may be expressed in the more conventional form of

$$\mathrm{Nu} = \mathrm{Re}\left(\frac{f}{2}\right)^{1/2}\phi\{\mathrm{Pr}\}$$

where Nu = hD/k is the *Nusselt number* and Pr = $c_p\mu/k$ is the *Prandtl number.* This result appears to be a valid asymptote for Re $\rightarrow \infty$ and a good approximation for even moderate values (>5000) for large values of Pr. Elimination of μ as well as D results in

$$\frac{h}{c_p(\tau_w\rho)^{1/2}} = A$$

or

$$\mathrm{Nu} = A\,\mathrm{Re}\,\mathrm{Pr}\left(\frac{f}{2}\right)^{1/2}$$

which appears to be an approximate asymptote for Re $\rightarrow \infty$ and Pr $\rightarrow 0$. Elimination of both c_p and ρ again yields the appropriate result for laminar flow, indicating that ρ rather than u_m is the meaningful variable to eliminate in this respect.

The numerical value of the coefficient A in the several expressions above depends on the mode of heating, a true variable, but one from which the purely functional expressions are independent. If j_w the heat flux density at the wall, and $T_w - T_m$, the temperature difference between the wall and the bulk of the fluid, were introduced as variables in place of $h \equiv j_w/(T_w - T_m)$, another group such as $c_p(T_w - T_m)\,(D\rho/\mu)^2$ or $\rho c_p(T_w - T_m)/\tau_w$ or $c_p(T_w - T_m)/u_m^2$, which represents the effect of viscous dissipation, would be obtained. This effect is usually but not always negligible. (See Chapter 4.)

Example 3.3: Free Convection from a Vertical Isothermal Plate

The behavior for this process may be postulated to be represented by

$$h = \phi\{g,\beta,T_w - T_\infty,x,\mu,\rho,c_p,k\}$$

where g is the acceleration due to gravity, β is the volumetric coefficient of expansion with temperature, T_∞ is the unperturbed temperature of the fluid, and x is the vertical distance along the plate. The corresponding tabulation is

	h	g	β	$T_w - T_\infty$	x	μ	ρ	c_p	k
M	1	0	0	0	0	1	1	0	1
L	0	1	0	0	1	–1	–3	2	1
θ	–3	–2	0	0	0	–1	0	–2	–3
T	–1	0	–1	1	0	0	0	–1	1

The minimal number of dimensionless groups indicated by both methods is 9 – 4 = 5. A satisfactory set of dimensionless groups, as found by any of the methods illustrated in Example 1 is

$$\frac{hx}{k} = \phi\left\{\frac{\rho^2 g x^3}{\mu^2},\frac{c_p\mu}{k},\beta(T_w - T_\infty),c_p(T_w - T_\infty)\left(\frac{\rho x}{\mu}\right)^2\right\}$$

It may be reasoned that the buoyant force which generates the convective motion must be proportional to $\rho g\beta(T_w - T_\infty)$, thus, g in the first term on the right-hand side must be multiplied by $\beta(T_w - T_\infty)$, resulting in

$$\frac{hx}{k} = \phi\left\{\frac{\rho^2 g\beta(T_w - T_\infty)x^3}{\mu^2},\frac{c_p\mu}{k},\beta(T_w - T_\infty),c_p(T_w - T_\infty)\left(\frac{\rho x}{\mu}\right)^2\right\}$$

The effect of expansion other than on the buoyancy is now represented by $\beta(T_w - T_\infty)$, and the effect of viscous dissipation by $c_p(T_w - T_\infty)(\rho x/\mu)^2$. Both effects are negligible for all practical circumstances. Hence, this expression may be reduced to

$$\frac{hx}{k} = \phi\left\{\frac{\rho^2 g\beta(T_w - T_\infty)x^3}{\mu^2}, \frac{c_p\mu}{k}\right\}$$

or

$$\mathrm{Nu}_x = \phi\{\mathrm{Gr}_x, \mathrm{Pr}\}$$

where $\mathrm{Nu}_x = hx/k$ and $\mathrm{Gr}_x = \rho^2 g\beta(T_w - T_\infty)x^3/\mu^2$ is the *Grashof number.*

Elimination of x speculatively now results in

$$\frac{hx}{k} = \left(\frac{\rho^2 g\beta(T_w - T_\infty)x^3}{\mu^2}\right)^{1/3}\phi\{\mathrm{Pr}\}$$

or

$$\mathrm{Nu}_x = \mathrm{Gr}_x^{1/3}\phi\{\mathrm{Pr}\}$$

This expression appears to be a valid asymptote for $\mathrm{Gr}_x \to \infty$ and a good approximation for the entire turbulent regime. Eliminating μ speculatively rather than x results in

$$\frac{hx}{k} = \phi\left\{\frac{\rho^2 c_p^2 g\beta(T_w - T_\infty)x^3}{k^2}\right\}$$

or

$$\mathrm{Nu}_x = \phi\{\mathrm{Gr}_x \mathrm{Pr}^2\}$$

The latter expression appears to be a valid asymptote for $\mathrm{Pr} \to 0$ for all Gr_x, that is, for both the laminar and the turbulent regimes. The development of a valid asymptote for large values of Pr requires more subtle reasoning. First $c_p\mu/k$ is rewritten as $\mu/\rho\alpha$ where $\alpha = k/\rho c_p$. Then ρ is eliminated speculatively except as it occurs in $\rho g\beta(T_w - T_\infty)$ and $k/\rho c_p$. The result is

$$\frac{hx}{k} = \phi\left\{\frac{c_p\rho^2 g\beta(T_w - T_\infty)x^3}{\mu k}\right\}$$

or

$$\mathrm{Nu}_x = \phi\{\mathrm{Ra}_x\}$$

where

$$\text{Ra}_x = \frac{c_p \rho^2 g \beta (T_w - T_\infty) x^3}{\mu k} = \text{Gr}_x \text{Pr}$$

is the *Rayleigh number.* The expression appears to be a valid asymptote for $\text{Pr} \to \infty$ and a reasonable approximation for even moderate values of Pr for all Gr_x, that is, for both the laminar and the turbulent regimes.

Eliminating x speculatively from the above expressions for small and large values of Pr results in

$$\text{Nu}_x = A\left(\text{Gr}_x \text{Pr}^2\right)^{1/3} = A\left(\text{Ra}_x \text{Pr}\right)^{1/3}$$

and

$$\text{Nu}_x = B\left(\text{Gr}_x \text{Pr}\right)^{1/3} = B\left(\text{Ra}_x\right)^{1/3}$$

The former appears to be a valid asymptote for $\text{Pr} \to 0$ and $\text{Gr}_x \to \infty$ and a reasonable approximation for very small values of Pr in the turbulent regime, while the latter is well confirmed as a valid asymptote for $\text{Pr} \to \infty$ and $\text{Gr}_x \to \infty$ and as a good approximation for moderate and large values of Pr over the entire turbulent regime. The expressions in terms of Gr_x are somewhat more complicated than those in terms of Ra_x, but are to be preferred since Gr_x is known to characterize the transition from laminar to turbulent motion in natural convection just as Re_D does in forced flow in a channel. The power of speculation combined with dimensional analysis is well demonstrated by this example in which valid asymptotes are thereby attained for several regimes.

Correlation of Experimental Data and Theoretical Values

Correlations of experimental data are generally developed in terms of dimensionless groups rather than in terms of the separate dimensional variables in the interests of compactness and in the hope of greater generality. For example, a complete set of graphical correlations for the heat transfer coefficient h of Example 3.2 above in terms of each of the six individual independent variables and physical properties might approach book length, whereas the dimensionless groupings both imply that a single plot with one parameter should be sufficient. Furthermore, the reduced expression for the turbulent regime implies that a plot of $\text{Nu}/\text{Re}\, f^{1/2}$ vs. Pr should demonstrate only a slight parametric dependence on Re or $\text{Re}\, f^{1/2}$. Of course, the availability of a separate correlation for f as a function of Re is implied.

Theoretical values, that is, ones obtained by numerical solution of a mathematical model in terms of either dimensional variables or dimensionless groups, are presumably free from imprecision. Even so, because of their discrete form, the construction of a correlation or correlations for such values may be essential for the same reasons as for experimental data.

Graphical correlations have the merit of revealing general trends, of providing a basis for evaluation of the choice of coordinates, and most of all of displaying visually the scatter of the individual experimental values about a curve representing a correlation or their behavior on the mean. (As mentioned in the previous subsection, the omission of a variable may give the false impression of experimental error in such a plot.) On the other hand, correlating equations are far more convenient as an input to a computer than is a graphical correlation. These two formats thus have distinct and complementary roles; both should generally be utilized. The merits and demerits of various graphical forms of correlations are discussed in detail by Churchill (1979), while the use of logarithmic and arithmetic coordinates, the effects of the appearance of a variable in both coordinates, and the effects of the distribution of error between the dependent and independent variable are further illustrated by Wilkie (1985).

Churchill and Usagi (1972; 1974) proposed general usage of the following expression for the formulation of correlating equations:

$$y^n\{x\} = y_0^n\{x\} + y_\infty^n\{x\} \tag{3.6}$$

where $y_o\{x\}$ and $y_\infty\{x\}$ denote asymptotes for small and large values of x, respectively, and n is an arbitrary exponent. For convenience and simplicity, Equation (3.6) may be rearranged in either of the following two forms:

$$(Y(x))^n = 1 + Z^n\{x\} \tag{3.7}$$

or

$$\left(\frac{Y\{x\}}{Z\{x\}}\right)^n = 1 + \frac{1}{Z^n\{x\}} \tag{3.8}$$

where $Y\{x\} \equiv y\{x\}/y_o\{x\}$ and $Z\{x\} \equiv y_\infty\{x\}/y_o\{x\}$. Equations (3.6), (3.7), and (3.9) are hereafter denoted collectively as the CUE (Churchill–Usagi equation). The principle merits of the CUE as a canonical expression for correlation are its simple form, generality, and minimal degree of explicit empiricism, namely, only that of the exponent n, since the asymptotes $y_o\{x\}$ and $y_\infty\{x\}$ are ordinarily known in advance from theoretical considerations or well-established correlations. Furthermore, as will be shown, the CUE is quite insensitive to the numerical value of n. Although the CUE is itself very simple in form, it is remarkably successful in representing closely very complex behavior, even including the dependence on secondary variables and parameters, by virtue of the introduction of such dependencies through $y_o\{x\}$ and $y_\infty\{x\}$. In the rare instances in which such dependencies are not represented in the asymptotes, n may be correlated as a function of the secondary variables and/or parameters. Although the CUE usually produces very close representations, it is empirical and not exact. In a few instances, numerical values of n have been derived or rationalized on theoretical grounds, but even then some degree of approximation is involved. Furthermore, the construction of a correlating expression in terms of the CUE is subject to the following severe limitations:

1. The asymptotes $y_o\{x\}$ and $y_\infty\{x\}$ must intersect once and only once;
2. The asymptotes $y_o\{x\}$ and $y_\infty\{x\}$ must be free of singularities. Even though a singularity occurs beyond the asserted range of the asymptote, it will persist and disrupt the prediction of the CUE, which is intended to encompass all values of the independent variable x; and
3. The asymptotes must both be upper or lower bounds.

In order to avoid or counter these limitations it may be necessary to modify or replace the asymptotes with others. Examples of this process are provided below. A different choice for the dependent variable may be an option in this respect. The suitable asymptotes for use in Equation (3.6) may not exist in the literature and therefore may need to be devised or constructed. See, for example, Churchill (1988b) for guidance in this respect. Integrals and derivatives of the CUE are generally awkward and inaccurate, and may include singularities not present or troublesome in the CUE itself. It is almost always preferable to develop a separate correlating equation for such quantities using derivatives or integrals of $y_o\{x\}$ and $y_\infty\{x\}$, simplified or modified as appropriate.

The Evaluation of *n*

Equation (3.6) may be rearranged as

$$n = \frac{\ln\left\{1+\left(\frac{y_\infty\{x\}}{y_0\{x\}}\right)^n\right\}}{\ln\left\{\frac{y\{x\}}{y_0\{x\}}\right\}} \tag{3.9}$$

and solved for n by iteration for any known value of $y\{x\}$, presuming that $y_o\{x\}$ and $y_\infty\{x\}$ are known. If $y\{x^*\}$ is known, where x^* represents the value of x at the point of intersection of the asymptotes, that is, for $y_o\{x\} = y_\infty\{x\}$, Equation (3.9) reduces to

$$n = \frac{\ln\{2\}}{\ln\left\{\frac{y\{x^*\}}{y_0\{x^*\}}\right\}} \tag{3.10}$$

and iterative determination of n is unnecessary.

A graphical and visual method of evaluation of n is illustrated in Figure 3.1 in which $Y\{Z\}$ is plotted vs. Z for $0 \le Z \le 1$ and $Y\{Z\}/Z$ vs. $1/Z$ for $0 \le 1/Z \le 1$ in arithmetic coordinates with n as a parameter. Values of $y\{x\}$ may be plotted in this form and the best overall value of n selected visually (as illustrated in Figure 3.2). A logarithmic plot of $Y\{Z\}$ vs. Z would have less sensitivity relative to the dependence on n. (See, for example, Figure 1 of Churchill and Usagi, 1972.) Figure 3.1 explains in part the success of the CUE. Although y and x may both vary from 0 to ∞, the composite variables plotted in Figure 3.1 are highly constrained in that the compound independent variables Z and $1/Z$ vary only between 0 and 1, while for $n \ge 1$, the compound dependent variables $Y\{Z\}$ and $Y\{Z\}/Z$ vary only from 1 to 2.

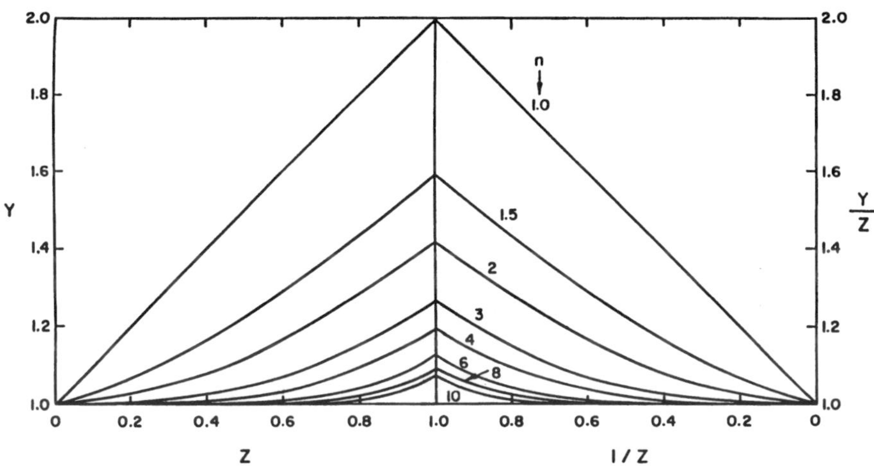

FIGURE 3.1 Arithmetic, split-coordinate plot of Equation 3.10. (From Churchill, S.W. and Usagi, R. *AIChE J.* 18(6), 1123, 1972. With permission from the American Institute of Chemical Engineers.)

Because of the relative insensitivity of the CUE to the numerical value of n, an integer or a ratio of two small integers may be chosen in the interest of simplicity and without significant loss of accuracy. For example, the maximum variance in Y (for $0 \le Z \le 1$) occurs at $Z = 1$ and increases only $100(2^{1/20} - 1) = 3.5\%$ if n is decreased from 5 to 4. If $y_o\{x\}$ and $y_\infty\{x\}$ are both lower bounds, n will be positive, and if they are both upper bounds, n will be negative. To avoid extending Figure 3.1 for negative values of n, $1/y\{x\}$ may simply be interpreted as the dependent variable.

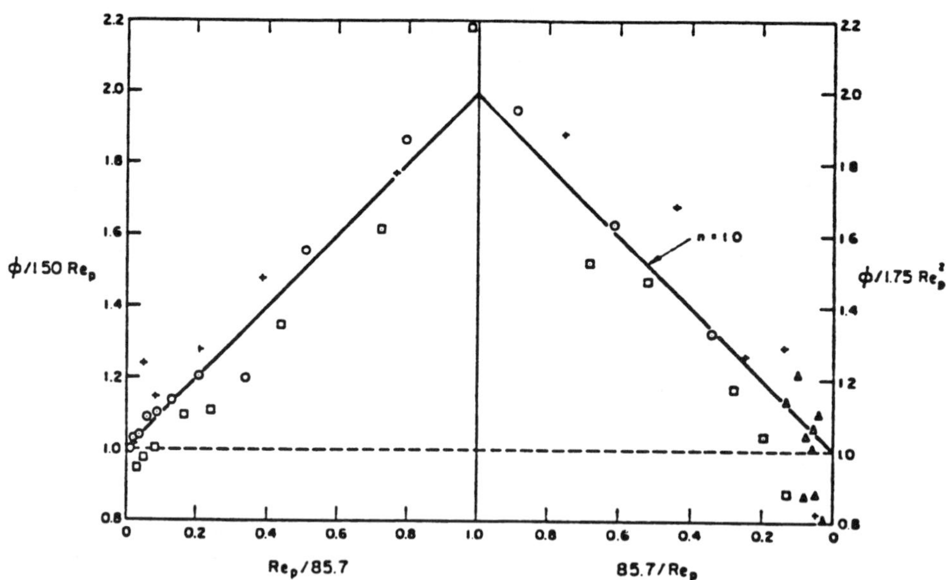

FIGURE 3.2 Arithmetic, split-coordinate plot of computed values and experimental data for laminar free convection from an isothermal vertical plate. (From Churchill, S.W. and Usagi, R. *AIChE J.* 18(6), 1124, 1972. With permission from the American Institute of Chemical Engineers.)

Intermediate Regimes

Equations (3.6), (3.7), and (3.8) imply a slow, smooth transition between $y_o\{x\}$ and $y_\infty\{x\}$ and, moreover, one that is symmetrical with respect to $x^*(Z = 1)$. Many physical systems demonstrate instead a relatively abrupt transition, as for example from laminar to turbulent flow in a channel or along a flat plate. The CUE may be applied serially as follows to represent such behavior if an expression $y_i\{x\}$ is postulated for the intermediate regime. First, the transition from the initial to the intermediate regime is represented by

$$y_1^n = y_0^n + y_i^n \tag{3.11}$$

Then the transition from this combined regime to the final regime by

$$y^m = y_1^m + y_\infty^m = \left(y_0^n + y_i^n\right)^{m/n} + y_\infty^m \tag{3.12}$$

Here, and throughout the balance of this subsection, in the interests of simplicity and clarity, the functional dependence of all the terms on x is implied rather written out explicitly. If y_o is a lower bound and y_i is implied to be one, y_1 and y_∞ must be upper bounds. Hence, n will then be positive and m negative. If y_o and y_i are upper bounds, y_1 and y_∞ must be lower bounds; then n will be negative and m positive. The reverse formulation starting with y_∞ and y_1 leads by the same procedure to

$$y^n = y_0^n + \left(y_i^m + y_\infty^m\right)^{n/m} \tag{3.13}$$

If the intersections of y_i with y_o and y_∞ are widely separated with respect to x, essentially the same pair of values for n and m will be determined for Equations (3.12) and (3.13), and the two representations for y will not differ significantly. On the other hand, if these intersections are close in terms of x, the pair of values of m and n may differ significantly and one representation may be quite superior to the other. In some instances a singularity in y_o or y_∞ may be tolerable in either Equation (3.12) or (3.13) because it is overwhelmed by the other terms. Equations (3.12) and (3.13) have one hidden flaw. For $x \to 0$, Equation (3.12) reduces to

$$y \to y_0 \left[1 + \left(\frac{y_\infty}{y_0}\right)^m\right]^{1/m} \tag{3.14}$$

If y_o is a lower bound, m is necessarily negative, and values of y less than y_o are predicted. If y_o/y_∞ is sufficiently small or if m is sufficiently large in magnitude, this discrepancy may be tolerable. If not, the following alternative expression may be formulated, again starting from Equation (3.11):

$$\left(y^n - y_0^n\right)^m = y_i^{nm} + \left(y_\infty^n - y_0^n\right)^m \tag{3.15}$$

Equation (3.15) is free from the flaw identified by means of Equation (3.14) and invokes no additional empiricism, but a singularity may occur at $y_\infty = y_o$, depending on the juxtapositions of y_o, y_i, and y_∞. Similar anomalies occur for Equation (3.13) and the corresponding analog of Equation (3.14), as well as for behavior for which $n < 0$ and $m > 0$. The preferable form among these four is best chosen by trying each of them.

One other problem with the application of the CUE for a separate transitional regime is the formulation of an expression for $y_i\{x\}$, which is ordinarily not known from theoretical considerations. Illustrations of the empirical determination of such expressions for particular cases may be found in Churchill and Usagi (1974), Churchill and Churchill (1975), and Churchill (1976; 1977), as well as in Example 3.5 below.

Example 3.4: The Pressure Gradient in Flow through a Packed Bed of Spheres

The pressure gradient at asymptotically low rates of flow (the creeping regime) can be represented by the Kozeny–Carman equation, $\Phi = 150\ \mathrm{Re}_p$, and at asymptotically high rates of flow (the inertial regime) by the Burke–Plummer equation, $\Phi = 1.75\ (\mathrm{Re}_p)^2$, where $\Phi = \rho\varepsilon^2 d_p(-dP_f/dx)\mu^2(1-\varepsilon)$, $\mathrm{Re}_p = d_p u_o \rho/\mu(1-\varepsilon)$, d_p = diameter of spherical particles, m, ε = void fraction of bed of spheres, dP_f/dx = dynamic pressure gradient (due to friction), Pa/m, and u_o = superficial velocity (in absence of the spheres), m/sec. For the origin of these two asymptotic expressions see Churchill (1988a). They both have a theoretical structure, but the numerical coefficients of 150 and 1.75 are basically empirical. These equations are both lower bounds and have one intersection. Experimental data are plotted in Figure 3.3, which has the form of Figure 3.1 with $Y = \Phi/150\ \mathrm{Re}_p$, $Y/Z = \Phi/(1.75\ \mathrm{Re}_p)^2$ and $Z = 1.75\ \mathrm{Re}_p^2/150\ \mathrm{Re}_p = \mathrm{Re}_p/85.7$. A value of $n = 1$ is seen to represent these data reasonably well on the mean, resulting in

$$\Phi = 150\mathrm{Re}_p + 1.75\left(\mathrm{Re}_p\right)^2$$

which was originally proposed as a correlating equation by Ergun (1952) on the conjecture that the volumetric fraction of the bed in "turbulent" flow is proportional to Re_p. The success of this expression in conventional coordinates is shown in Figure 3.4. The scatter, which is quite evident in the arithmetic split coordinates of Figure 3.3, is strongly suppressed in a visual sense in the logarithmic coordinates of Figure 3.4.

Example 3.5: The Friction Factor for Commercial Pipes for All Conditions

The serial application of the CUE is illustrated here by the construction of a correlating equation for both smooth and rough pipes in the turbulent regime followed by combination of that expression with ones for the laminar and transitional regimes.

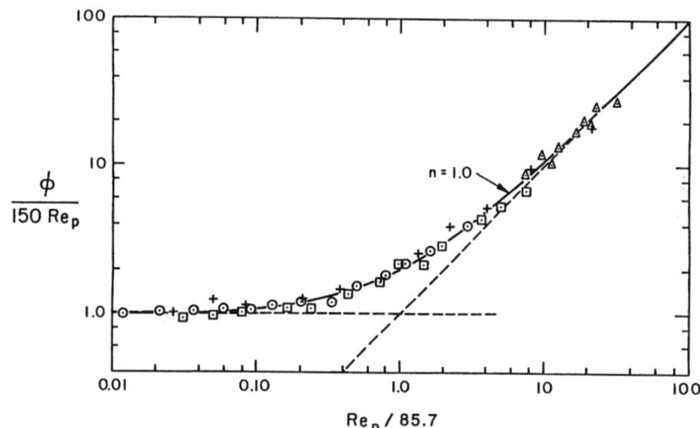

FIGURE 3.3 Arithmetic, split-coordinate plot of experimental data for the pressure drop in flow through a packed bed of spheres. (From Churchill, S.W. and Usagi, R. *AIChE J.* 18(6), 1123, 1972. With permission from the American Institute of Chemical Engineers.)

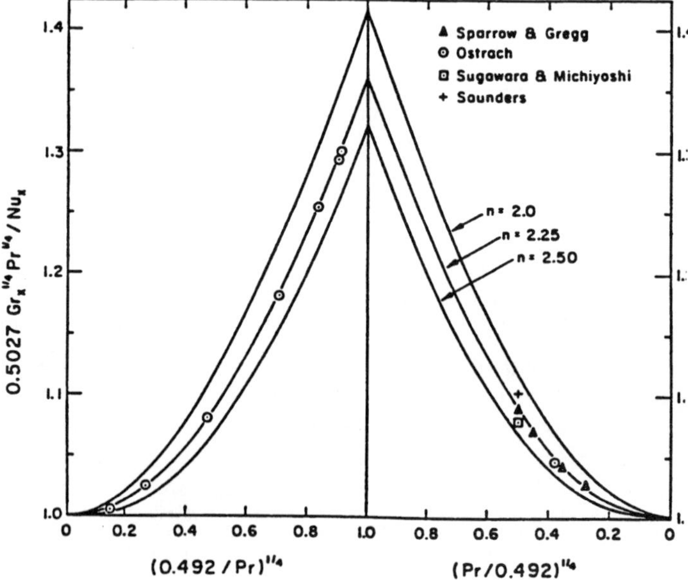

FIGURE 3.4 Logarithmic correlation of experimental data for the pressure drop in flow through a packed bed of spheres. (From Churchill, S.W. and Usagi, R. *AIChE J.* 18(6), 1123, 1972. With permission from the American Institute of Chemical Engineers.)

The Turbulent Regime. The Fanning friction factor, f_F, for turbulent flow in a smooth round pipe for asymptotically large rates of flow (say $Re_D > 5000$) may be represented closely by the empirical expression:

$$\left(\frac{2}{f_F}\right)^{1/2} = 0.256 + 2.5 \ln\left\{\left(\frac{f_F}{2}\right)^{1/2} Re_D\right\}$$

A corresponding empirical representation for naturally rough pipe is

$$\left(\frac{2}{f_F}\right)^{1/2} = 3.26 + 2.5\ln\left\{\frac{D}{e}\right\}$$

Direct combination of these two expressions in the form of the CUE does not produce a satisfactory correlating equation, but their combination in the following rearranged forms:

$$e^{(1/2.5)(2/f_F)^{1/2}} = 1.108\left(\frac{f_F}{2}\right)^{1/2}\mathrm{Re}_L$$

and

$$e^{(1/2.5)(2/f_F)^{1/2}} = 3.68\left(\frac{D}{e}\right)$$

with $n = -1$ results in, after the reverse rearrangement,

$$\left(\frac{2}{f_F}\right)^{1/2} = 0.256 + 2.5\ln\left\{\frac{\left(\frac{f_F}{2}\right)^{1/2}\mathrm{Re}_D}{1 + 0.3012\left(\frac{e}{D}\right)\left(\frac{f_F}{2}\right)^{1/2}\mathrm{Re}_D}\right\}$$

The exact equivalent of this expression in structure but with the slightly modified numerical coefficients of 0.300, 2.46, and 0.304 was postulated by Colebrook (1938–1939) to represent his own experimental data. The coefficients of the expression given here are presumed to be more accurate, but the difference in the predictions of f_F with the two sets of coefficients is within the band of uncertainty of the experimental data. The turbulent regime of the "friction-factor" plot in most current textbooks and handbooks is simply a graphical representation of the Colebrook equation. Experimental values are not included in such plots since e, the effective roughness of commercial pipes, is simply a correlating factor that forces agreement with the Colebrook equation. Values of e for various types of pipe in various services are usually provided in an accompanying table, that thereby constitutes an integral part of the correlation.

The Laminar Region. The Fanning friction factor in the laminar regime of a round pipe ($\mathrm{Re}_d < 1800$) is represented exactly by the following theoretical expression known as Poiseuille's law: $f_F = 16/\mathrm{Re}_D$. This equation may be rearranged as follows for convenience in combination with that for turbulent flow:

$$\left(\frac{2}{f_F}\right)^{1/2} = \frac{\mathrm{Re}_D(f_F/2)^{1/2}}{8}$$

The Transitional Regime. Experimental data as well as semitheoretical computed values for the limiting behavior in the transition may be represented closely by $(f_F/2) = (\mathrm{Re}_D/37500)^2$. This expression may be rewritten, in terms of $(2/f_F)^{1/2}$ and $\mathrm{Re}_D(f_F/2)^{1/2}$, as follows:

$$\left(\frac{f_F}{2}\right)^{1/2} = \left(\frac{37500}{\mathrm{Re}_D(f_F/2)^{1/2}}\right)^{1/2}$$

Overall Correlation. The following correlating equation for all $\mathrm{Re}_D(f_F/2)^{1/2}$ and e/D may now be constructed by the combination of the expressions for the turbulent and transition regimes in the form of the CUE with $n = 8$, and then that expression and that for the laminar regime with $n = -12$, both components being chosen on the basis of experimental data and predicted values for the full regime of transition:

$$\left(\frac{2}{f_F}\right)^{1/2} = \left[\left(\frac{8}{\mathrm{Re}_D(f_F/2)^{1/2}}\right)^{12} + \left(\left[\frac{37500}{\mathrm{Re}_D(f_F/2)^{1/2}}\right]^4 + \left|2.5\ln\left\{\frac{1.108\,\mathrm{Re}_D(f_F/2)^{1/2}}{1+0.3012\left(\frac{e}{a}\right)\mathrm{Re}_D(f_F/2)^{1/2}}\right\}\right|^8\right)^{-3/2}\right]^{-1/12}$$

The absolute value signs are only included for aesthetic reasons; the negative values of the logarithmic term for very small values of $\mathrm{Re}_D(f_F/2)^{1/2}$ do not affect the numerical value of $(2/f_F)^{1/2}$ in the regime in which they occur. This overall expression appears to have a complicated structure, but it may readily be recognized to reduce to its component parts when the corresponding term is large with respect to the other two. It is insensitive to the numerical values of the two arbitrary exponents. For example, doubling their values would have almost no effect on the predictions of $(f_F/2)^{1/2}$. The principal uncertainty is associated with the expression for the transition regime, but the overall effect of the corresponding term is very small. The uncertainties associated with this correlating equation are common to most graphical correlations and algebraic expressions for the friction factor, and are presumed to be fairly limited in magnitude and to be associated primarily with the postulated value of e. Although the overall expression is explicit in $\mathrm{Re}_D(f_F/2)^{1/2}$ rather than Re_D, the latter quantity may readily be obtained simply by multiplying the postulated value of $\mathrm{Re}_D(f_F/2)^{1/2}$ by the computed values of $(2/f_F)^{1/2}$.

References

Buckingham, E. 1914. On physically similar systems; illustrations of the use of dimensional equations. *Phys. Rev., Ser. 2,* 4(4):345–375.

Churchill, S.W. 1976. A comprehensive correlating equation for forced convection from plates. *AIChE J.* 22(2):264–268.

Churchill, S.W. 1977. Comprehensive correlating equation for heat, mass and momentum transfer in fully developed flow in smooth tubes. *Ind. Eng. Chem. Fundam.* 16(1):109–116.

Churchill, S.W. 1979. *The Interpretation and Use of Rate Data. The Rate Process Concept,* rev. printing, Hemisphere Publishing Corp., Washington, D.C.

Churchill, S.W. 1981. The use of speculation and analysis in the development of correlations. *Chem. Eng. Commun.* 9:19–38.

Churchill, S.W. 1988a. Flow through porous media, Chapter 19 in *Laminar Flows. The Practical Use of Theory,* pp. 501–538, Butterworths, Boston.

Churchill, S.W. 1988b. Derivation, selection, evaluation and use of asymptotes. *Chem. Eng. Technol.* 11:63–72.

Churchill, S.W. and Churchill, R.U. 1975. A general model for the effective viscosity of pseudoplastic and dilatant fluids. *Rheol. Acta.* 14:404–409.

Churchill, S.W. and Usagi, R. 1972. A general expression for the correlation of rates of transfer and other phenomena. *AIChE J.* 18(6):1121–1128.

Churchill, S.W. and Usagi, R. 1974. A standardized procedure for the production of correlations in the form of a common empirical equation. *Ind. Eng. Chem. Fundam.* 13(1):39–44.

Colebrook, C.R. 1938–1939. Turbulent flow in pipes with particular reference to the transition region between the smooth and rough pipe laws. *J. Inst. Civ. Eng.* 11(5024):133–156.

Ergun, S. 1952. Fluid flow through packed beds. *Chem. Eng. Prog.* 48(2):81–96.

Hellums, J.D. and Churchill, S.W. 1964. Simplifications of the mathematical description of boundary and initial value problems. *AIChE J.* 10(1):110–114.

Wilkie, D. 1985. The correlation of engineering data reconsidered. *Int. J. Heat Fluid Flow.* 8(2):99–103.

Zlokarnik, M. 1991. *Dimensional Analysis and Scale-Up in Chemical Engineering.* Springer-Verlag, Berlin.

4

Hydraulics of Pipe Systems

J. Paul Tullis

Utah State University

4 Hydraulics of Pipe Systems 47
Basic Computations • Pipe Design • Valve Selection • Pump Selection • Other Considerations

Hydraulics of Pipe Systems

Basic Computations

Equations

Solving fluid flow problems involves the application of one or more of the three basic equations: continuity, momentum, and energy. These three basic tools are developed from the law of conservation of mass, Newton's second law of motion, and the first law of thermodynamics.

The simplest form of the continuity equation is for one-dimensional incompressible steady flow in a conduit. Applying continuity between any two sections gives

$$A_1 V_1 = A_2 V_2 = Q \tag{4.1}$$

For a variable density the equation can be written

$$\rho_1 A_1 V_1 = \rho_2 A_2 V_2 = \dot{m} \tag{4.2}$$

in which A is the cross-sectional area of the pipe, V is the mean velocity at that same location, Q is the flow rate, ρ is the fluid density, and $\dot{m}$ is the mass flow rate. The equations are valid for any rigid conduit as long as there is no addition or loss of liquid between the sections.

For steady state pipe flow, the momentum equation relates the net force in a given direction (F_x) acting on a control volume (a section of the fluid inside the pipe), to the net momentum flux through the control volume.

0-8493-0055-X/00/$0.00+$.50

$$F_x = \rho_2 A_2 V_2 V_{2x} - \rho_1 A_1 V_1 V_{1x} \tag{4.3}$$

For incompressible flow this equation can be reduced to

$$F_x = \rho Q\left(V_{2x} - V_{1x}\right) \tag{4.4.}$$

These equations can easily be applied to a three-dimensional flow problem by adding equations in the y and z directions.

A general form of the energy equation applicable to incompressible pipe or duct flow

$$\frac{P_1}{\gamma} + Z_1 + \frac{V_1^2}{2g} = \frac{P_2}{\gamma} + Z_2 + \frac{V_2^2}{2g} - H_p + H_t + H_f \tag{4.5}$$

The units are energy per unit weight of liquid: ft · lb/lb or N · m/N which reduce to ft or m. The first three terms are pressure head (P/γ), elevation head (Z) (above some datum), and velocity head ($V^2/2g$). The last three terms on the right side of the equation are the total dynamic head added by a pump (H_p) or removed by a turbine (H_t) and the friction plus minor head losses (H_f). The sum of the first three terms in Equation 4.5 is defined as the total head, and the sum of the pressure and elevation heads is referred to as the piezometric head.

The purpose of this section is to determine the pressure changes resulting from incompressible flow in pipe systems. Since pipes of circular cross sections are most common in engineering application, the analysis in this section will be performed for circular geometry. However, the results can be generalized for a pipe of noncircular geometry by substituting for the diameter D in any of the equations, the hydraulic diameter, D_h, defined as

$$D_h = 4 \times \frac{\text{the cross sectional area}}{\text{the wetted perimeter}}$$

The analysis in this section can also be applied to gases and vapors, provided the Mach number in the duct does not exceed 0.3. For greater values of the Mach number, the compressibility effect becomes significant and the reader is referred to Section 7 on compressible flow.

Fluid Friction

The calculation of friction loss in pipes and ducts depends on whether the flow is laminar or turbulent. The Reynolds number is the ratio of inertia forces to viscous forces and is a convenient parameter for predicting if a flow condition will be laminar or turbulent. It is defined as

$$\mathrm{Re}_D = \frac{\rho V D}{\mu} = \frac{VD}{\nu} \tag{4.6}$$

in which V is the mean flow velocity, D diameter, ρ fluid density, μ dynamic viscosity, and ν kinematic viscosity.

Friction loss (H_f) depends on pipe diameter (d), length (L), roughness (e), fluid density (ρ) or specific weight (γ), viscosity (ν), and flow velocity (V). Dimensional analysis can be used to provide a functional relationship between the friction loss H_f, pipe dimensions, fluid properties, and flow parameters. The resulting equation is called the Darcy–Weisbach equation:

$$H_f = \frac{fLV^2}{2gd} = \frac{fLQ^2}{1.23gD^5} \tag{4.7}$$

The friction factor f is a measure of pipe roughness. It has been evaluated experimentally for numerous pipes. The data wee used to create the Moody friction factor chart shown as Figure 4.1. For Re < 2000, the flow in a pipe will be laminar and f is only a function of Re_D. It can be calculated by

$$f = \frac{64}{Re_D} \tag{4.8}$$

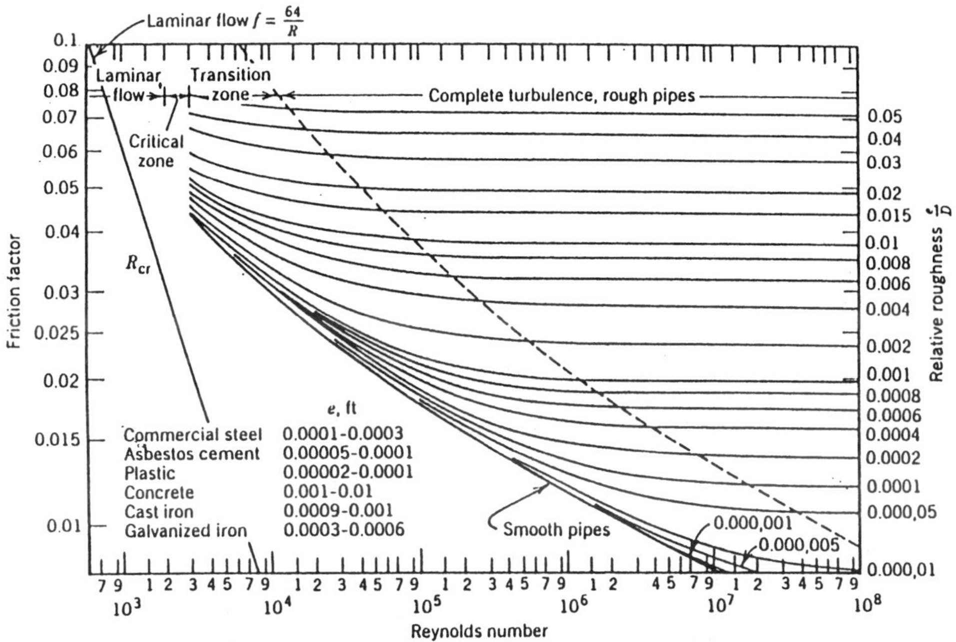

FIGURE 4.1 The Moody diagram.

At Reynolds numbers between about 2000 and 4000 the flow is unstable as a result of the onset of turbulence (critical zone in Figure 4.1). In this range, friction loss calculations are difficult because it is impossible to determine a unique value of f. For Re > 4000 the flow becomes turbulent and f is a function of both Re and relative pipe roughness (e/d). At high Re, f eventually depends only on e/d; defining the region referred to as fully turbulent flow. This is the region in Figure 4.1 where the lines for different e/d become horizontal (e is the equivalent roughness height and d pipe diameter). The Re_D at which this occurs depends on the pipe roughness. Laminar flow in pipes is unusual. For example, for water flowing in a 0.3-m-diameter pipe, the velocity would have to be below 0.02 m/sec for laminar flow to exist. Therefore, most practical pipe flow problems are in the turbulent region.

Using the Moody chart in Figure 4.1 to get f requires that Re and e/d be known. Calculating Re is direct if the water temperature, velocity, and pipe diameter are known. The problem is obtaining a good value for e. Typical values of e are listed in Figure 4.1. These values should be considered as guides only and not used if more–exact values can be obtained from the pipe supplier.

Since roughness may vary with time due to buildup of solid deposits or organic growths, f is also time dependent. Manufacturing tolerances also cause variations in the pipe diameter and surface roughness. Because of these factors, the friction factor for any pipe can only be approximated. A designer is required to use good engineering judgment in selecting a design value for f so that proper allowance is made for these uncertainties.

For noncircular pipes, the only change in the friction loss equation is the use of an equivalent diameter — based on the hydraulic radius (R), i.e., $d = 4R$ — in place of the circular pipe diameter d. R is the ratio of the flow area to the wetter perimeter.

Wood (1966) developed equations which can be used in place of the Moody diagram to estimate f for Re > 10^4 and $10^{-5} < k < 0.04$ ($k = e/d$).

$$f = a + b\text{Re}^{-c} \tag{4.9}$$

$$a = 0.094k^{0.225} + 0.53k, \quad b = 88k^{0.44}, \quad c = 1.62k^{0.134}$$

The practical problem is still obtaining a reliable value for e. It cannot be directly measured but must be determined from friction loss tests of the pipe.

An exact solution using the Darcy–Weisbach equation can require a trial-and-error solution because of the dependency of f on Re if either the flow or pipe diameter are not known. A typical approach to solving this problem is to estimate a reasonable fluid velocity to calculate Re and obtain f from the Moody chart or Equation (4.9). Next, calculate a new velocity and repeat until the solution converges. Converging on a solution is greatly simplified with programmable calculators and a variety of software available for computer.

For long gravity flow pipelines, the starting point in selecting the pipe diameter is to determine the smallest pipe that can pass the required flow without friction loss exceeding the available head. For pumped systems, the selection must be based on an economic analysis that compares the pipe cost with the cost of building and operating the pumping plant.

Local Losses

Flow through valves, orifices, elbows, transitions, etc. causes flow separation which results in the generation and dissipation of turbulent eddies. For short systems containing many bends, valves, tees, etc. local or minor losses can exceed friction losses. The head loss h_l associated with the dissipation caused by a minor loss is proportional to the velocity head and can be accounted for as a minor or local loss using the following equation.

$$h_l = K_l \frac{Q^2}{\left(2gA_m^2\right)} \tag{4.10}$$

in which K_l is the minor loss coefficient and A_m is the area of the pipe at the inlet to the local loss. The loss coefficient K_l is analogous to fL/d in Equation 4.7.

The summation of all friction and local losses in a pipe system can be expressed as

$$H_f = h_f + h_l \tag{4.11}$$

$$H_f = \left[\sum\left(\frac{fL}{2gdA_p^2}\right) + \sum\left(\frac{K_l}{2gA_m^2}\right)\right]Q^2 = CQ^2 \tag{4.12}$$

in which

$$C = \sum\left(\frac{fL}{2gdA_p^2}\right) + \sum\left(\frac{K_l}{2gA_m^2}\right) \tag{4.13}$$

It is important to use the correct pipe diameter for each pipe section and local loss.

In the past some have expressed the local losses as an equivalent pipe length: $L/d = K_l/f$. It simply represents the length of pipe that produces the same head loss as the local or minor loss. This is a simple, but not a completely accurate method of including local losses. The problem with this approach is that since the friction coefficient varies from pipe to pipe, the equivalent length will not have a unique value. When local losses are truly minor, this problem becomes academic because the error only influences losses which make up a small percentage of the total. For cases where accurate evaluation of all losses is important, it is recommended that the minor loss coefficients K_l be used rather than an equivalent length.

The challenging part of making minor loss calculations is obtaining reliable values of K_l. The final results cannot be any more accurate than the input data. If the pipe is long, the friction losses may be large compared with the minor losses and approximate values of K_l will be sufficient. However, for short systems with many pipe fittings, the local losses can represent a significant portion of the total system losses, and they should be accurately determined. Numerous factors influence K_l. For example, for elbows, K_l is influenced by the shape of the conduit (rectangular vs. circular), by the radius of the bend, the bend angle, the Reynolds number, and the length of the outlet pipe. For dividing or combining tees or Y-branches, the percent division of flow and the change in pipe diameter must also be included when estimating K_l. One factor which is important for systems where local losses are significant is the interaction between components placed close together. Depending on the type, orientation, and spacing of the components, the total loss coefficient may be greater or less than the simple sum of the individual K_l values.

Comparing the magnitude of $\Sigma(fL/2gA^2)$ to $\Sigma(K_l/2gA_m^2)$ will determine how much care should be given to the selection of the K_l values. Typical values of K_l are listed in Table 4.1 (Tullis, 1989). When more comprehensive information on loss coefficients is needed, the reader is referred to Miller (1990).

Pipe Design

Pipe Materials

Materials commonly used for pressure pipe transporting liquids are ductile iron, concrete, steel, fiberglass, PVC, and polyolefin. Specifications have been developed by national committees for each of these pipe materials. The specifications discuss external loads, internal design pressure, available sizes, quality of materials, installation practices, and information regarding linings. Standards are available from the following organizations:

American Water Works Association (AWWA)
American Society for Testing and Materials (ASTM)
American National Standards Institute (ANSI)
Canadian Standards Association (CSA)
Federal Specifications (FED)
Plastic Pipe Institute (PPI)

In addition, manuals and other standards have been published by various manufacturers and manufacturer's associations. All of these specifications and standards should be used to guide the selection of pipe material. ASCE (1992) contains a description of each of these pipe materials and a list of the specifications for the various organizations which apply to each material. It also discusses the various pipe-lining materials available for corrosion protection.

For air- and low-pressure liquid applications one can use unreinforced concrete, corrugated steel, smooth sheet metal, spiral rib (sheet metal), and HDPE (high-density polyethylene) pipe. The choice of a material for a given application depends on pipe size, pressure requirements, resistance to collapse from internal vacuums, external loads, resistance to internal and external corrosion, ease of handling and installing, useful life, and economics.

TABLE 4.1 Minor Loss Coefficients

Item	K_l Typical Value	Typical Range
Pipe inlets		
Inward projecting pipe	0.78	0.5–0.9
Sharp corner-flush	0.50	—
Slightly rounded	0.20	0.04–0.5
Bell mouth	0.04	0.03–0.1
Expansions[a]	$(1 - A_1/A_2)^2$ (based on V_1)	
Contractions[b]	$(1/C_c - 1)^2$ (based on V_2)	

A_2/A_1	0.1	0.2	0.3	0.4	0.5	0.6	0.7	0.8	0.9
C_c	0.624	0.632	0.643	0.659	0.681	0.712	0.755	0.813	0.892

Item	K_l Typical Value	Typical Range
Bends[c]		
Short radius, $r/d = 1$		
90	—	0.24
45	—	0.1
30	—	0.06
Long radius, $r/d = 1.5$		
90	—	0.19
45	—	0.09
30	—	0.06
Mitered (one miter)		
90	1.1	—
60	0.50	0.40–0.59
45	0.3	0.35–0.44
30	0.15	0.11–0.19
Tees	[c]	—
Diffusers	[c]	—
Valves		
Check valve	0.8	0.5–1.5
Swing check	1.0	0.29–2.2
Tilt disk	1.2	0.27–2.62
Lift	4.6	0.85–9.1
Double door	1.32	1.0–1.8
Full-open gate	0.15	0.1–0.3
Full-open butterfly	0.2	0.2–0.6
Full-open globe	4.0	3–10

[a] See Streeter and Wylie, 1975, p. 304.
[b] See Streeter and Wylie, 1975, p. 305.
[c] See Miller, 1990.
[d] See Kalsi Engineering and Tullis Engineering Consultants, 1993.

Pressure Class Guidelines

Procedures for selecting the pressure class of pipe vary with the type of pipe material. Guidelines for different types of materials are available from AWWA, ASTM, ANSI, CSA, FED, PPI and from the pipe manufacturers. These specifications should be obtained and studied for the pipe materials being considered.

The primary factors governing the selection of a pipe pressure class are (1) the maximum steady state operating pressure, (2) surge and transient pressures, (3) external earth loads and live loads, (4) variation of pipe properties with temperature or long-time loading effects, and (5) damage that could result from handling, shipping, and installing or reduction in strength due to chemical attack or other aging factors. The influence of the first three items can be quantified, but the last two are very subjective and are generally accounted for with a safety factor which is the ratio of the burst pressure to the rated pressure.

There is no standard procedure on how large the safety factor should be or on how the safety factor should be applied. Some may feel that it is large enough to account for all of the uncertainties. Past failures of pipelines designed using this assumption prove that it is not always a reliable approach. The procedure recommended by the author is to select a pipe pressure class based on the internal design pressure (IDP) defined as

$$\text{IDP} = \left(P_{\max} + P_s\right)\text{SF} \tag{4.14}$$

in which P_{max} is the maximum steady state operating pressure, P_s is the surge or water hammer pressure, and SF is the safety factor applied to take care of the unknowns (items 3 to 5) just enumerated. A safety factor between 3 and 4 is typical.

The maximum steady state operating pressure (P_{max}) in a gravity flow system is usually the difference between the maximum reservoir elevation and the lowest elevation of the pipe. For a pumped system it is usually the pump shutoff head calculated based on the lowest elevation of the pipe.

Surge and transient pressures depend on the specific pipe system design and operation. Accurately determining P_s requires analyzing the system using modern computer techniques. The most commonly used method is the "Method of Characteristics" (Tullis, 1989; Wylie and Streeter, 1993). Some of the design standards give general guidelines to predict P_s that can be used if a detailed transient analysis is not made. However, transients are complex enough that simple "rules of thumb" are seldom accurate enough. Transients are discussed again in a later subsection.

Selection of wall thickness for larger pipes is often more dependent on collapse pressure and handling loads than it is on burst pressure. A thin-wall, large-diameter pipe may be adequate for resisting relatively high internal pressures but may collapse under negative internal pressure or, if the pipe is buried, the soil and groundwater pressure plus live loads may be sufficient to cause collapse even if the pressure inside the pipe is positive.

External Loads

There are situations where the external load is the controlling factor determining if the pipe will collapse. The magnitude of the external load depends on the diameter of the pipe, the pipe material, the ovality (out of roundness) of the pipe cross section, the trench width, the depth of cover, the specific weight of the soil, the degree of soil saturation, the type of backfill material, the method used to backfill, the degree of compaction, and live loads. The earth load increases with width and depth of the trench, and the live load reduces with depth of cover. The cumulative effect of all these sources of external loading requires considerable study and analysis.

There are no simple guidelines for evaluating external pipe loads. Because of the complexity of this analysis, the default is to assume that the safety factor is adequate to account for external loads as well as the other factors already mentioned. One should not allow the safety factor to replace engineering judgment and calculations. One option to partially compensate for the lack of a detailed analysis is to use a higher-pressure class of pipe in areas where there will be large live loads or where the earth loading is unusually high. One should consider the cost of a pipe failure caused by external loads compared with the cost of using a thicker pipe or the cost of performing a detailed analysis. Those interested in the details of performing calculations of earth loading should be Spranger and Handy, 1973.

Limiting Velocities

There are concerns about upper and lower velocity limits. If the velocity is too low, problems may develop due to settling of suspended solids and air being trapped at high points and along the crown of the pipe. The safe lower velocity limit to avoid collecting air and sediment depends on the amount and type of sediment and on the pipe diameter and pipe profile. Velocities greater than about 1 m/sec (3 ft/sec) are usually sufficient to move trapped air to air release valves and keep the sediment in suspension.

Problems associated with high velocities are (1) erosion of the pipe wall or liner (especially if coarse suspended sediment is present), (2) cavitation at control valves and other restrictions, (3) increased pumping costs, (4) removal of air at air release valves, (5) increased operator size and concern about valve shaft failures due to excessive flow torques, and (6) an increased risk of hydraulic transients. Each of these should be considered before making the final pipe diameter selection. A typical upper velocity for many applications if 6 m/sec (20 ft/sec). However, with proper pipe design and analysis (of the preceding six conditions), plus proper valve selection, much higher velocities can be tolerated. A typical upper velocity limit for standard pipes and valves is about 6 m/sec (20 ft/sec). However, with proper design and analysis, much higher velocities can be tolerated.

Valve Selection

Valves serve a variety of functions. Some function as isolation or block valves that are either full open or closed. Control valves are used to regulate flow or pressure and must operate over a wide range of valve openings. Check valves prevent reverse flow, and air valves release air during initial filling and air that is collected during operation and admit air when the pipe is drained.

Control Valves

For many flow control applications it is desirable to select a valve that has linear control characteristics. This means that if you close the valve 10%, the flow reduces about 10%. Unfortunately, this is seldom possible since the ability of a valve to control flow depends as much on the system as it does on the design of the valve. The same valve that operates linearly in one system may not in another.

Selecting the proper flow control valve should consider the following criteria:

1. The valve should not produce excessive pressure drop when full open.
2. The valve should control over at least 50% of its movement.
3. At maximum flow, the operating torque must not exceed the capacity of the operator or valve shaft and connections.
4. The valve should not be subjected to excessive cavitation.
5. Pressure transients should not exceed the safe limits of the system.
6. Some valves should not be operated at very small openings. Other valves should be operated near full open.

Controllability. To demonstrate the relationship between a valve and system, consider a butterfly valve that will be used to control the flow between two reservoirs with an elevation difference of ΔZ. System A is a short pipe (0.3 m dia., 100 m long, $\Delta Z = 10$ m) where pipe friction is small $fL/2gdA_p^2 = 46.9$, and System B is a long pipe (0.3 m dia., 10,000 m long, $\Delta Z = 200$ m) with high friction $fL/2gdA_p^2 = 4690$. Initially, assume that the same butterfly valve will be used in both pipes and it will be the same size as the pipe. The flow can be calculated using the energy equation (Equation 4.5) and the system loss equation (Equation (4.12):

$$Q = \sqrt{\frac{\Delta Z}{\left[\sum\left(\frac{fL}{2gdA_p^2}\right) + \sum\left(\frac{K_l}{2gA_m^2}\right)\right]}} \tag{4.15}$$

For the valve, assume that the K_l full open is 0.2 and at 50% open it is 9.0. Correspondingly, $K_l/2gA_m^2$ = 1.905 and 85.7. For System A, the flow with the valve full open will be 0.453 m³/sec and at 50% open 0.275 m³/sec, a reduction of 39%. Repeating these calculations over the full range of valve openings would show that the flow for System A reduces almost linearly as the valve closes.

For System B, the flow with the valve full open will be 0.206 m³/sec and at 50% open 0.205 m³/sec, a reduction of less than 1%. For System B the valve does not control until the valve loss, expressed by $K_l/2gA_m^2$ becomes a significant part of the friction term (4690). The same valve in System B will not start to control the flow until it has closed more than 50%. A line-size butterfly valve is obviously not a good choice for a control valve in System B. One solution to this problem is to use a smaller valve. If the butterfly valve installed in System B was half the pipe diameter, it would control the flow over more of the stroke of the valve.

The range of opening over which the valve controls the flow also has a significant effect on the safe closure time for control valves. Transient pressures are created when there is a sudden change in the flow. Most valve operators close the valve at a constant speed. If the valve does not control until it is more than 50% closed, over half of the closing time is wasted and the effective valve closure time is less than half the total closing time. This will increase the magnitude of the transients that will be generated.

Torque. To be sure that the valve shaft, connections, and operator are properly sized, the maximum torque or thrust must be known. If the maximum force exceeds operator capacity, it will not be able to open and close the valve under extreme flow conditions. If the shaft and connectors are underdesigned, the valve may fail and slam shut causing a severe transient.

For quarter-turn valves, the force required to operate a valve consists of seating, bearing, and packing friction, hydrodynamic (flow) forces, and inertial forces. These forces are best determined experimentally. A key step in applying experimental torque information is the determination of the flow condition creating maximum torque. This requires that the system be analyzed for all possible operating conditions and valve openings. For a given size and type of valve, the flow torque depends on the torque coefficient (which is dependent on the specific valve design) and the pressure drop which, in turn, depends on the flow. In short systems where there is little friction loss and high velocities, a quarter-turn valve will see maximum torques at large openings where the flow is high. In long systems with high reservoir heads and smaller velocities, the same valve will see maximum torque at small openings where the pressure drop is high.

One situation where it is easy to overlook the condition causing maximum torque is with parallel pumps. Each pump normally will have a discharge control valve. The maximum system flow occurs with all three pumps operating. However, the flow and the torque for any of the pump discharge valves is maximum for only one pump operating. One specific example (Tullis, 1989) showed that the torque on a butterfly valve was three times higher when one pump was operating compared with three pumps operating in parallel.

Cavitation. Cavitation is frequently an important consideration in selection and operation of control valves. It is necessary to determine if cavitation will exist, evaluate its intensity, and estimate its effect on the system and environment. Cavitation can cause noise, vibration, and erosion damage and can decrease performance. The analysis should consider the full range of operation. Some valves cavitate worst at small openings and others cavitate heavily near full open. It depends on both the system and the valve design. If cavitation is ignored in the design and selection of the valves, repairs and replacement of the valves may be necessary. Information for making a complete cavitation analysis is beyond the scope of this section. Detailed information on the process to design for cavitation is contained in Tullis (1989; 1993).

The first step in a cavitation analysis is selecting the acceptable level of cavitation. Experimental data are available for four limits: incipient (light, intermittent noise), critical (light, steady noise), incipient damage (pitting damage begins), and choking (very heavy damage and performance drops off). Limited cavitation data are available for each of these limits (Tullis, 1989; 1993). Choosing a cavitation limit depends on several factors related to the operating requirements, expected life, location of the device, details of the design, and economics. For long-term operation of a control valve in a system where noise can be tolerated, the valve should never operate beyond incipient damage. In systems where noise is objectionable, critical cavitation would be a better operating limit.

Using a choking cavitation as a design limit is often misused. It is generally appropriate as a design limit for valves that only operate for short periods of time, such as a pressure relief valve. The intensity of cavitation and the corresponding noise vibration and erosion damage at the valve are at their maximum just before a valve chokes. If the valve operates beyond choking (sometimes referred to as supercavitation), the collapse of the vapor cavities occurs remote from the valve. Little damage is likely to occur at the valve, but farther downstream serious vibration and material erosion problems can occur.

If the cavitation analysis indicates that the valve, orifice, or other device will be operating at a cavitation level greater than can be tolerated, various techniques can be used to limit the level of cavitation. One is to select a different type of valve. Recent developments in valve design have produced a new generation of valves that are more resistant to cavitation. Most of them operate on the principle of dropping the pressure in stages. They usually have multiple paths with numerous sharp turns or orifices in series. Two limitations to these valves are that they often have high pressure drops (even when full open), and they are only usable with clean fluids.

A similar approach is to place multiple conventional valves in series or a valve in series with orifice plates. Proper spacing of valves and orifices placed in series is important. The spacing between valves depends upon the type. For butterfly valves it is necessary to have between five and eight diameters of pipe between valves to prevent flutter of the leaf of the downstream valve and to obtain the normal pressure drop at each valve. For globe, cone, and other types of valves, it is possible to bolt them flange to flange and have satisfactory operation.

For some applications another way to suppress cavitation is to use a free-discharge valve that is vented so cavitation cannot occur. There are valves specifically designed for this application. Some conventional valves can also be used for free discharge, if they can be adequately vented.

Cavitation damage can be suppressed by plating critical areas of the pipe and valve with cavitation-resistant materials. Based on tests using a magnetostriction device, data show that there is a wide variation in the resistance of the various types of material. Limited testing has been done on the erosion resistance of different materials and coating to cavitation in flowing systems. The available data show that there is less variation in the damage resistance of materials in actual flowing systems. However, experience has shown the plating parts of the valve with the right material will extend valve life.

Injecting air to suppress cavitation is a technique which has been used for many years with varying degrees of success. The most common mistake is placing the air injection port in the wrong location so the air does not get to the cavitation zone. If an adequate amount of air is injected into the proper region, the noise, vibrations, and erosion damage can be significantly reduced. The air provides a cushioning effect reducing the noise, vibration, and erosion damage. If the system can tolerate some air being injected, aeration is usually the cheapest and best remedy for cavitation.

Transients. Transient pressures can occur during filling and flushing air from the line, while operating valves, and when starting or stopping pumps. If adequate design provisions and operational procedures are not established, the transient pressure can easily exceed the safe operating pressure of the pipe. A system should be analyzed to determine the type and magnitudes of possible hydraulic transients. The basic cause is rapid changes in velocity. The larger the incremental velocity change and the faster that change takes place, the greater will be the transient pressure. If the piping system is not designed to withstand the high transient pressures, or if controls are not included to limit the pressure, rupture of the pipe or damage to equipment can result.

All pipelines experience transients. Whether or not the transient creates operational problems or pipe failure depends upon its magnitude and the ability of the pipes and mechanical equipment to tolerate high pressures without damage. For example, an unreinforced concrete pipeline may have a transient pressure head allowance of only a meter above its operating pressure before damage can occur. For such situations even slow closing of control valves or minor interruptions of flow due to any cause may create sufficient transient pressures to rupture the pipeline. In contrast, steel and plastic pipes can take relatively high transient pressures without failure.

Transients caused by slow velocity changes, such as the rise and fall of the water level in a surge tank, are called surges. Surge analysis, or "rigid column theory" involves mathematical or numerical solution of simple ordinary differential equations. The compressibility of the fluid and the elasticity of the conduit are ignored, and the entire column of fluid is assumed to move as a rigid body.

When changes in velocity occur rapidly, both the compressibility of the liquid and the elasticity of the pipe must be included in the analysis. This procedure is often called "elastic" or "waterhammer" analysis and involves tracking acoustic pressure waves through the pipe. The solution requires solving partial differential equations.

An equation predicting the head rise ΔH caused by a sudden change of velocity $\Delta V = V_2 - V_1$ can be derived by applying the unsteady momentum equation to a control volume of a section of the pipe where the change of flow is occurring. Consider a partial valve closure which instantly reduces the velocity by an amount ΔV. Reduction of the velocity can only be accomplished by an increase in the pressure upstream from the valve. This creates a pressure wave of magnitude ΔH which travels up the pipe at the acoustic velocity **a**. The increased pressure compresses the liquid and expands the pipe. The transient head rise due to an incremental change in velocity is

$$\Delta H = -\mathbf{a}\Delta V/g, \quad \text{for } \mathbf{a} \gg \Delta V \tag{4.16}$$

This equation is easy to use for multiple incremental changes of velocity as long as the first wave has not been reflected back to the point of origin.

The derivation of Equation (4.16) is based on an assumption of an instant velocity change. For a valve closing at the end of the pipe, instant closure actually refers to a finite time. It is the longest time that a valve can be closed and still cause a pressure rise equal to that of an instant closure. Normally, it is equal to $2L/a$ sec (which is the time required for the first pressure wave to travel to and from the other end of the pipe of length L); the head rise at the valve will be the same as if the valve were closed instantly. The $2L/a$ time is therefore often the instant closure time.

For a valve at the end of a long pipeline, the instant closure time can be considerably greater than $2L/a$. This is because when the friction loss coefficient fL/d is much greater than the loss coefficient for the valve K_l, the valve can be closed a long way before the flow changes. This dead time must be added to the $2L/a$ time to identify the actual instant closure time. To avoid the maximum potential transient pressure rise, the valve must be closed much slower than the instant closure time.

Computational techniques for estimating transient pressures caused by unsteady flow in pipelines are too complex to be done with simple hand calculations. The solution involves solving partial differential equations based on the equations of motion and continuity. These equations are normally solved by the method of characteristics. This technique transforms the equations into total differential equations. After integration, the equations can be solved numerically by finite differences. This analysis provides equations that can be used to predict the flow and head at any interior pipe section at any time (Tullis, 1989; Wiley and Streeter, 1993).

To complete the analysis, equations describing the boundary conditions are required. Typical boundary conditions are the connection of a pipe to a reservoir, a valve, changes in pipe diameter or material, pipe junctions, etc. Friction loss is included in the development of the basic equations and minor losses are handled as boundary conditions. The analysis properly models friction and the propagation and reflections of the pressure wave. It can also be used for surge calculations.

It is recommended that every pipe system should have at least a cursory transient analysis performed to identify the possibility of serious transients and decide whether or not a detailed analysis is necessary. If an analysis indicates that transients are a problem, the types of solutions available to the engineer include

1. Increasing the closing time of control valves.
2. Using a smaller valve to provide better control.
3. Designing special facilities for filling, flushing, and removing air from pipelines.
4. Increasing the pressure class of the pipeline.
5. Limiting the flow velocity.
6. Using pressure relief valves, surge tanks, air chambers, etc.

Check Valves

Selecting the wrong type or size of check valve can result in poor performance, severe transients, and frequent repairs (Kalsi, 1993). Proper check valve selection requires understanding the characteristics of the various types of check valves and analyzing how they will function as a part of the system in which they will be installed. A check valve that operates satisfactorily in one system may be totally inadequate in another. Each valve type has unique characteristics that give it advantages or disadvantages compared with the others. The characteristics of check valves that describe their hydraulic performance and which should be considered in the selection process include

1. Opening characteristics, i.e., velocity vs. disk position data.
2. Velocity required to fully open and firmly backseat the disk.
3. Pressure drop at maximum flow.
4. Stability of the disk at partial openings.
5. Sensitivity of disk flutter to upstream disturbances.
6. Speed of valve closure compared with the rate of flow reversal of the system.

Disk stability varies with flow rate, disk position, and upstream disturbances and is an important factor in determining the useful life of a check valve. For most applications it is preferable to size the check valve so that the disk is fully open and firmly backseated at normal flow rates. One of the worst design errors is to oversize a check valve that is located just downstream from a disturbance such as a pump, elbow, or control valve. If the disk does not fully open, it will be subjected to severe motion that will accelerate wear. To avoid this problem, it may be necessary to select a check valve that is smaller than the pipe size.

The transient pressure rise generated at check valve closure is another important consideration. The pressure rise is a function of how fast the valve disk closes compared with how fast the flow in the system reverses. The speed that the flow in a system reverses depends on the system. In systems where rapid flow reversals occur, the disk can slam shut causing a pressure transient (Thorley, 1989).

The closing speed of a valve is determined by the mass of the disk, the forces closing the disk, and the distance of travel from full open to closed. Fast closing valves have the following properties: the disk (including all moving parts) is lightweight, closure is assisted by springs, and the full stroke of the disk is short. Swing check valves are the slowest-closing valves because they violate all three of these criteria; i.e., they have heavy disks, no springs, and long disk travel. The nozzle check valve is one of the fastest-closing valves because the closing element is light, is spring loaded, and has a short stroke. The silent, duo, double door, and lift check valves with springs are similar to nozzle valves in their closing times, mainly because of the closing force of the spring.

Systems where rapid flow reversals occur include parallel pumps, where one pump is stopped while the others are still operating, and systems that have air chambers or surge tanks close to the check valve. For these systems there is a high-energy source downstream from the check valve to cause the flow to quickly reverse. As the disk nears its seat, it starts to restrict the reverse flow. This builds up the pressure, accelerates the disk, and slams it into the seat. Results of laboratory experiments, field tests, and computer simulations show that dramatic reductions in the transient pressures at disk closure can be achieved by replacing a slow-closing swing check valve with a fast-acting check valve. For example, in a system containing parallel pumps where the transient was generated by stopping one of the pumps, the peak transient pressure was reduced from 745 to 76 kPa when a swing check was replaced with a nozzle check valve. Such a change improved performance and significantly reduced maintenance.

Air Valves

There are three types of automatic air valves: (1) air/vacuum valves, (2) air release valves, and (3) combination valves. The air/vacuum valve is designed for releasing air while the pipe is being filled and for admitting air when the pipe is being drained. The valve must be large enough that it can admit and expel large quantities of air at a low pressure differential. The outlet orifice is generally the same diameter as the inlet pipe.

These valves typically contain a float, which rises and closes the orifice as the valve body fills with water. Once the line is pressurized, this type of valve cannot reopen to remove air that may subsequently accumulate until the pressure becomes negative, allowing the float to drop. If the pressure becomes negative during a transient or while draining, the float drops and admits air into the line. For thin-walled pipes that can collapse under internal vacuums, the air/vacuum valves should be sized for a full pipe break at the lowest pipe elevation. The vacuum valve must supply an air flow equal to the maximum drainage rate of the water from the pipe break and at an internal pipe pressure above the pipe collapse pressure.

The critical factor in sizing air/vacuum valves is usually the air flow rate to protect the pipe from a full pipe break. Since a pipe is filled much slower than it would drain during a full break, the selected valve will be sized so that the air is expelled during filling without pressurizing the pipe. Sizing charts are provided by manufacturers.

1Locating air valves in a piping system depends on the pipe profile, pipe length, and flow rates. Preferably, pipes should be laid to grade with valves placed at the high points or at intervals if there are no high points. One should use engineering judgment when defining a high point. If the pipe has numerous high points that are close together, or if the high points are not pronounced, it will not be necessary to have an air valve at each high point. If the liquid flow velocity is above about 1 m/sec (3 ft/sec), the flowing water can move the entrained air past intermediate high points to a downstream air valve. Releasing the air through an air valve prevents any sizable air pockets under high pressure from forming in the pipe. Trapped air under high pressure is extremely dangerous.

Velocity of the flow during filling is important. A safe way to fill a pipe is to limit the initial fill rate to an average flow velocity of about 0.3 m/sec (1 ft/sec) until most of the air is released and the air/vacuum valves close. The next step is to flush the system at about 1 m/sec (3 ft/sec), at a low system pressure, to flush the remaining air to an air release valve. It is important that the system not be pressurized until the air has been removed. Allowing large quantities of air under high pressure to accumulate and move through the pipe can generate severe transients. This is especially true if the compressed air is allowed to pass through a control valve or manual air release valve. When pressurized air flows through a partially open valve, the sudden acceleration and deceleration of the air and liquid can generate high pressure transients.

Pump Selection

Optimizing the life of a water supply system requires proper selection, operation, and maintenance of the pumps. During the selection process, the designer must be concerned about matching the pump performance to the system requirements and must anticipate problems that will be encountered when the pumps are started or stopped and when the pipe is filled and drained. The design should also consider the effect of variations in flow requirements, and also anticipate problems that will be encountered due to increased future demands and details of the installation of the pumps.

Selecting a pump for a particular service requires matching the system requirements to the capabilities of the pump. The process consists of developing a system equation by applying the energy equation to evaluate the pumping head required to overcome the elevation difference, friction, and minor losses. For a pump supplying water between two reservoirs, the pump head required to produce a given discharge can be expressed as

$$Hp = \Delta Z + H_f \tag{4.17}$$

or

$$Hp = \Delta Z + CQ^2 \tag{4.18}$$

in which the constant C is defined by Equation (4.13).

Figure 4.2 shows a system curve for a pipe having an 82-m elevation lift and moderate friction losses. If the elevation of either reservoir is a variable, then there is not a single curve but a family of curves corresponding to differential reservoir elevations.

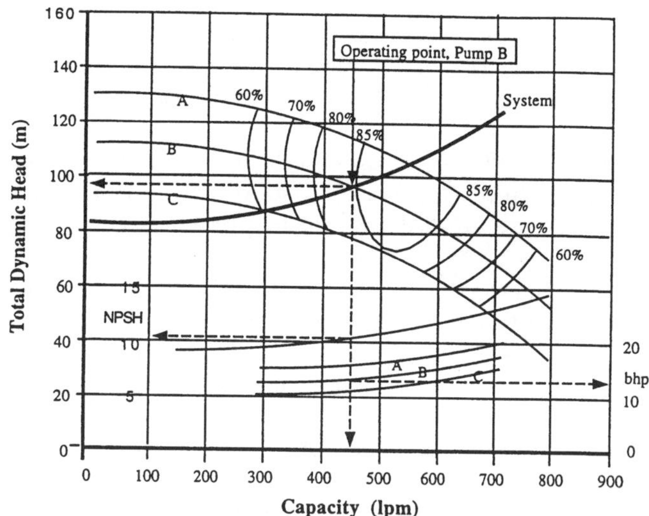

FIGURE 4.2 Pump selection for a single pump.

The three pump curves shown in Figure 4.2 represent different impeller diameters. The intersections of the system curve with the pump curves identify the flow that each pump will supply if installed in that system. For this example both A and B pumps would be a good choice because they both operate at or near their best efficiency range. Figure 4.2 shows the head and flow that the B pump will produce when operating in that system are 97 m and 450 L/m. The net positive suction head (NPSH) and brake horsepower (bhp) are obtained as shown in the figure.

The selection process is more complex when the system demand varies, either due to variations in the water surface elevation or to changing flow requirements. If the system must operate over a range of reservoir elevations, the pump should be selected so that the system curve, based on the mean (or the most frequently encountered) water level, intersects the pump curve near the midpoint of the best efficiency range. If the water level variation is not too great, the pump may not be able to operate efficiently over the complete flow range.

The problem of pump selection also becomes more difficult when planning for future demands or if the pumps are required to supply a varying flow. If the flow range is large, multiple pumps or a variable-speed drive may be needed. Recent developments in variable-frequency drives for pumps make them a viable alternative for systems with varying flows. Selection of multiple pumps and the decision about installing them in parallel or in series depend on the amount of friction in the system. Parallel installations are most effective for low-friction systems. Series pumps work best in high-friction systems.

For parallel pump operation the combined two pump curve is constructed by adding the flow of each pump. Such a curve is shown in Figure 4.3 (labeled 2 pumps). The intersection of the two-pump curve with the system curve identifies the combined flow for the two pumps. The pump efficiency for each pump is determined by projecting horizontally to the left to intersect the single-pump curve. For this example, a C pump, when operating by itself, will be have an efficiency of 83%. With two pumps operating, the efficiency of each will be about 72%. For the two pumps to operate in the most efficient way, the selection should be made so the system curve intersects the single-pump curve to the right of its best efficiency point.

Starting a pump with the pipeline empty will result in filling at a very rapid rate because initially there is little friction to build backpressure. As a result, the pump will operate at a flow well above the design flow. This may cause the pump to cavitate, but the more serious problem is the possibility of high pressures generated by the rapid filling of the pipe. Provisions should be made to control the rate of filling to a safe rate. Start-up transients are often controlled by starting the pump against

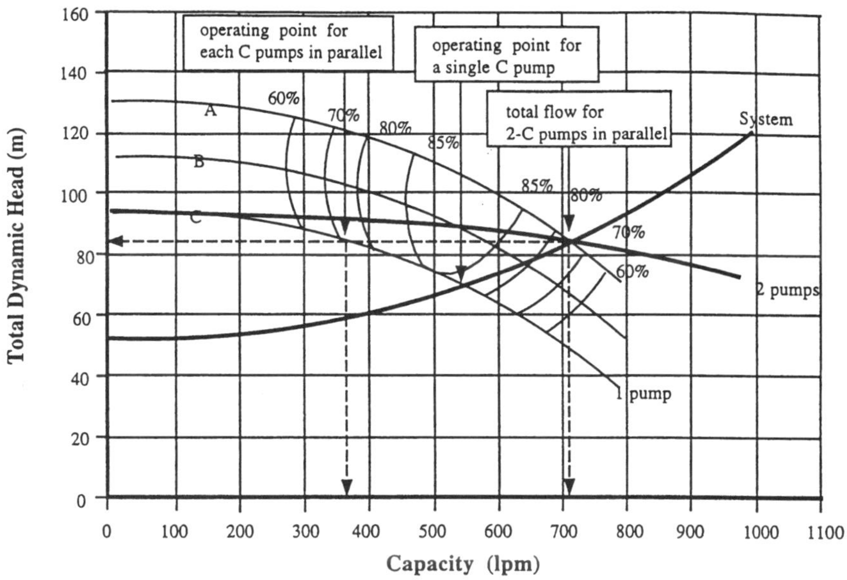

FIGURE 4.3 Selection of parallel pumps.

a partially open discharge valve located near the pump and using a bypass line around the pump. This allows the system to be filled slowly and safely. If the pipe remains full and no air is trapped, after the initial filling, subsequent start-up of the pumps generally does not create any serious problem. Adequate air release valves should be installed to release the air under low pressure.

For some systems, stopping the pump, either intentionally or accidentally, can generate high pressures that can damage the pipe and controls. If the design process does not consider these potential problems, the system may not function trouble free. Downtime and maintenance costs may be high. Not all systems will experience start-up and shutdown problems, but the design should at least consider the possibility. The problem is more severe for pipelines that have a large elevation change and multiple high points. The magnitude of the transient is related to the length and profile of the pipeline, the pump characteristics, the magnitude of the elevation change, and the type of check valve used. The downsurge caused by stopping the pump can cause column separation and high pressures due to flow reversals and closure of the check valves. Surge-protection equipment can be added to such systems to prevent damage and excessive maintenance.

Another operational problem occurs with parallel pumps. Each pump must have a check valve to prevent reverse flow. When one of the pumps is turned off, the flow reverses almost immediately in that line because of the high manifold pressure supplied by the operating pumps. This causes the check valve to close. If a slow-closing check valve is installed, the flow can attain a high reverse velocity before the valve closes, generating high pressure transients.

Numerous mechanical devices and techniques have been used to suppress pump shutdown transients. These include increasing the rotational inertia of the pump, use of surge tanks or air chambers near the pump, pressure relief valves, vacuum-breaking valves, and surge-anticipating valves. Selection of the proper transient control device will improve reliability, extend the economic life of the system, and reduce maintenance. Failure to complete a transient analysis and include the required controls will have the opposite effect. A system is only as good as it is designed to be.

Other Considerations

Feasibility Study

Designing pipelines, especially long transmission lines, involves more than just determining the required type and size of pipe. A starting point for major projects is usually a feasibility study which involves social, environmental, political, and legal issues, as well as an economic evaluation of the engineering alternatives developed during the preliminary design. The preliminary design should identify the scope of the project and all major features that influence the cost or viability. Since local laws, social values, and environmental concerns vary significantly between geographic areas, the engineer must be aware of the problems unique to the area.

Choices that affect the economics of the project include alternative pipe routes, amount of storage and its effect on reliability and controllability of flow, choice of pipe material, diameter and pressure class, provision for future demands, etc. In making decisions one must consider both the engineering and economic advantages of the alternatives. Reliability, safety, maintenance, operating, and replacement costs must all be given their proper value. The analysis should consider (1) expected life of the pipe, which is a function of the type of pipe material and the use of linings or protective coatings; (2) economic life, meaning how long the pipe will supply the demand; (3) planning for future demand; (4) pumping cost vs. pipe cost; and (5) provisions for storage.

During the feasibility study only a general design has been completed so a detailed analysis of all hydraulic problems and their solutions is not available. Even so, it is necessary to anticipate the need for special facilities or equipment and problems such as safe filling, provisions for draining, cavitation at control valves, and transient problems caused by valve or pump operation. Provisions should be made for the cost of the detailed analysis, design, and construction costs required to control special operational problems. Attention should also be given to costs associated with winterizing, stream crossings, highways crossing, special geologic or topographic problems, and any other items that would have a significant influence on the cost, reliability, or safety of the project.

Storage

The purposes of storage tanks and intermediate reservoirs include (1) to supply water when there is a temporary interruption of flow from the supply, (2) to provide supplemental water during peak periods, (3) to sectionalize the pipe to reduce mean and transient pressures, (4) to maintain pressure (elevated storage), and (5) to simplify control. Storage also has a significant impact on the control structures, pumping plants, and general operation of the pipeline. If there is adequate storage, large fluctuations in demand can be tolerated. Any mismatch in supply and demand is made up for by an increase or decrease in storage, and valves in the transmission main will require only infrequent adjustments to maintain storage. Pumps can be activated by level controls at the storage tank and not by fluctuations in demand so they can operate for long periods near their design point.

If there is no storage, the system may have to provide continuous fine adjustment of the flow to provide the required flow within safe pressure limits. For gravity systems this may require automatic pressure- or flow-regulating valves. For pumped systems, the variations in flow can cause constant-speed centrifugal pumps to operate both below and above their design point where power consumption is high, efficiency is low, and where there is more chance of operational problems. Selection of a variable-frequency drive can avoid these problems. The selection of multiple pumps vs. a variable-speed pump is primarily a economic decision.

Thrust Blocks

Any time there is a change of pipe alignment, an unbalanced force is developed. The force required to restrain the pipe can be calculated with the two-dimensional, steady state momentum equation. For buried pipelines, this force can be transmitted to the soil with a thrust block. Determining the size of the block and, consequently, the bearing surface area depends on pipe diameter, fluid pressure, deflection angle of the pipe, and bearing capacity of the soil. A convenient monograph for sizing thrust blocks was published in the *Civil Engineering* in 1969 (Morrison, 1969).

References

ASCE. 1992. *Pressure Pipeline Design for Water and Wastewater.* Prepared by the Committee on Pipeline Planning of the Pipeline Division of the American Society of Civil Engineers, New York.

Kalsi Engineering and Tullis Engineering Consultants. 1993. *Application Guide for Check Valves in Nuclear Power Plants,* Revision 1, NP-5479. Prepared for Nuclear Maintenance Applications Center, Charlotte, NC.

Miller, D.S. 1990. *Internal Flow Systems — Design and Performance Prediction,* 2nd ed. Gulf Publishing Company, Houston.

Morrison, E.B. 1969. Monograph for the design of thrust blocks. *Civil Eng.,* 39, June, 55–51.

Spanger, M.G. and Handy, R.L. 1973. *Soil Engineering,* 3rd ed. Intext Eductational Publishers, New York, Chap. 25 and 26.

Streeter, V.L. and Wylie, E.B. 1975. *Fluid Mechanics,* 6th ed. McGraw-Hill, New York, 752 pp.

Thorley, A.R.D. 1989. Check valve behavior under transient flow conditions: a state-of-the-art review. *ASME,* 111, Vol. 2, June. *J. Fluids Engineering: Transactions of the ASME,* pp. 173–183.

Tullis, J.P. 1989. *Hydraulics of Pipelines — Pumps, Valves, Cavitation, Transients,* John Wiley and Sons, New York.

Tullis, J.P. 1993. Cavitation Guide for Control Valves, NUREG/CR-6031, U.S. Nuclear Regulatory Commission, Washington, D.C.

Wood, D.J. 1966. An explicit friction factor relationship. *Civil Eng.,* 36, December, 60–61.

Wylie, E.B. and Streeter, V.L. 1993. *Fluid Transients in Systems,* Prentice-Hall, Englewood Cliffs, N.J.

Further Information

Bean, H.S., ed. 1971. *Fluid Meters, Their Theory and Application,* 6th ed. The American Society of Mechanical Engineers, New York.

Handbook of PVC-Design and Construction. 1979. Uni-Bell Plastic Pipe Association, Dallas.

King, H.W. 1954. *Handbook of Hydraulics — For the Solution of Hydraulic Problems,* 4th ed. Revised by E.F. Brater. McGraw-Hill, New York.

Stephenson, D. 1981. *Pipeline Design for Water Engineers,* 2nd ed. Elsevier Scientific Publishing Company, Distributed in the U.S. by Gulf Publishing Company, Houston.

Stutsman, R.D. 1993. Steel Penstocks, ASCE Manuals and Reports on Engineering Practice No. 79, Energy Division, American Society of Civil Engineers, New York.

Watkins, R.K. and Spangler, M.G., Some Characteristics of the Modulus of Passive Resistance of Soil: A Study in Similitude. Highway Research Board Proceedings Vol. 37, 1958, pp. 576–583.

Tullis, J.P. 1996. Valves, in Dorf, R.C., Ed., *The Engineering Handbook,* CRC Press, Boca Raton, FL.

5

Open Channel Flow

Frank M. White
University of Rhode Island

5 Open Channel Flow 65
Definition • Uniform Flow • Critical Flow • Hydraulic Jump • Weirs • Gradually Varied Flow

Open Channel Flow

Definition

The term *open channel flow* denotes the gravity-driven flow of a liquid with a free surface. Technically, we may study any flowing liquid and any gas interface. In practice, the vast majority of open channel flows concern water flowing beneath atmospheric air in artificial or natural channels.

The geometry of an arbitrary channel is shown in Figure 5.1. The area A is for the water cross section only, and b is its top width. The wetted perimeter P covers only the bottom and sides, as shown, not the surface (whose air resistance is neglected). The water depth at any location is y, and the channel slope is θ, often denoted as $S_o = \sin\theta$. All of these parameters may vary with distance x along the channel. In unsteady flow (not discussed here) they may also vary with time.

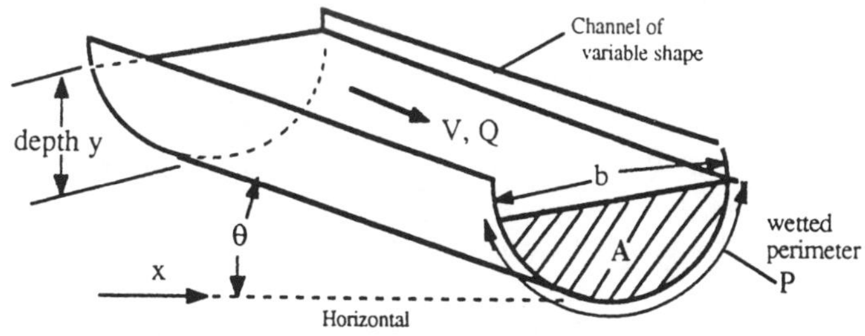

FIGURE 5.1 Definition sketch for an open channel.

0-8493-0055-X/00/$0.00+$.50

Uniform Flow

A simple reference condition, called *uniform flow*, occurs in a long straight prismatic channel of constant slope S_o. There is no acceleration, and the water flows at constant depth with fluid weight exactly balancing the wetted wall shear force: $\rho gLA \sin\theta = \tau_w PL$, where L is the channel length. Thus, $\tau_w = \rho g R_h S_o$, where $R_h = A/P$ is called the *hydraulic radius* of the channel. If we relate wall shear stress to the Darcy friction factor f, $\tau_w = (f/8)\rho V^2$, we obtain the basic uniform flow open channel relation:

$$\text{Uniform flow:}\quad V = \sqrt{\frac{8g}{f}}\sqrt{R_h S_o},\quad \text{where}\quad \sqrt{\frac{8g}{f}} = C = \text{Chézy coefficient} \tag{5.1}$$

Antoine Chézy first derived this formula in 1769. It is satisfactory to base f upon the pipe-flow Moody diagram (Figure 4.1) using the hydraulic diameter, $D_h = 4R_h$, as a length scale. That is, $f = fcn\,(VD_h/\nu, \varepsilon/D_h)$ from the Moody chart. In ordinary practice, however, engineers assume fully rough, high-Reynolds-number flow and use Robert Manning's century-old correlation:

$$C \approx \frac{\zeta}{n} R_h^{1/6},\ \text{or}\ V_{\text{uniform}} \approx \frac{\zeta}{n} R_h^{2/3} S_o^{1/2}\ \text{and}\ Q = VA \tag{5.2}$$

where ζ is a conversion factor equal to 1.0 in SI units and 1.486 in English units. The quantity n is Manning's roughness parameter, with typical values, along with the associated roughness heights ε, listed in Table 5.1.

Critical Flow

Since the surface is always atmospheric, pressure head is not important in open channel flows. Total energy E relates only to velocity and elevation:

$$\text{Specific energy } E = y + \frac{V^2}{2g} = y + \frac{Q^2}{2gA^2}$$

At a given volume flow rate Q, the energy passes through a minimum at a condition called *critical flow*, where $dE/dy = 0$, or $dA/dy = b = gA^3/Q^2$:

$$A_{\text{crit}} = \left(\frac{bQ^2}{g}\right)^{1/3} \qquad V_{\text{crit}} = \frac{Q}{A_{\text{crit}}} = \left(\frac{gA_{\text{crit}}}{b}\right)^{1/2} \tag{5.3}$$

where b is the top-surface width as in Figure 5.1. The velocity V_{crit} equals the speed of propagation of a surface wave along the channel. Thus, we may define the Froude number Fr of a channel flow, for any cross section, as $Fr = V/V_{\text{crit}}$. The three regimes of channel flow are

$$\text{Fr} < 1\text{: subcritical flow;}\quad \text{Fr} = 1\text{: critical flow;}\quad \text{Fr} > 1\text{: supercritical flow}$$

There are many similarities between Froude number in channel flow and Mach number in variable-area duct flow (see Section 6).

For a rectangular duct, $A = by$, we obtain the simplified formulas

$$V_{\text{crit}} = \sqrt{gy} \qquad \text{Fr} = \frac{V}{\sqrt{gy}} \tag{5.4}$$

TABLE 5.1 Average Roughness Parameters for Various Channel Surfaces

	n	Average Roughness Height ε	
		ft	mm
Artificial lined channels			
Glass	0.010 ± 0.002	0.0011	0.3
Brass	0.011 ± 0.002	0.0019	0.6
Steel; smooth	0.012 ± 0.002	0.0032	1.0
Painted	0.014 ± 0.003	0.0080	2.4
Riveted	0.015 ± 0.002	0.012	3.7
Cast iron	0.013 ± 0.003	0.0051	1.6
Cement; finished	0.012 ± 0.002	0.0032	1.0
Unfinished	0.014 ± 0.002	0.0080	2.4
Planed wood	0.012 ± 0.002	0.0032	1.0
Clay tile	0.014 ± 0.003	0.0080	2.4
Brickwork	0.015 ± 0.002	0.012	3.7
Asphalt	0.016 ± 0.003	0.018	5.4
Corrugated metal	0.022 ± 0.005	0.12	37
Rubble masonry	0.025 ± 0.005	0.26	80
Excavated earth channels			
Clean	0.022 ± 0.004	0.12	37
Gravelly	0.025 ± 0.005	0.26	80
Weedy	0.030 ± 0.005	0.8	240
Stony; cobbles	0.035 ± 0.010	1.5	500
Natural channels			
Clean and straight	0.030 ± 0.005	0.8	240
Sluggish, deep pools	0.040 ± 0.010	3	900
Major rivers	0.035 ± 0.010	1.5	500
Floodplains			
Pasture, farmland	0.035 ± 0.010	1.5	500
Light brush	0.05 ± 0.02	6	2000
Heavy brush	0.075 ± 0.025	15	5000
Trees	0.15 ± 0.05	?	?

independent of the width of the channel.

Example 5.1

Water (ρ = 998 kg/m³, μ = 0.001 kg/m · sec) flows uniformly down a half-full brick 1-m-diameter circular channel sloping at 1°. Estimate (a) Q; and (b) the Froude number.

Solution 5.1 (a). First compute the geometric properties of a half-full circular channel:

$$A=\frac{\pi}{8}(1\text{ m})^2=0.393\text{ m}^2;\quad P=\frac{\pi}{2}(1\text{ m})=1.57\text{ m};\quad R=\frac{A}{P}=\frac{0.393}{1.57}=0.25\text{ m}$$

From Table 5.1, for brickwork, $n \approx 0.015$. Then, Manning's formula, Equation (5.2) predicts

$$V=\frac{\zeta}{n}R_h^{1/6}S_o^{1/2}=\frac{1.0}{0.015}(0.25)^{1/6}(\sin 1°)^{1/2}\approx 3.49\frac{\text{m}}{\text{sec}};\quad Q=3.49(0.393)\approx \mathbf{1.37\frac{m^3}{sec}}\qquad \textit{Solution 5.1(a)}$$

The uncertainty in this result is about ±10%. The flow rate is quite large (21,800 gal/min) because 1°, although seemingly small, is a substantial slope for a water channel.

One can also use the Moody chart. with $V \approx 49$ m/sec, compute Re = $\rho V D_h/\mu \approx 49$ E6 and $\varepsilon/D_h \approx$ 0.0037, then compute $f \approx 0.0278$ from the Moody chart. Equation (5.1) then predicts

$$V = \sqrt{\frac{8g}{f} R_h S_o} = \sqrt{\frac{8(9.81)}{0.0278}(0.25)(\sin 1°)} \approx 3.51 \frac{\text{m}}{\text{sec}}; \quad Q = VA \approx \mathbf{1.38} \frac{\mathbf{m^3}}{\mathbf{sec}}$$

Solution 5.1 (b). With Q known from part (a), compute the critical conditions from Equation (5.3):

$$A_{\text{crit}} = \left(\frac{bQ^2}{g}\right)^{1/3} = \left[\frac{1.0(1.37)^2}{9.81}\right]^{1/3} = 0.576 \text{ m}^2, \quad V_{\text{crit}} = \frac{Q}{A_{\text{crit}}} = \frac{1.37}{0.576} = 2.38 \frac{\text{m}}{\text{sec}}$$

Hence

$$\text{Fr} = \frac{V}{V_{\text{crit}}} = \frac{3.49}{2.38} \approx \mathbf{1.47} \text{ (supercritical)} \quad \textit{Solution 5.1(b)}$$

Again the uncertainty is approximately ±10%, primarily because of the need to estimate the brick roughness.

Hydraulic Jump

In gas dynamics (Section 6), a supersonic gas flow may pass through a thin normal shock and exit as a subsonic flow at higher pressure and temperature. By analogy, a supercritical open channel flow may pass through a *hydraulic jump* and exit as a subcritical flow at greater depth, as in Figure 5.2. Application of continuity and momentum to a jump in a rectangular channel yields

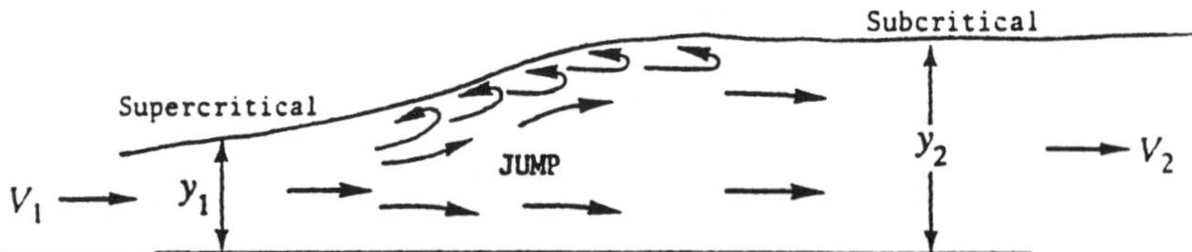

FIGURE 5.2 A two-dimensional hydraulic jump.

$$V_2 = V_1 \frac{y_1}{y_2} \quad y_2 = \frac{y_1}{2}\left[-1 + \sqrt{1 + 8\text{Fr}_1^2}\right] \text{ where } \text{Fr}_1 = \frac{V_1}{\sqrt{gy_1}} > 1 \tag{5.5}$$

Both the normal shock and the hydraulic jump are dissipative processes: the entropy increases and the effective energy decreases. For a rectangular jump,

$$\Delta E = E_1 - E_2 = \frac{(y_2 - y_1)^3}{4y_1y_2} > 0 \tag{5.6}$$

For strong jumps, this loss in energy can be up to 85% of E_1. The second law of thermodynamics requires $\Delta E > 0$ and $y_2 > y_1$ or, equivalently, $\text{Fr}_1 > 1$,

Note from Figure 5.2 that a hydraulic jump is not thin. Its total length is approximately four times the downstream depth. Jumps also occur in nonrectangular channels, and the theory is much more algebraically laborious.

Weirs

If an open channel flow encounters a significant obstruction, it will undergo rapidly varied changes which are difficult to model analytically but can be correlated with experiment. An example is the *weir* in Figure 5.3 (colloquially called a *dam*), which forces the flow to deflect over the top. If $L << Y$, the weir is termed *sharp-crested*; if $L = O(Y)$ it is *broad-crested*. Small details, such as the upper front corner radius or the crest roughness, may be significant. The crest is assumed level and of width b into the paper.

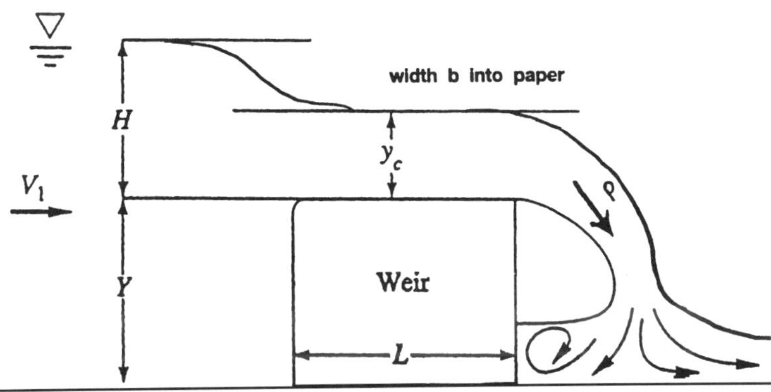

FIGURE 5.3 Geometry and notation for flow over a weir.

If there is a free overfall, as in Figure 5.3, the flow accelerates from subcritical upstream to critical over the crest to supercritical in the overfall. There is no flow when the excess upstream depth $H = 0$. A simple Bernoulli-type analysis predicts that the flow rate Q over a wide weir is approximately proportional to $bg^{1/2}H^{3/2}$. An appropriate correlation is thus

$$Q_{\text{weir}} = C_d b g^{1/2} H^{3/2}, \quad \text{where } C_d = \text{dimensionless weir coefficient} \tag{5.7}$$

If the upstream flow is turbulent, the weir coefficient depends only upon geometry, and Reynolds number effects are negligible. If the weir has sidewalls and is narrow, replace width b by $(b - 0.1H)$.

Two recommended empirical correlations for Equation (5.7) are as follows:

$$\begin{aligned} &\text{Sharp-crested:} \quad C_d \approx 0.564 + 0.0846\frac{H}{Y} \quad \text{for} \quad \frac{L}{Y} < 0.07 \\ &\text{Broad-crested:} \quad C_d \approx 0.462 \quad \text{for} \quad 0.08 < \frac{H}{L} < 0.33 \end{aligned} \tag{5.8}$$

These data are for wide weirs with a sharp upper corner in front. Many other weir geometries are discussed in the references for this section. Of particular interest is the sharp-edged vee-notch weir, which has no length scale b. If 2θ is the total included angle of the notch, the recommended correlation is

$$\text{Vee-notch, angle } 2\theta: \quad Q \approx 0.44 \tan\theta\, g^{1/2} H^{5/2} \quad \text{for} \quad 10° < \theta \leq 50° \tag{5.9}$$

The vee-notch is more sensitive at low flow rates (large H for a small Q) and thus is popular in laboratory measurements of channel flow rates.

A weir in the field will tend to spring free and form a natural *nappe*, or air cavity, as in Figure 5.3. Narrow weirs, with sidewalls, may need to be aerated artificially to form a nappe and keep the flow from sliding down the face of the weir. The correlations above assume nappe formation.

Gradually Varied Flow

Return to Figure 5.1 and suppose that (y, A, b, P, S_o) are all variable functions of horizontal position x. If these parameters are slowly changing, with no hydraulic jumps, the flow is termed *gradually varied* and satisfies a simple one-dimensional first-order differential equation if Q = constant:

$$\frac{dy}{dx} \approx \frac{S_o - S}{1 - \frac{V^2 b}{gA}}, \quad \text{where } V = \frac{Q}{A} \quad \text{and} \quad S = \frac{f}{D_h}\frac{V^2}{2g} = \frac{n^2 V^2}{\zeta^2 R_h^{4/3}} \tag{5.10}$$

The conversion factor $\zeta^2 = 1.0$ for SI units and 2.208 for English units. If flow rate, bottom slope, channel geometry, and surface roughness are known, we may solve for $y(x)$ for any given initial condition $y = y_o$ at $x = x_o$. The solution is computed by any common numerical method, e.g., Runge–Kutta.

Recall from Equation (5.3) that the term $V^2b/(gA) \equiv \text{Fr}^2$, so the sign of the denominator in Equation (5.10) depends upon whether the flow is sub- or supercritical. The mathematical behavior of Equation (5.10) differs also. If Fr is near unity, the change dy/dx will be very large, which probably violates the basic assumption of "gradual" variation.

For a given flow rate and local bottom slope, two reference depths are useful and may be computed in advance:

(a) The *normal* depth y_n for which Equation (5.2) yields the flow rate:
(b) The *critical* depth y_c for which Equation (5.3) yields the flow rate.

Comparison of these two, and their relation to the actual local depth y, specifies the type of solution curve being computed. The five bottom-slope regimes (mild M, critical C, steep S, horizontal H, and adverse A) create 12 different solution curves, as illustrated in Figure 5.4. All of these may be readily generated by a computer solution of Equation 5.10. The following example illustrates a typical solution to a gradually varied flow problem.

Example 5.2

Water, flowing at 2.5 m³/sec in a rectangular gravelly earth channel 2 m wide, encounters a broad-crested weir 1.5 m high. Using gradually varied theory, estimate the water depth profile back to 1 km upstream of the weir. The bottom slope is 0.1°.

Solution. We are given Q, $Y = 1.5$ m, and $b = 2$ m. We may calculate excess water level H at the weir (see Figure 5.3) from Equations (5.7) and (5.8):

$$Q = 2.5\frac{\text{m}^3}{\text{sec}} = C_d b_{\text{eff}} g^{1/2} H^{3/2} = 0.462(2.0 - 0.1H)(9.81)^{1/2} H^{3/2}, \quad \text{solve for } H \approx \mathbf{0.94}\text{ m}$$

Since the weir is not too wide, we have subtracted 0.1 H from b as recommended. The weir serves as a "control structure" which sets the water depth just upstream. This is our initial condition for gradually varied theory: $y(0) = Y + H = 1.5 + 0.94 \approx 2.44$ m at $x = 0$. Before solving Equation (5.10), we find the normal and critical depths to get a feel for the problem:

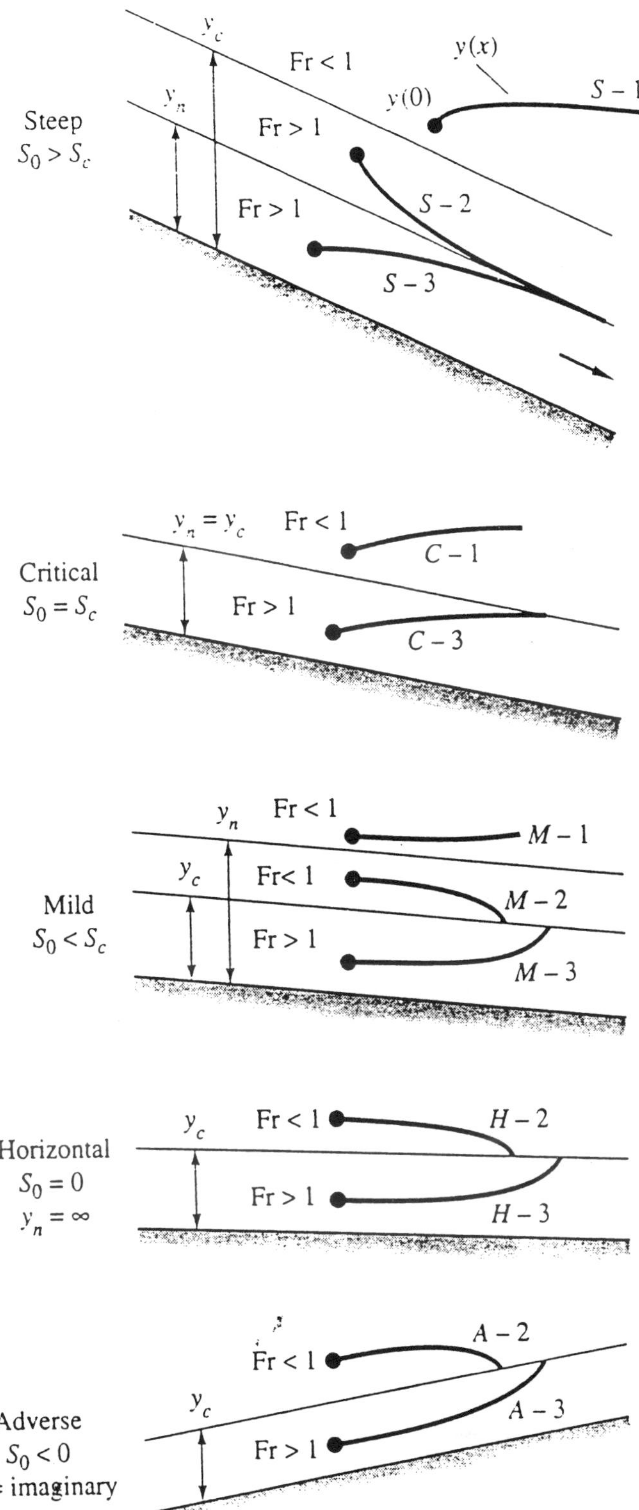

FIGURE 5.4 Classification of solution curves for gradually varied flow.

Normal depth: $Q = 2.5\frac{\text{m}^3}{\text{sec}} = \frac{1.0}{0.025}(2.0y_n)\left(\frac{2.0y_n}{2.0+2y_n}\right)^{3/2}\sqrt{\sin 0.1°}$, solve $y_n \approx \mathbf{1.14}$ m

Critical depth: $A_c = 2.0y_c - \left(\frac{bQ^2}{g}\right)^{1/3} = \left[\frac{2.0(2.5)^2}{9.81}\right]^{1/3}$, solve $y_c \approx \mathbf{0.54}$ m

We have taken $n \approx 0.025$ for gravelly earth, from Table 5.1. Since $y(0) > y_n > y_c$, we are on a mild slope $M-1$ "backwater" curve, as in Figure 5.4. For our data, Equation (5.10) becomes

$$\frac{dy}{dx} \approx \frac{S_o - n^2Q^2/\left(\zeta^2 A^2 R_h^{4/3}\right)}{1 - Q^2 b/\left(gA^3\right)}$$

where $Q = 2.5$, $b = 2$, $\zeta = 1$, $A = 2y$, $S_o = \sin 0.1°$, $R_h = 2y/(2 + 2y)$, $g = 9.81$, $y(0) = 2.44$ at $x = 0$.

Integrate numerically backward, that is, for $\Delta x < 0$, until $x = -1$ km $= -1000$ m. The complete solution curve is shown in Figure 5.5. The water depth decreases upstream and is approximately $y \approx 1.31$ m at $x = -1000$ m. If slope and channel width remain constant, the water depth asymptotically approaches the normal depth y_n far upstream.

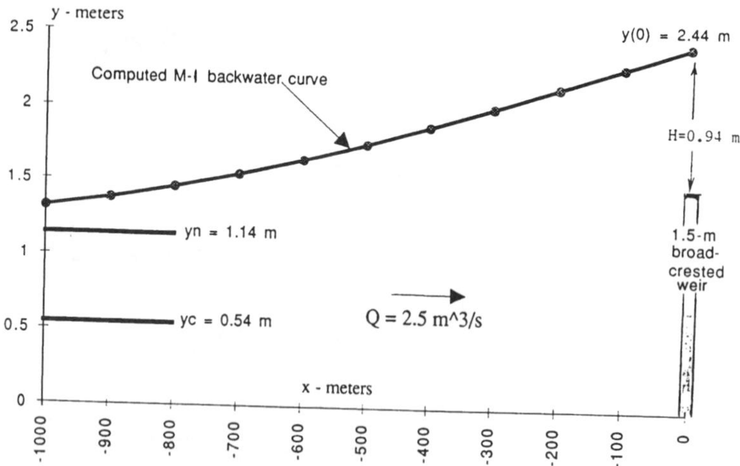

FIGURE 5.5 Backwater solution curve for Example 5.2. from White, F., Fluid Mechanics, 3rd ed., 1994; reproduced with permission of McGraw-Hill, Inc.

Nomenclature

English symbols

A = water cross section area
b = channel upper-surface width
C = Chézy coefficient, Equation (5.1)
C_d = weir discharge coefficient, Equation (5.7)
D_h = hydraulic diameter, $= 4R_h$
E = specific energy, $= y + V^2/2g$
f = Moody friction factor
Fr = Froude number, $= V/V_{crit}$

g = acceleration of gravity
H = excess water level above weir, Figure 5.3
L = weir length, Figure 5.3
n = Manning roughness factor, Table 5.1
P = wetted perimeter
Q = volume flow rate
R_h = hydraulic radius, = A/P
S = frictional slope, Equation (5.10)
S_o = bottom slope
V = average velocity
x = horizontal distance along the channel
y = water depth
Y = weir height, Figure (5.3)

Greek Symbols

ε = wall roughness height, Table 5.1
ρ = fluid density
μ = fluid viscosity
ν = fluid kinematic viscosity, = μ/ρ
ζ = conversion factor, = 1.0 (SI) and 1.486 (English)

Subscripts

c,crit = critical, at Fr = 1
n = normal, in uniform flow

References

Ackers, P. et al. 1978. *Weirs and Flumes for Flow Measurement*, John Wiley, New York.

Bos, M.G. 1985. *Long-Throated Flumes and Broad-Crested Weirs,* Martinus Nijhoff (Kluwer), Dordrecht, The Netherlands.

Bos, M.G., Replogle, J.A., and Clemmens, A.J. 1984. *Flow-Measuring Flumes for Open Channel Systems,* John Wiley, New York.

Brater, E.F. 1976. *Handbook of Hydraulics,* 6th ed., McGraw-Hill, New York.

Chow, V.T. 1959. *Open Channel Hydraulics,* McGraw-Hill, New York.

French, R.H. 1985. *Open Channel Hydraulics,* McGraw-Hill, New York.

Henderson, F.M. 1966. *Open Channel Flow,* Macmillan, New York.

Sellin, R.H.J. 1970. *Flow in Channels*, Gordon & Breach, London.

Spitzer, D.W. (Ed.). 1991. *Flow Measurement: Practical Guides for Measurement and Control*, Instrument Society of America, Research Triangle Park, NC.

6

External Incompressible Flows

Alan T. McDonald
Purdue University

6 External Incompressible Flows 75
Introduction and Scope • Boundary Layers • Drag • Lift •
Boundary Layer Control • Computation vs. Experiment

External Incompressible Flows

Introduction and Scope

Potential flow theory (Section 2) treats an incompressible *ideal fluid* with zero viscosity. There are no shear stresses; pressure is the only stress acting on a fluid particle. Potential flow theory predicts no drag force when an object moves through a fluid, which obviously is not correct, because all real fluids are viscous and cause drag forces. The objective of this section is to consider the behavior of viscous, incompressible fluids flowing over objects.

A number of phenomena that occur in external flow at high Reynolds number over an object are shown in Figure 6.1. The freestream flow divides at the stagnation point and flows around the object. Fluid at the object surface takes on the velocity of the body as a result of the no-slip condition. Boundary layers form on the upper and lower surfaces of the body; flow in the boundary layers is initially laminar, then **transition** to turbulent flow may occur (points "T").

Boundary layers thickening on the surfaces cause only a slight displacement of the streamlines of the external flow (their thickness is greatly exaggerated in the figure). **Separation** may occur in the region of increasing pressure on the rear of the body (points "S"); after separation boundary layer fluid no longer remains in contact with the surface. Fluid that was in the boundary layers forms the viscous *wake* behind the object.

0-8493-0055-X/00/$0.00+$.50

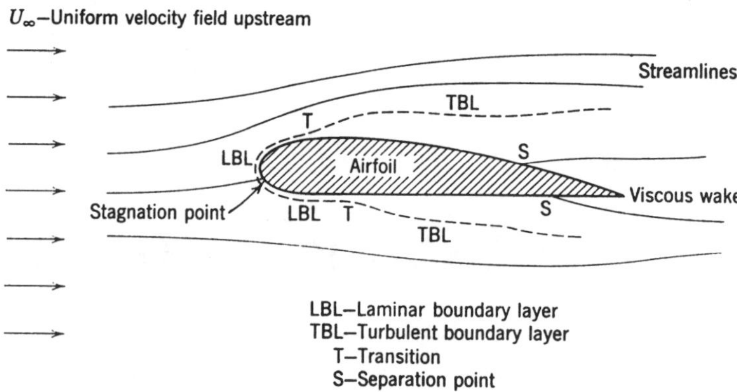

FIGURE 6.1 Viscous flow around an airfoil (boundary layer thickness exaggerated for clarity).

The Bernoulli equation is valid for steady, incompressible flow without viscous effects. It may be used to predict pressure variations outside the boundary layer. Stagnation pressure is constant in the uniform inviscid flow far from an object, and the Bernoulli equation reduces to

$$p_\infty + \frac{1}{2}\rho V^2 = \text{constant} \tag{6.1}$$

where p_∞ is pressure far upstream, ρ is density, and V is velocity. Therefore, the local pressure can be determined if the local freestream velocity, U, is known.

Boundary Layers

The Boundary Layer Concept

The **boundary layer** is the thin region near the surface of a body in which viscous effects are important. By recognizing that viscous effects are concentrated near the surface of an object, Prandtl showed that only the Euler equations for inviscid flow need be solved in the region outside the boundary layer. Inside the boundary layer, the elliptic Navier-Stokes equations are simplified to boundary layer equations with parabolic form that are easier to solve. The thin boundary layer has negligible pressure variation across it; pressure from the freestream is impressed upon the boundary layer.

Basic characteristics of all laminar and turbulent boundary layers are shown in the developing flow over a flat plate in a semi-infinite fluid. Because the boundary layer is thin, there is negligible disturbance of the inviscid flow outside the boundary layer, and the **pressure gradient** along the surface is close to zero. Transition from laminar to turbulent boundary layer flow on a flat plate occurs when Reynolds number based on x exceeds $\mathrm{Re}_x = 500{,}000$. Transition may occur earlier if the surface is rough, pressure increases in the flow direction, or separation occurs. Following transition, the turbulent boundary layer thickens more rapidly than the laminar boundary layer as a result of increased shear stress at the body surface.

Boundary Layer Thickness Definitions

Boundary layer disturbance thickness, δ, is usually defined as the distance, y, from the surface to the point where the velocity within the boundary layer, u, is within 1% of the local freestream velocity, U. As shown in Figure 6.2, the boundary layer velocity profile merges smoothly and asymptotically into the freestream, making δ difficult to measure. For this reason and for their physical significance, we define two integral measures of boundary layer thickness. Displacement thickness, δ^*, is defined as

$$\frac{\delta^*}{\delta} = \int_0^\infty \left(1 - \frac{u}{U}\right) d\left(\frac{y}{\delta}\right) \tag{6.2}$$

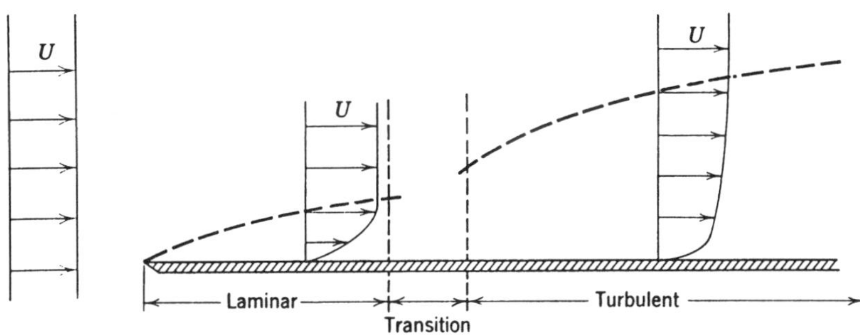

FIGURE 6.2 Boundary layer on a flat plate (vertical thickness exaggerated for clarity).

Physically, δ^* is the distance the solid boundary would have to be displaced into the freestream in a frictionless flow to produce the mass flow deficit caused by the viscous boundary layer. Momentum thickness, θ, is defined as

$$\frac{\theta}{\delta} = \int_0^\infty \frac{u}{U}\left(1 - \frac{u}{U}\right) d\left(\frac{y}{\delta}\right) \tag{6.3}$$

Physically, θ is the thickness of a fluid layer, having velocity U, for which the momentum flux is the same as the deficit in momentum flux within the boundary layer (momentum flux is momentum per unit time passing a cross section).

Because δ^* and θ are defined in terms of integrals for which the integrand vanishes in the freestream, they are easier to evaluate experimentally than disturbance thickness δ.

Exact Solution of the Laminar Flat-Plate Boundary Layer

Blasius obtained an exact solution for laminar boundary layer flow on a flat plate. He assumed a thin boundary layer to simplify the streamwise momentum equation. He also assumed *similar* velocity profiles in the boundary layer, so that when written as $u/U = f(y/\delta)$, velocity profiles do not vary with x. He used a similarity variable to reduce the partial differential equations of motion and continuity to a single third-order ordinary differential equation.

Blasius used numerical methods to solve the ordinary differential equation. Unfortunately, the velocity profile must be expressed in tabular form. The principal results of the Blasius solution may be expressed as

$$\frac{\delta}{x} = \frac{5}{\sqrt{\mathrm{Re}_x}} \tag{6.4}$$

and

$$C_f = \frac{\tau_w}{\frac{1}{2}\rho U^2} = \frac{0.664}{\sqrt{\mathrm{Re}_x}} \tag{6.5}$$

These results characterize the laminar boundary layer on a flat plate; they show that laminar boundary layer thickness varies as $x^{1/2}$ and wall shear stress varies as $1/x^{1/2}$.

Approximate Solutions

The Blasius solution cannot be expressed in closed form and is limited to laminar flow. Therefore, approximate methods that give solutions for both laminar and turbulent flow in closed form are desirable. One such method is the *momentum integral equation* (MIE), which may be developed by integrating the boundary layer equation across the boundary layer or by applying the streamwise momentum equation to a differential control volume (Fox and McDonald, 1992). The result is the ordinary differential equation

$$\frac{d\theta}{dx} = \frac{\tau_w}{\rho U^2} - \left(\frac{\delta^*}{\theta} + 2\right)\frac{\theta}{U}\frac{dU}{dx} \tag{6.6}$$

The first term on the right side of Equation (6.6) contains the influence of wall shear stress. Since τ_w is always positive, it always causes θ to increase. The second term on the right side contains the pressure gradient, which can have either sign. Therefore, the effect of the pressure gradient can be to either increase or decrease the rate of growth of boundary layer thickness.

Equation (6.6) is an ordinary differential equation that can be solved for θ as a function of x on a flat plate (zero pressure gradient), provided a reasonable shape is assumed for the boundary layer velocity profile and shear stress is expressed in terms of the other variables. Results for laminar and turbulent flat-plate boundary layer flows are discussed below.

Laminar Boundary Layers. A reasonable approximation to the laminar boundary layer velocity profile is to express u as a polynomial in y. The resulting solutions for δ and τ_w have the same dependence on x as the exact Blasius solution. Numerical results are presented in Table 6.1. Comparing the approximate and exact solutions shows remarkable agreement in view of the approximations used in the analysis. The trends are predicted correctly and the approximate values are within 10% of the exact values.

TABLE 6.1 Exact and Approximate Solutions for Laminar Boundary Layer Flow over a Flat Plate at Zero Incidence

Velocity Distribution $\frac{u}{U} = f\left(\frac{y}{\delta}\right) = f(\eta)$	$\frac{\theta}{\delta}$	$\frac{\delta^*}{\delta}$	$a = \frac{\delta}{x}\sqrt{\mathrm{Re}_x}$	$b = C_f\sqrt{\mathrm{Re}_x}$
$f(\eta) = 2\eta - \eta^2$	2/15	1/3	5.48	0.730
$f(\eta) = 3/2\ \eta - 1/2\ \eta^3$	39/280	3/8	4.64	0.647
$f(\eta) = \sin(\pi/2\ \eta)$	$(4 - \pi)/2\pi$	$(\pi - 2)/\pi$	4.80	0.654
Exact	0.133	0.344	5.00	0.664

Turbulent Boundary Layers. The turbulent velocity profile may be expressed well using a power law, $u/U = (y/\delta)^{1/n}$, where n is an integer between 6 and 10 (frequently 7 is chosen). For turbulent flow it is not possible to express shear stress directly in terms of a simple velocity profile; an empirical correlation is required. Using a pipe flow data correlation gives

$$\frac{\delta}{x} = \frac{0.382}{\mathrm{Re}_x^{1/5}} \tag{6.7}$$

and

$$C_f = \frac{\tau_w}{\frac{1}{2}\rho U^2} = \frac{0.0594}{\mathrm{Re}_x^{1/5}} \tag{6.8}$$

These results characterize the turbulent boundary layer on a flat plate. They show that turbulent boundary layer thickness varies as $x^{4/5}$ and wall shear stress varies as $1/x^{1/5}$.

Approximate results for laminar and turbulent boundary layers are compared in Table 6.2. At a Reynolds number of 1 million, wall shear stress for the turbulent boundary layer is nearly six times as large as for the laminar layer. For a turbulent boundary layer, thickness increases five times faster with distance along the surface than for a laminar layer. These approximate results give a physical feel for relative magnitudes in the two cases.

TABLE 6.2 Thickness and Skin Friction Coefficient for Laminar and Turbulent Boundary Layers on a Flat Plate

Reynolds	Boundary Layer Thickness/x		Skin Friction Coefficient		Turbulent/Laminar Ratio	
Number	Laminar BL	Turbulent BL	Laminar BL	Turbulent BL	BL Thickness	Skin Friction
2E + 05	0.0112	0.0333	0.00148	0.00517	2.97	3.48
5E + 05	0.00707	0.0277	0.000939	0.00431	3.92	4.58
1E + 06	0.00500	0.0241	0.000664	0.00375	4.82	5.64
2E + 06	0.00354	0.0210	0.000470	0.00326	5.93	6.95
5E + 06	0.00224	0.0175	0.000297	0.00272	7.81	9.15
1E + 07	0.00158	0.0152	0.000210	0.00236	9.62	11.3
2E + 07	0.00112	0.0132	0.000148	0.00206	11.8	13.9
5E + 07	0.000707	0.0110	0.0000939	0.00171	15.6	18.3

Note: BL = boundary layer.

The MIE cannot be solved in closed form for flows with nonzero pressure gradients. However, the role of the pressure gradient can be understood qualitatively by studying the MIE.

Effect of Pressure Gradient

Boundary layer flow with favorable, zero, and adverse pressure gradients is depicted schematically in Figure 6.3. (Assume a thin boundary layer, so flow on the lower surface behaves as external flow on a surface, with the pressure gradient impressed on the boundary layer.) The pressure gradient is favorable when $\partial p/\partial x < 0$, zero when $\partial p/\partial x = 0$, and adverse when $\partial p/\partial x > 0$, as indicated for Regions 1, 2, and 3.

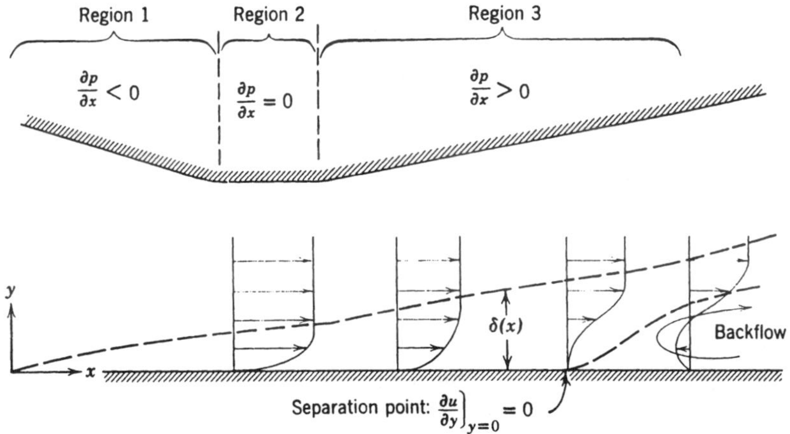

FIGURE 6.3 Boundary layer flow with presssure gradient (thickness exaggerated for clarity).

Viscous shear always causes a net retarding force on any fluid particle within the boundary layer. For zero pressure gradient, shear forces alone can never bring the particle to rest. (Recall that for laminar and turbulent boundary layers the shear stress varied as $1/x^{1/2}$ and $1/x^{1/5}$, respectively; shear stress never becomes zero for finite x.) Since shear stress is given by $\tau_w = \mu \; \partial u/\partial y)_{y=0}$, the velocity gradient cannot be zero. Therefore, flow cannot separate in a zero pressure gradient; shear stresses alone can never cause flow separation.

In the favorable pressure gradient of Region 1, pressure forces tend to maintain the motion of the particle, so flow cannot separate. In the adverse pressure gradient of Region 3, pressure forces oppose the motion of a fluid particle. An adverse pressure gradient is a necessary condition for flow separation.

Velocity profiles for laminar and turbulent boundary layers are shown in Figure 6.2. It is easy to see that the turbulent velocity profile has much more momentum than the laminar profile. Therefore, the turbulent velocity profile can resist separation in an adverse pressure gradient better than the laminar profile.

The freestream velocity distribution must be known before the MIE can be applied. We obtain a first approximation by applying potential flow theory to calculate the flow field around the object. Much effort has been devoted to calculation of velocity distributions over objects of known shape (the "direct" problem) and to determination of shapes to produce a desired pressure distribution (the "inverse" problem). Detailed discussion of such calculation schemes is beyond the scope of this section; the state of the art continues to progress rapidly.

Drag

Any object immersed in a viscous fluid flow experiences a net force from the shear stresses and pressure differences caused by the fluid motion. *Drag* is the force component parallel to, and *lift* is the force component perpendicular to, the flow direction. *Streamlining* is the art of shaping a body to reduce fluid dynamic drag. Airfoils (hydrofoils) are designed to produce lift in air (water); they are streamlined to reduce drag and thus to attain high lift–drag ratios.

In general, lift and drag cannot be predicted analytically for flows with separation, but progress continues on computational fluid dynamics methods. For many engineering purposes, drag and lift forces are calculated from experimentally derived coefficients, discussed below.

Drag coefficient is defined as

$$C_D = \frac{F_D}{\frac{1}{2}\rho V^2 A} \tag{6.9}$$

where $^1/_2\rho V^2$ is dynamic pressure and A is the area upon which the coefficient is based. Common practice is to base drag coefficients on projected *frontal area* (Fox and McDonald, 1992).

Similitude was treated in Section 3. In general, the drag coefficient may be expressed as a function of Reynolds number, Mach number, Froude number, relative roughness, submergence divided by length, and so forth. In this section we consider neither high-speed flow nor free-surface effects, so we will consider only Reynolds number and roughness effects on drag coefficient.

Friction Drag

The total friction drag force acting on a plane surface aligned with the flow direction can be found by integrating the shear stress distribution along the surface. The drag coefficient for this case is defined as friction force divided by dynamic pressure and *wetted area* in contact with the fluid. Since shear stress is a function of Reynolds number, so is drag coefficient (see Figure 6.4). In Figure 6.4, transition occurs at $Re_x = 500{,}000$; the dashed line represents the drag coefficient at larger Reynolds numbers. A number of empirical correlations may be used to model the variation in C_D shown in Figure 6.4 (Schlichting, 1979).

Extending the laminar boundary layer line to higher Reynolds numbers shows that it is beneficial to delay transition to the highest possible Reynolds number. Some results are presented in Table 6.3; drag is reduced more than 50% by extending laminar boundary layer flow to $Re_L = 10^6$.

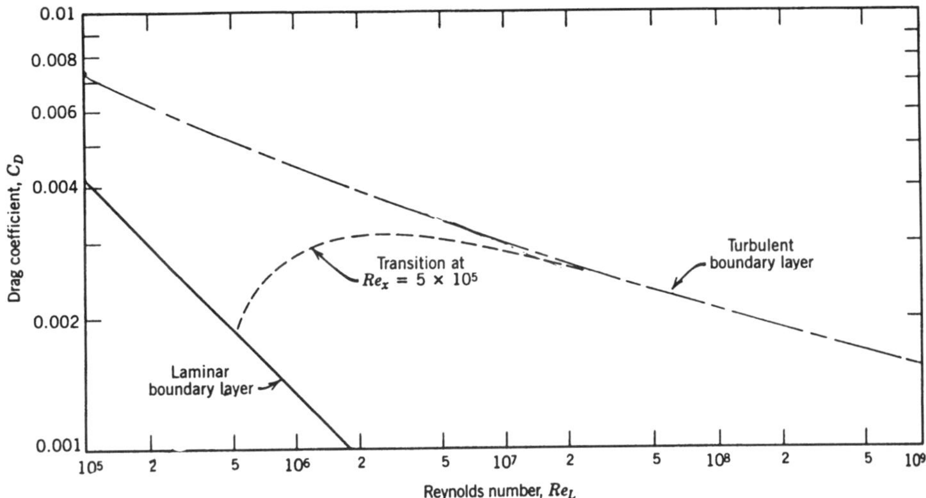

FIGURE 6.4 Drag coefficient vs. Reynolds number for a smooth flat plate parallel to the flow.

TABLE 6.3 Drag Coefficients for Laminar, Turbulent, and Transition Boundary Layers on a Flat Plate

Reynolds Number	Drag Coefficient			Laminar/ Transition	% Drag Reduction
	Laminar BL	Turbulent BL	Transition		
2E + 05	0.00297	0.00615	—	—	—
5E + 05	0.00188	0.00511	0.00189	—	—
1E + 06	0.00133	0.00447	0.00286	0.464	53.6
2E + 06	0.000939	0.00394	0.00314	0.300	70.0
5E + 06	0.000594	0.00336	0.00304	0.195	80.5
1E + 07	0.000420	0.00300	0.00284	0.148	85.2
2E + 07	0.000297	0.00269	0.00261	0.114	88.6
5E + 07	0.000188	0.00235	0.00232	0.081	9.19

Note: BL = Boundary layer.

Pressure Drag

A thin flat surface normal to the flow has no area parallel to the flow direction. Therefore, there can be no friction force parallel to the flow; all drag is caused by pressure forces. Drag coefficients for objects with sharp edges tend to be independent of Reynolds number (for Re > 1000), because the separation points are fixed by the geometry of the object. Drag coefficients for selected objects are shown in Table 6.4.

Rounding the edges that face the flow reduces drag markedly. Compare the drag coefficients for the hemisphere and C-section shapes facing into and away from the flow. Also note that the drag coefficient for a two-dimensional object (long square cylinder) is about twice that for the corresponding three-dimensional object (square cylinder with $b/h = 1$).

Friction and Pressure Drag: Bluff Bodies

Both friction and pressure forces contribute to the drag of *bluff bodies* (see Shapiro, 1960, for a good discussion of the mechanisms of drag). As an example, consider the drag coefficient for a smooth sphere shown in Figure 6.5. Transition from laminar to turbulent flow in the boundary layers on the forward portion of the sphere causes a dramatic dip in drag coefficient at the *critical Reynolds number* ($Re_D \approx 2 \times 10^5$). The turbulent boundary layer is better able to resist the adverse pressure

TABLE 6.4 Drag Coefficient Data for Selected Objects (Re > 1000)

Object	Diagram		$C_D(\mathrm{RE}^* \gtrsim 10^3)$
Square prism	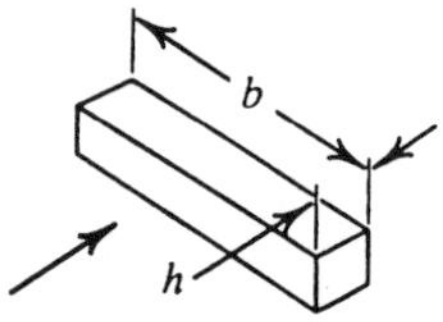	$b/h = \infty$	2.05
		$b/h = 1$	1.05
Disk	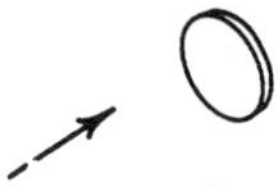		1.17
Ring			1.20[b]
Hemisphere (open end facing flow)	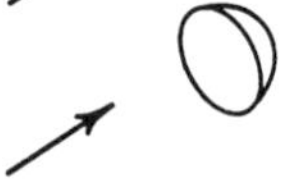		1.42
Hemisphere (open end facing downstream)			0.38
C-section (open side facing flow)	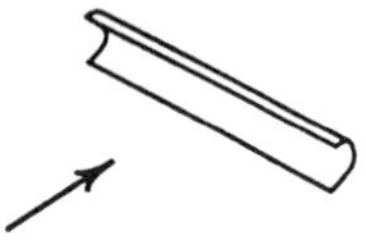		2.30
C-section (open side facing downstream)	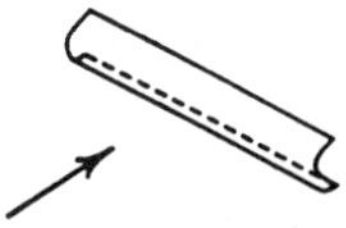		1.20

[a] Data from Hoerner, 1965.
[b] Based on ring area.

gradient on the rear of the sphere, so separation is delayed and the wake is smaller, causing less pressure drag.

Surface roughness (or freestream disturbances) can reduce the critical Reynolds number. Dimples on a golf ball cause the boundary layer to become turbulent and, therefore, lower the drag coefficient in the range of speeds encountered in a drive.

Streamlining

Streamlining is adding a faired tail section to reduce the extent of separated flow on the downstream portion of an object (at high Reynolds number where pressure forces dominate drag). The adverse pressure gradient is taken over a longer distance, delaying separation. However, adding a faired tail increases surface area, causing skin friction drag to increase. Thus, streamlining must be optimized for each shape.

Front contours are of principal importance in road vehicle design; the angle of the back glass also is important (in most cases the entire rear end cannot be made long enough to control separation and reduce drag significantly).

Lift

Lift coefficient is defined as

$$C_L = \frac{F_L}{\frac{1}{2}\rho V^2 A} \tag{6.10}$$

Note that lift coefficient is based on projected *planform area.*

Airfoils

Airfoils are shaped to produce lift efficiently by accelerating flow over the upper surface to produce a low-pressure region. Because the flow must again decelerate, inevitably there must be a region of adverse pressure gradient near the rear of the upper surface (pressure distributions are shown clearly in Hazen, 1965).

Lift and drag coefficients for airfoil sections depend on Reynolds number and *angle of attack* between the chord line and the undisturbed flow direction. The *chord line* is the straight line joining the leading and trailing edges of the airfoil (Abbott and von Doenhoff, 1959).

As the angle of attack is increased, the minimum pressure point moves forward on the upper surface and the minimum pressure becomes lower. This increases the adverse pressure gradient. At some angle of attack, the adverse pressure gradient is strong enough to cause the boundary layer to separate completely from the upper surface, causing the airfoil to *stall.* The separated flow alters the pressure distribution, reducing lift sharply.

Increasing the angle of attack also causes the the drag coefficient to increase. At some angle of attack below stall the ratio of lift to drag, the *lift–drag* ratio, reaches a maximum value.

Drag Due to Lift

For wings (airfoils of finite span), lift and drag also are functions of aspect ratio. Lift is reduced and drag increased compared with infinite span, because end effects cause the lift vector to rotate rearward. For a given geometric angle of attack, this reduces effective angle of attack, reducing lift. The additional component of lift acting in the flow direction increases drag; the increase in drag due to lift is called *induced drag.*

The effective aspect ratio includes the effect of planform shape. When written in terms of effective aspect ratio, the drag of a finite-span wing is

$$C_D = C_{D,\infty} + \frac{C_L^2}{\pi ar} \tag{6.11}$$

where ar is effective aspect ratio and the subscript ∞ refers to the infinite section drag coefficient at C_L. For further details consult the references.

The lift coefficient must increase to support aircraft weight as speed is reduced. Therefore, induced drag can increase rapidly at low flight speeds. For this reason, minimum allowable flight speeds for commercial aircraft are closely controlled by the FAA.

Boundary Layer Control

The major part of the drag on an airfoil or wing is caused by skin friction. Therefore, it is important to maintain laminar flow in the boundary layers as far aft as possible; laminar flow sections are designed to do this. It also is important to prevent flow separation and to achieve high lift to reduce takeoff and landing speeds. These topics fall under the general heading of boundary layer control.

Profile Shaping

Boundary layer transition on a conventional airfoil section occurs almost immediately after the minimum pressure at about 25% chord aft the leading edge. Transition can be delayed by shaping the profile to maintain a favorable pressure gradient over more of its length. The U.S. National Advisory Committee

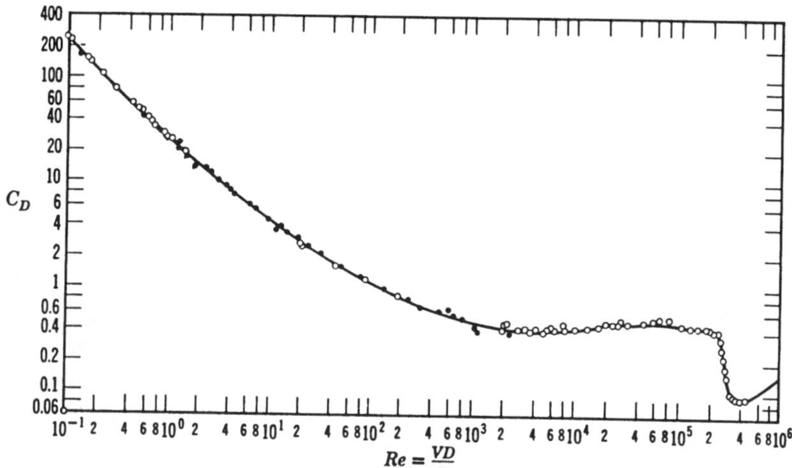

FIGURE 6.5 Drag coefficient vs. Reynolds number for a smooth sphere. (From Schlichting, H. 1979. *Boundary Layer Theory,* 7th ed., McGraw-Hill, New York. With permission.)

for Aeronautics (NACA) developed several series of profiles that delayed transition to 60 or 65% of chord, reducing drag coefficients (in the design range) 60% compared with conventional sections of the same thickness ratio (Abbott and von Doenhoff, 1959).

Flaps and Slats

Flaps are movable sections near the trailing edge of a wing. They extend and/or deflect to increase wing area and/or increase wing camber (curvature), to provide higher lift than the clean wing. Many aircraft also are fitted with leading edge slats which open to expose a slot from the pressure side of the wing to the upper surface. The open slat increases the effective radius of the leading edge, improving maximum lift coefficient. The slot allows energized air from the pressure surface to flow into the low-pressure region atop the wing, energizing the boundary layers and delaying separation and stall.

Suction and Blowing

Suction removes low-energy fluid from the boundary layer, reducing the tendency for early separation. Blowing via high-speed jets directed along the surface reenergizes low-speed boundary layer fluid. The objective of both approaches is to delay separation, thus increasing the maximum lift coefficient the wing can achieve. Powered systems add weight and complexity; they also require bleed air from the engine compressor, reducing thrust or power output.

Moving Surfaces

Many schemes have been proposed to utilize moving surfaces for boundary layer control. Motion in the direction of flow reduces skin friction, and thus the tendency to separate; motion against the flow has the opposite effect. The aerodynamic behavior of sports balls — baseballs, golf balls, and tennis balls — depends significantly on aerodynamic side force (lift, down force, or side force) produced by spin. These effects are discussed at length in Fox and McDonald (1992) and its references.

Computation vs. Experiment

Experiments cannot yet be replaced completely by analysis. Progress in modeling, numerical techniques, and computer power continues to be made, but the role of the experimentalist likely will remain important for the foreseeable future.

Computational Fluid Dynamics (CFD)

Computation of fluid flow requires accurate mathematical modeling of flow physics and accurate numerical procedures to solve the equations. The basic equations for laminar boundary layer flow are well known. For turbulent boundary layers generally it is not possible to resolve the solution space into sufficiently small cells to allow direct numerical simulation. Instead, empirical models for the turbulent stresses must be used. Advances in computer memory storage capacity and speed (e.g., through use of massively parallel processing) continue to increase the resolution that can be achieved.

A second source of error in CFD work results from the numerical procedures required to solve the equations. Even if the equations are exact, approximations must be made to discretize and solve them using finite-difference or finite-volume methods. Whichever is chosen, the solver must guard against introducing numerical instability, round-off errors, and numerical diffusion (Hoffman, 1992).

Role of the Wind Tunnel

Traditionally, wind tunnel experiments have been conducted to verify the design and performance of components and complete aircraft. Design verification of a modern aircraft may require expensive scale models, several thousand hours of wind tunnel time at many thousands of dollars an hour, and additional full-scale flight testing.

New wind tunnel facilities continue to be built and old ones refurbished. This indicates a need for continued experimental work in developing and optimizing aircraft configurations.

Many experiments are designed to produce baseline data to validate computer codes. Such systematic experimental data can help to identify the strengths and weaknesses of computational methods.

CFD tends to become only indicative of trends when massive zones of flow separation are present. Takeoff and landing configurations of conventional aircraft, with landing gear, high-lift devices, and flaps extended, tend to need final experimental confirmation and optimization. Many studies of vertical takeoff and vectored thrust aircraft require testing in wind tunnels.

Defining Terms

Boundary layer: Thin layer of fluid adjacent to a surface where viscous effects are important; viscous effects are negligible outside the boundary layer.

Drag coefficient: Force in the flow direction exerted on an object by the fluid flowing around it, divided by dynamic pressure and area.

Lift coefficient: Force perpendicular to the flow direction exerted on an object by the fluid flowing around it, divided by dynamic pressure and area.

Pressure gradient: Variation in pressure along the surface of an object. For a *favorable* pressure gradient, pressure *decreases* in the flow direction; for an *adverse* pressure gradient, pressure *increases* in the flow direction.

Separation: Phenomenon that occurs when fluid layers adjacent to a solid surface are brought to rest and boundary layers depart from the surface contour, forming a low-pressure *wake* region. Separation can occur only in an *adverse pressure gradient.*

Transition: Change from laminar to turbulent flow within the boundary layer. The location depends on distance over which the boundary layer has developed, pressure gradient, surface roughness, freestream disturbances, and heat transfer.

References

Abbott, I.H. and von Doenhoff, A.E. 1959. *Theory of Wing Sections, Including a Summary of Airfoil Data.* Dover, New York.

Fox, R.W. and McDonald, A.T. 1992. *Introduction to Fluid Mechanics,* 4th ed. John Wiley & Sons, New York.

Hazen, D.C. 1965. *Boundary Layer Control,* film developed by the National Committee for Fluid Mechanics Films (NCFMF) and available on videotape from Encyclopaedia Britannica Educational Corporation, Chicago.

Hoerner, S.F. 1965. *Fluid-Dynamic Drag,* 2nd ed. Published by the author, Midland Park, NJ.

Hoffman, J.D. 1992. *Numerical Methods for Engineers and Scientists.* McGraw-Hill, New York.

Schlichting, H. 1979. *Boundary-Layer Theory,* 7th ed. McGraw-Hill, New York.

Shapiro, A.H. 1960. *The Fluid Dynamics of Drag,* film developed by the National Committee for Fluid Mechanics Film (NCFMF) and available on videotape from Encyclopaedia Britannica Educational Corporation, Chicago.

Further Information

A comprehensive source of basic information is the *Handbook of Fluid Dynamics,* edited by Victor L. Streeter (McGraw-Hill, New York, 1960).

Timely reviews of important topics are published in the *Annual Review of Fluid Mechanics* series (Annual Reviews, Inc., Palo Alto, CA.). Each volume contains a cumulative index.

ASME (American Society of Mechanical Engineers, New York, NY) publishes the *Journal of Fluids Engineering* quarterly. *JFE* contains fluid machinery and other engineering applications of fluid mechanics.

The monthly *AIAA Journal* and bimonthly *Journal of Aircraft* (American Institute for Aeronautics and Astronautics, New York) treat aerospace applications of fluid mechanics.

7

Compressible Flow

Ajay Kumar

NASA Langley Research Center

7 Compressible Flow 87
Introduction • One-Dimensional Flow • Normal Shock Wave • One-Dimensional Flow with Heat Addition • Quasi-One-Dimensional Flow • Two-Dimensional Supersonic Flow

Compressible Flow

Introduction

This section deals with compressible flow. Only one- or two-dimensional steady, inviscid flows under perfect gas assumption are considered. Readers are referred to other sources of information for unsteady effects, viscous effects, and three-dimensional flows.

The term *compressible flow* is routinely used to define variable density flow which is in contrast to incompressible flow, where the density is assumed to be constant throughout. In many cases, these density variations are principally caused by the pressure changes from one point to another. Physically, the *compressibility* can be defined as the fractional change in volume of the gas element per unit change in pressure. It is a property of the gas and, in general, can be defined as

$$\tau = \frac{1}{\rho}\frac{d\rho}{dp}$$

where τ is the compressibility of the gas, ρ is the density, and p is the pressure being exerted on the gas. A more precise definition of compressibility is obtained if we take into account the thermal and frictional losses. If during the compression the temperature of the gas is held constant, it is called the isothermal compressibility and can be written as

$$\tau_T = \frac{1}{\rho}\left(\frac{\partial \rho}{\partial p}\right)_T$$

However, if the compression process is reversible, it is called the isentropic compressibility and can be written as

0-8493-0055-X/00/$0.00+$.50

$$\tau_s = \frac{1}{\rho}\left(\frac{\partial \rho}{\partial p}\right)_s$$

Gases in general have high compressibility (τ_T for air is 10^{-5} m^2/N at 1 atm) as compared with liquids (τ_T for water is 5×10^{-10} m^2/N at 1 atm).

Compressibility is a very important parameter in the analysis of compressible flow and is closely related to the *speed of sound*, *a*, which is the velocity of propagation of small pressure disturbances and is defined as

$$a^2 = \left(\frac{\partial p}{\partial \rho}\right)_s \quad \text{or} \quad a = \sqrt{\left(\frac{\partial p}{\partial \rho}\right)_s}$$

In an isentropic process of a perfect gas, the pressure and density are related as

$$\frac{p}{\rho^\gamma} = \text{constant}$$

Using this relation along with the perfect gas relation $p = \rho RT$, we can show that for a perfect gas

$$a = \sqrt{\gamma RT} = \sqrt{\frac{\gamma p}{\rho}}$$

where γ is the ratio of specific heats at constant pressure and constant volume, R is the gas constant, and T is the temperature. For air under normal conditions, γ is 1.4 and R is 287 m^2/sec^2 K so that the speed of sound for air becomes $a = 20.045\sqrt{T}$ m/sec where T is in kelvin.

Another important parameter in compressible flows is the *Mach number, M,* which is defined as the ratio of the gas velocity to the speed of sound or

$$M = \frac{V}{a}$$

where V is the velocity of gas. Depending upon the Mach number of the flow, we can define the following flow regimes:

$M \ll 1$ Incompressible flow

$M < 1$ Subsonic flow

$M \approx 1$ Transonic flow

$M > 1$ Supersonic flow

$M \gg 1$ Hypersonic flow

Subsonic through hypersonic flows are compressible in nature. In these flows, the velocity is appreciable compared with the speed of sound, and the fractional changes in pressure, temperature, and density are all of significant magnitude. We will restrict ourselves in this section to subsonic through flows only.

Before we move on to study these flows, let us define one more term. Let us consider a gas with static pressure p and temperature T, traveling at some velocity V and corresponding Mach number M. If this gas is brought isentropically to stagnation or zero velocity, the pressure and temperature which the gas achieves are defined as *stagnation pressure* p_0 and *stagnation temperature* T_0 (also called total pressure and total temperature). The speed of sound at stagnation conditions is called the *stagnation speed of sound* and is denoted as a_0.

One-Dimensional Flow

In one-dimensional flow, the flow properties vary only in one coordinate direction. Figure 7.1 shows two streamtubes in a flow. In a *truly one-dimensional flow* illustrated in Figure 7.1(a), the flow variables are a function of x only and the area of the stream tube is constant. On the other hand, Figure 7.1(b) shows a flow where the area of the stream tube is also a function of x but the flow variables are still a function of x only. This flow is defined as the *quasi-one-dimensional flow.* We will first discuss the truly one-dimensional flow.

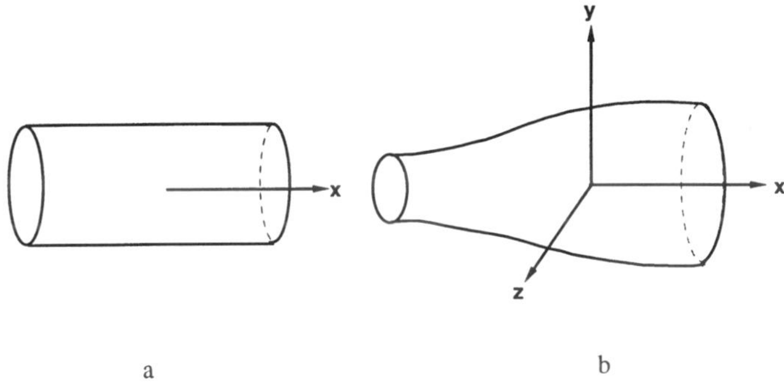

FIGURE 7.1 (a) One-dimensional flow; (b) quasi-one-dimensional flow.

In a steady, truly one-dimensional flow, conservation of mass, momentum, and energy leads to the following simple algebraic equations.

$$\begin{aligned} \rho u &= \text{constan} \\ p + \rho u^2 &= \text{constan} \\ h + \frac{u^2}{2} + q &= \text{constan} \end{aligned} \tag{7.1}$$

where q is the heat added per unit mass of the gas. These equations neglect body forces, viscous stresses, and heat transfer due to thermal conduction and diffusion. These relations given by Equation 7.1, when applied at points 1 and 2 in a flow with no heat addition, become

$$\begin{aligned} \rho_1 u_1 &= \rho_2 u_2 \\ p_1 + \rho_1 u_1^2 &= p_2 + \rho_2 u_2^2 \\ h_1 + \frac{u_1^2}{2} &= h_2 + \frac{u_2^2}{2} \end{aligned} \tag{7.2}$$

The energy equation for a calorically perfect gas, where $h = c_p T$, becomes

$$c_p T_1 + \frac{u_1^2}{2} = c_p T_2 + \frac{u_2^2}{2}$$

Using $c_p = \gamma R/(\gamma - 1)$ and $a^2 = \gamma RT$, the above equation can be written as

$$\frac{a_1^2}{\gamma - 1} + \frac{u_1^2}{2} = \frac{a_2^2}{\gamma - 1} + \frac{u_2^2}{2} \tag{7.3}$$

Since Equation (7.3) is written for no heat addition, it holds for an adiabatic flow. If the energy equation is applied to the stagnation conditions, it can be written as

$$c_p T + \frac{u^2}{2} = c_p T_0$$

$$\frac{T_0}{T} = 1 + \frac{\gamma - 1}{2} M^2 \tag{7.4}$$

It is worth mentioning that in arriving at Equation (7.4), only adiabatic flow condition is used whereas stagnation conditions are defined as those where the gas is brought to rest isentropically. Therefore, the definition of stagnation temperature is less restrictive than the general definition of stagnation conditions. According to the general definition of isentropic flow, it is a reversible adiabatic flow. This definition is needed for the definition of stagnation pressure and density. For an isentropic flow,

$$\frac{p_0}{p} = \left(\frac{\rho_0}{\rho}\right)^{\gamma} = \left(\frac{T_0}{T}\right)^{\gamma/(\gamma-1)} \tag{7.5}$$

From Equations 7.4 and 7.5, we can write

$$\frac{p_0}{p} = \left(1 + \frac{\gamma - 1}{2} M^2\right)^{\gamma/(\gamma-1)} \tag{7.6}$$

$$\frac{\rho_0}{\rho} = \left(1 + \frac{\gamma - 1}{2} M^2\right)^{1/(\gamma-1)} \tag{7.7}$$

Values of stagnation conditions are tabulated in Anderson (1982) as a function of M for $\gamma = 1.4$.

Normal Shock Wave

A shock wave is a very thin region (of the order of a few molecular mean free paths) across which the static pressure, temperature, and density increase whereas the velocity decreases. If the shock wave is perpendicular to the flow, it is called a *normal shock wave.* The flow is supersonic ahead of the normal shock wave and subsonic behind it. Figure 7.2 shows the flow conditions across a normal shock wave which is treated as a discontinuity. Since there is no heat added or removed, the flow across the shock wave is adiabatic. By using Equations 7.2 the normal shock equations can be written as

$$\rho_1 u_1 = \rho_2 u_2$$

$$p_1 + \rho_1 u_1^2 = p_2 + \rho_2 u_2^2 \tag{7.8}$$

$$h_1 + \frac{u_1^2}{2} = h_2 + \frac{u_2^2}{2}$$

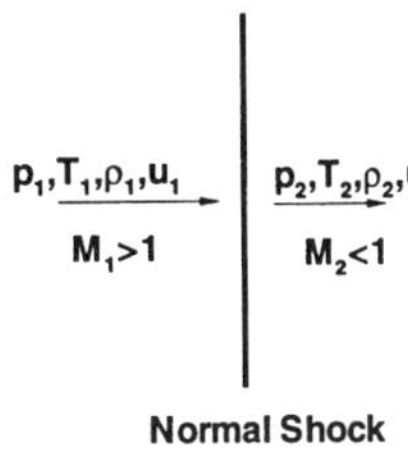

FIGURE 7.2 Flow conditions across a normal shock.

Equations (7.8) are applicable to a general type of flow; however, for a calorically perfect gas, we can use the relations $p = \rho RT$ and $h = c_pT$ to derive a number of equations relating flow conditions downstream of the normal shock to those at upstream. These equations (also known as Rankine–Hugoniot relations) are

$$\frac{p_2}{p_1} = 1 + \frac{2\gamma}{\gamma+1}\left(M_1^2 - 1\right)$$

$$\frac{\rho_2}{\rho_1} = \frac{u_1}{u_2} = \frac{(\gamma+1)M_1^2}{2+(\gamma-1)M_1^2} \tag{7.9}$$

$$\frac{T_2}{T_1} = \frac{h_2}{h_1} = \left[1 + \frac{2\gamma}{\gamma+1}\left(M_1^2 - 1\right)\right]\left[\frac{2+(\gamma-1)M_1^2}{(\gamma+1)M_1^2}\right]$$

$$M_2^2 = \frac{1 + \dfrac{\gamma-1}{2}M_1^2}{\gamma M_1^2 - \dfrac{\gamma-1}{2}}$$

Again, the values of p_2/p_1, ρ_2/ρ_1, T_2/T_1, etc. are tabulated in Anderson (1982) as a function of M_1 for $\gamma = 1.4$. Let us examine some limiting cases. As $M_1 \to 1$, Equations 7.9 yield $M_2 \to 1$, $p_2/p_1 \to 1$, $\rho_2/\rho_1 \to 1$, and $T_2/T_1 \to 1$. This is the case of an extremely weak normal shock across which no finite changes occur. This is the same as the sound wave. On the other hand, as $M_1 \to \infty$, Equations (7.9) yield

$$M_2 \to \sqrt{\frac{\gamma-1}{2\gamma}} = 0.378; \quad \frac{\rho_2}{\rho_1} \to \frac{\gamma+1}{\gamma-1} = 6; \quad \frac{p_2}{p_1} \to \infty; \quad \frac{T_2}{T_1} \to \infty$$

However, the calorically perfect gas assumption no longer remains valid as $M_1 \to \infty$.

Let us now examine why the flow ahead of a normal shock wave must be supersonic even though Equations (7.8) hold for $M_1 < 1$ as well as $M_1 > 1$. From the second law of thermodynamics, the entropy change across the normal shock can be written as

$$s_2 - s_1 = c_p \ln\frac{T_2}{T_1} - R\ln\frac{p_2}{p_1}$$

By using Equations (7.9) it becomes

$$s_2 - s_1 = c_p \ln\left\{\left[1 + \frac{2\gamma}{\gamma+1}\left(M_1^2 - 1\right)\right]\left[\frac{2 + (\gamma-1)M_1^2}{(\gamma+1)M_1^2}\right]\right\} - R\ln\left[1 + \frac{2\gamma}{\gamma+1}\left(M_1^2 - 1\right)\right] \quad (7.10)$$

Equation (7.10) shows that the entropy change across the normal shock is also a function of M_1 only. Using Equation (7.10) we see that

$$\begin{aligned} s_2 - s_1 &= 0 \quad \text{for} \quad M_1 = 1 \\ &< 0 \quad \text{for} \quad M_1 < 1 \\ &> 0 \quad \text{for} \quad M_1 > 1 \end{aligned}$$

Since it is necessary that $s_2 - s_1 \geqslant 0$ from the second law, $M_1 \geq 1$. This, in turn, requires that $p_2/p_1 \geqslant 1$, $\rho_2/\rho_1 \geqslant 1$, $T_2/T_1 \geqslant 1$, and $M_2 \leqslant 1$.

We now examine how the stagnation conditions change across a normal shock wave. For a calorically perfect gas, the energy equation in Equations (7.9) gives

$$c_p T_{01} = c_p T_{02} \quad \text{or} \quad T_{01} = T_{02}$$

In other words, the total temperature remains constant across a stationary normal shock wave.

Let us now apply the entropy change relation across the shock using the stagnation conditions.

$$s_2 - s_1 = c_p \ln\frac{T_{02}}{T_{01}} - R\ln\frac{p_{02}}{p_{01}}$$

Note that entropy at stagnation conditions is the same as at the static conditions since to arrive at stagnation conditions, the gas is brought to rest isentropically. Since $T_{02} = T_{01}$,

$$s_2 - s_1 = -R\ln\frac{p_{02}}{p_{01}}$$

$$\frac{p_{02}}{p_{01}} = e^{-(s_2 - s_1)/R} \quad (7.11)$$

Since $s_2 > s_1$ across the normal shockwave, Equation (7.11) gives $P_{02} < P_{01}$ or, in other words, the total pressure decreases across a shock wave.

One-Dimensional Flow with Heat Addition

Consider one-dimensional flow through a control volume as shown in Figure 7.3. Flow conditions going into this control volume are designated by 1 and coming out by 2. A specified amount of heat per unit mass, q, is added to the control volume. The governing equations relating conditions 1 and 2 can be written as

$$\rho_1 u_1 = \rho_2 u_2$$

$$p_1 + \rho_1 u_1^2 = p_2 + \rho_2 u_2^2 \tag{7.12}$$

$$h_1 + \frac{u_1^2}{2} + q = h_2 + \frac{u_2^2}{2}$$

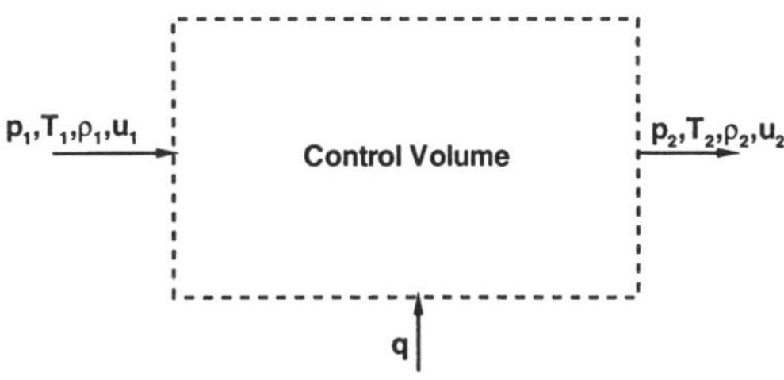

FIGURE 7.3 One-dimensional control volume with heat addition.

The following relations can be derived from Equation (7.12) for a calorically perfect gas

$$q = c_p\left(T_{02} - T_{01}\right) \tag{7.13}$$

$$\frac{p_2}{p_1} = \frac{1+\gamma M_1^2}{1+\gamma M_2^2} \tag{7.14}$$

$$\frac{T_2}{T_1} = \left(\frac{1+\gamma M_1^2}{1+\gamma M_2^2}\right)^2 \left(\frac{M_2}{M_1}\right)^2 \tag{7.15}$$

$$\frac{\rho_2}{\rho_1} = \left(\frac{1+\gamma M_2^2}{1+\gamma M_1^2}\right)^2 \left(\frac{M_1}{M_2}\right)^2 \tag{7.16}$$

Equation (7.13) indicates that the effect of heat addition is to directly change the stagnation temperature T_0 of the flow. Table 7.1 shows some physical trends which can be obtained with heat addition to subsonic and supersonic flow. With heat extraction the trends in Table 7.1 are reversed.

TABLE 7.1 Effect of Heat Addition on Subsonic and Supersonic Flow

	$M_1 < 1$	$M_1 > 1$
M_2	Increases	Decreases
p_2	Decreases	Increases
T_2	Increases for $M_1 < \gamma^{-1/2}$ and decreases for $M_1 > \gamma^{-1/2}$	Increases
u_2	Increases	Decreases
T_{02}	Increases	Increases
p_{02}	Decreases	Decreases

Figure 7.4 shows a plot between enthalpy and entropy, also known as the Mollier diagram, for one-dimensional flow with heat addition. This curve is called the Rayleigh curve and is drawn for a set of given initial conditions. Each point on this curve corresponds to a different amount of heat added or removed. It is seen from this curve that heat addition always drives the Mach numbers toward 1. For a

certain amount of heat addition, the flow will become sonic. For this condition, the flow is said to be *choked.* Any further increase in heat addition is not possible without adjustment in initial conditions. For example, if more heat is added in region 1, which is initially supersonic, than allowed for attaining Mach 1 in region 2, then a normal shock will form inside the control volume which will suddenly change the conditions in region 1 to subsonic. Similarly, in case of an initially subsonic flow corresponding to region 1′, any heat addition beyond that is needed to attain Mach 1 in region 2, the conditions in region 1′ will adjust to a lower subsonic Mach number through a series of pressure waves.

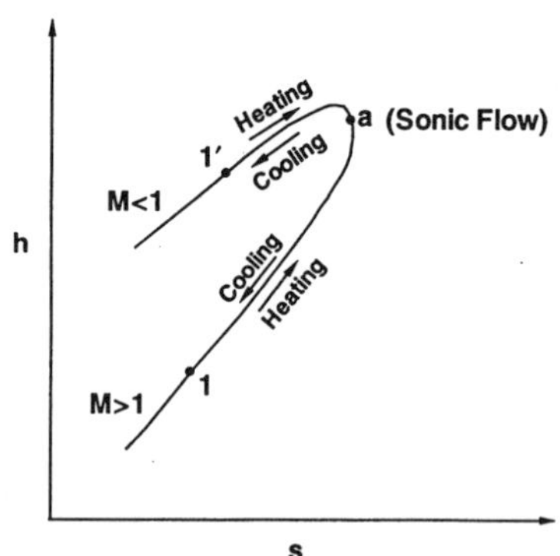

FIGURE 7.4 The Rayleigh curve.

Similar to the preceding heat addition or extraction relationships, we can also develop relationships for one-dimensional steady, adiabatic flow but with frictional effects due to viscosity. In this case, the momentum equation gets modified for frictional shear stress. For details, readers are referred to Anderson (1982).

Quasi-One-Dimensional Flow

In quasi-one-dimensional flow, in addition to flow conditions, the area of duct also changes with x. The governing equations for quasi-one-dimensional flow can be written in a differential form as follows using an infinitesimal control volume shown in Figure 7.5.

$$d(\rho u A) = 0 \tag{7.17}$$

$$dp + \rho u \, du = 0 \tag{7.18}$$

$$dh + u \, du = 0 \tag{7.19}$$

Equation 7.17 can be written as

$$\frac{d\rho}{\rho} + \frac{du}{u} + \frac{dA}{A} = 0 \tag{7.20}$$

which can be further written as follows for an isentropic flow:

$$\frac{dA}{A} = \left(M^2 - 1\right)\frac{du}{u} \tag{7.21}$$

Some very useful physical insight can be obtained from this area–velocity relation.

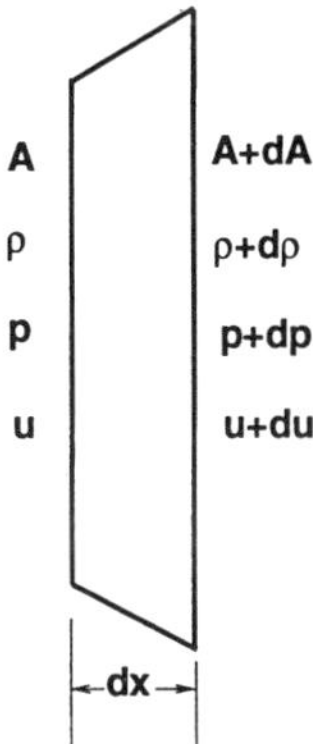

FIGURE 7.5 Control volume for quasi-one-dimensional flow.

- For subsonic flow ($0 \le M < 1$), an increase in area results in decrease in velocity, and vice versa.
- For supersonic flow ($M > 1$), an increase in area results in increase in velocity, and vice versa.
- For sonic flow ($M = 1$), $dA/A = 0$, which corresponds to a minimum or maximum in the area distribution, but it can be shown that a minimum in area is the only physical solution.

Figure 7.6 shows the preceding results in a schematic form.

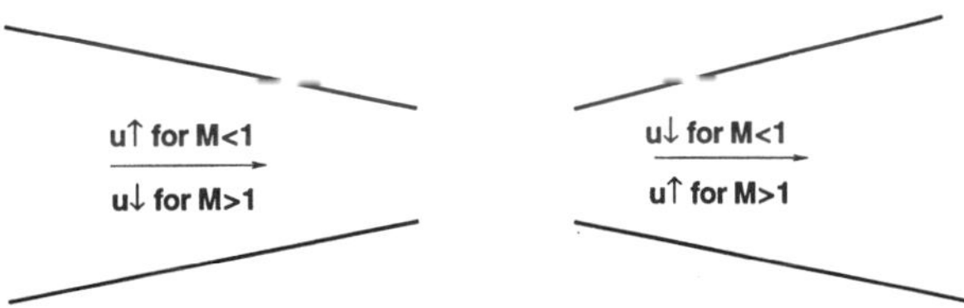

FIGURE 7.6 Compressible flow in converging and diverging ducts.

It is obvious from this discussion that for a gas to go isentropically from subsonic to supersonic, and vice versa, it must flow through a convergent–divergent nozzle, also known as the de Laval nozzle. The minimum area of the nozzle at which the flow becomes sonic is called the throat. This physical observation forms the basis of designing supersonic wind tunnels shown schematically in Figure 7.7. In

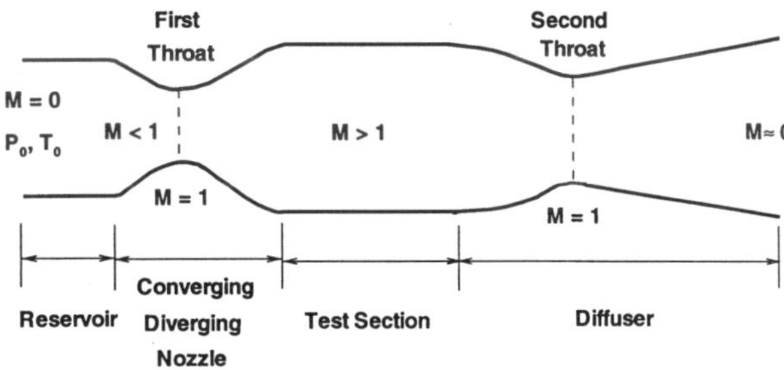

FIGURE 7.7 Schematic of a typical supersonic wind tunnel.

general, in a supersonic wind tunnel, a stagnant gas is first expanded to the desired supersonic Mach number. The supersonic flow enters the test section where it passes over a model being tested. The flow then is slowed down by compressing it through a second convergent–divergent nozzle, also known as a diffuser, before it is exhausted to the atmosphere.

Now, using the equations for quasi-one-dimensional flow and the isentropic flow conditions, we can derive a relation for the area ratio that is needed to accelerate or decelerate the gas to sonic conditions. Denoting the sonic conditions by an asterisk, we can write $u^* = a^*$. The area is denoted as A^*, and it is obviously the minimum area for the throat of the nozzle. From Equation (7.17) we have

$$\rho u A = \rho^* u^* A^*$$

$$\frac{A}{A^*} = \frac{\rho^* u^*}{\rho u} = \frac{\rho^*}{\rho_0}\frac{\rho_0}{\rho}\frac{u^*}{u} \tag{7.22}$$

Under isentropic conditons,

$$\frac{\rho_0}{\rho} = \left(1 + \frac{\gamma - 1}{2}M^2\right)^{1/(\gamma-1)} \tag{7.23}$$

$$\frac{\rho_0}{\rho^*} = \left(1 + \frac{\gamma - 1}{2}\right)^{1/(\gamma-1)} = \left(\frac{\gamma + 1}{2}\right)^{1/(\gamma-1)} \tag{7.24}$$

Also, $u^*/u = a^*/u$. Let us define a Mach number $M^* = u/a^*$. M^* is known as the *characteristic Mach number* and it is related to the local Mach number by the following relation:

$$M^{*2} = \frac{\frac{\gamma + 1}{2}M^2}{1 + \frac{\gamma - 1}{2}M^2} \tag{7.25}$$

Using Equations (7.23) through (7.25) in Equation (7.22) we can write

$$\left(\frac{A}{A^*}\right)^2 = \frac{1}{M^2}\left[\left(\frac{2}{\gamma + 1}\right)\left(1 + \frac{\gamma - 1}{2}M^2\right)\right]^{(\gamma+1)/(\gamma-1)} \tag{7.26}$$

Equation (7.26) is called the area Mach number relation. Figure 7.8 shows a plot of A/A^* against Mach number. A/A^* is always ≥ 1 for physically viable solutions.

The area Mach number relation says that for a given Mach number, there is only one area ratio A/A^*. This is a very useful relation and is frequently used to design convergent–divergent nozzles to produce a desired Mach number. Values of A/A^* are tabulated as a function of M in Anderson (1982).

Equation (7.26) can also be written in terms of pressure as follows:

$$\frac{A}{A^*} = \frac{\left[1 - \left(\frac{p}{p_0}\right)^{(\gamma-1)/\gamma}\right]^{1/2}\left(\frac{p}{p_0}\right)^{1/\gamma}}{\left(\frac{\gamma - 1}{2}\right)^{1/2}\left(\frac{2}{\gamma + 1}\right)^{(\gamma+1)/2(\gamma-1)}} \tag{7.27}$$

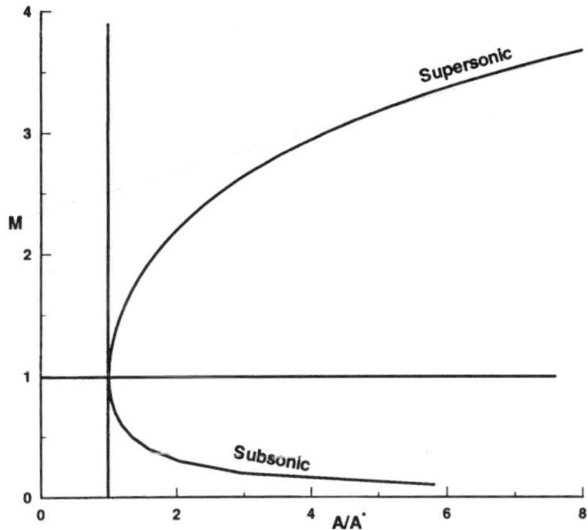

FIGURE 7.8 Variation of area ratio A/A* as a function of Mach number for a quasi-one-dimensional flow.

Nozzle Flow

Using the area relations, we can now plot the distributions of Mach number and pressure along a nozzle. Figure 7.9 shows pressure and Mach number distributions along a given nozzle and the wave configurations for several exit pressures. For curves a and b, the flow stays subsonic throughout and the exit pressure controls the flow in the entire nozzle. On curve c, the throat has just become sonic, and so the pressure at the throat, and upstream of it, can decrease no further. There is another exit pressure corresponding to curve j ($p_j < p_c$) for which a supersonic isentropic solution exists. But if the pressure lies between p_c and p_j, there is no isentropic solution possible. For example, for an exit pressure p_d, a shock will form in the nozzle at location s which will raise the pressure to $p_{d'}$ and turn the flow subsonic. The pressure will then rise to p_d as the subsonic flow goes through an increasing area nozzle. The location, s, depends on the exit pressure. Various possible situations are shown in Figure 7.9. It is clear that if the exit pressure is equal to or below p_f, the flow within the nozzle is fully supersonic. This is the principle used in designing supersonic wind tunnels by operating from a high-pressure reservoir or into a vacuum receiver, or both.

Diffuser

If a nozzle discharges directly into the receiver, the minimum pressure ratio for full supersonic flow in the test section is

$$\left(\frac{p_0}{p_E}\right)_{\min} = \frac{p_0}{p_f}$$

where p_f is the value of p_E at which the normal shock stands right at the nozzle exit. However, by adding an additional diverging section, known as a diffuser, downstream of the test section as shown in Figure 7.10 it is possible to operate the tunnel at a lower pressure ratio than p_0/p_f. This happens because the diffuser can now decelerate the subsonic flow downstream of the shock isentropically to a stagnation pressure p_0'. The pressure ratio required then is the ratio of stagnation pressures across a normal shock wave at the test section Mach number. In practice, the diffuser gives lower than expected recovery as a result of viscous losses caused by the interaction of shock wave and the boundary layer which are neglected here.

The operation of supersonic wind tunnels can be made even more efficient; i.e., they can be operated at even lower pressure ratios than p_0/p_0', by using the approach shown in Figure 7.7 where the diffuser has a second throat. It can slow down the flow to subsonic Mach numbers isentropically and, ideally,

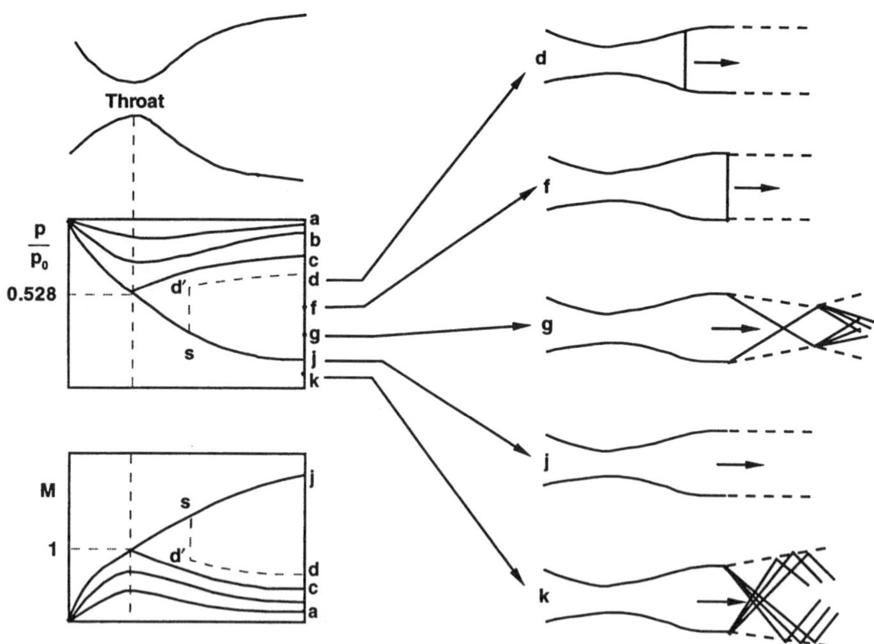

FIGURE 7.9 Effect of exit pressure on flow through a nozzle.

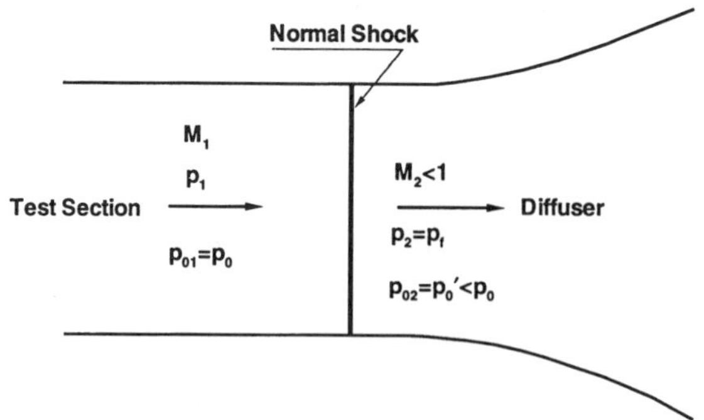

FIGURE 7.10 Normal shock diffuser.

can provide complete recovery, giving $p_0' = p_0$. However, due to other considerations, such as the starting process of the wind tunnel and viscous effects, it is not realized in real life.

Two-Dimensional Supersonic Flow

When supersonic flow goes over a wedge or an expansion corner, it goes through an oblique shock or expansion waves, respectively, to adjust to the change in surface geometry. Figure 7.11 shows the two flow situations. In Figure 7.11(a) an oblique shock abruptly turns the flow parallel to the wedge surface. The Mach number behind the shock is less than ahead of it, whereas the pressure, temperature, and density increase. In the case of an expansion corner, oblique expansion waves smoothly turn the flow to become parallel to the surface downstream of the expansion corner. In this case, the Mach number increases, but the pressure, temperature, and density decrease as the flow goes through the expansion corner. Oblique shocks and expansion waves occur in two- and three-dimensional supersonic flows. In this section, we will restrict ourselves to steady, two-dimensional supersonic flows only.

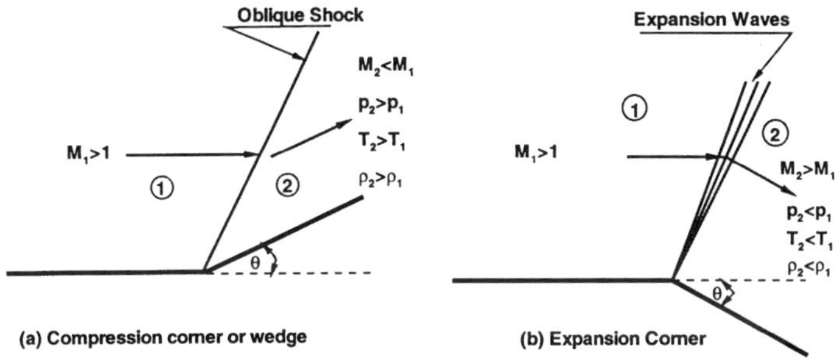

FIGURE 7.11 Supersonic flow over a corner.

Oblique Shock Waves

The oblique shock can be treated in the same way as the normal shock by accounting for the additional velocity component. If a uniform velocity v is superimposed on the flow field of the normal shock, the resultant velocity ahead of the shock can be adjusted to any flow direction by adjusting the magnitude and direction of v. If v is taken parallel to the shock wave, as shown in Figure 7.12, the resultant velocity ahead of the shock is $w_1 = \sqrt{u_1^2 + v_1^2}$ and its direction from the shock is given by $\beta = \tan^{-1}(u_1/v)$. On the downstream side of the shock, since u_2 is less than u_1, the flow always turns toward the shock. The magnitude of u_2 can be determined by the normal shock relations corresponding to velocity u_1 and the magnitude of v is such that the flow downstream of the shock turns parallel to the surface. Since imposition of a uniform velocity does not affect the pressure, temperature, etc., we can use normal shock relations with Mach number replaced in them to correspond to velocity u_1 or u_1/a_1, which is nothing but $M_1 \sin \beta$. Thus, oblique shock relations become

$$\frac{p_2}{p_1} = 1 + \frac{2\gamma}{\gamma+1}\left(M_1^2 \sin^2\beta - 1\right) \tag{7.28}$$

$$\frac{\rho_2}{\rho_1} = \frac{(\gamma+1)M_1^2 \sin^2\beta}{(\gamma-1)M_1^2 \sin^2\beta + 2} \tag{7.29}$$

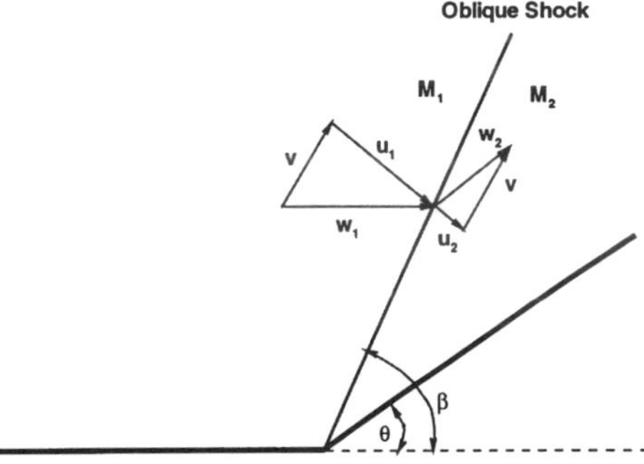

FIGURE 7.12 Oblique shock on a wedge.

$$\frac{T_2}{T_1} = \frac{a_2^2}{a_1^2} = \left[1 + \frac{2\gamma}{\gamma+1}\left(M_1^2 \sin^2\beta - 1\right)\right]\left[\frac{2 + (\gamma-1)M_1^2 \sin^2\beta}{(\gamma+1)M_1^2 \sin^2\beta}\right] \tag{7.30}$$

The Mach number M_2 ($= w_2/a_2$) can be obtained by using a Mach number corresponding to velocity u_2 ($= w_2 \sin(\beta - \theta)$) in the normal shock relation for the Mach number. In other words,

$$M_2^2 \sin^2(\beta - \theta) = \frac{1 + \frac{\gamma - 1}{2} M_1^2 \sin^2 \beta}{\gamma M_1^2 \sin^2 \beta - \frac{\gamma - 1}{2}} \tag{7.31}$$

To derive a relation between the wedge angle θ and the wave angle β, we have from Figure 7.12

$$\tan\beta = \frac{u_1}{v} \quad \text{and} \quad \tan(\beta - \theta) = \frac{u_2}{v}$$

so that

$$\frac{\tan(\beta - \theta)}{\tan\beta} = \frac{u_2}{u_1} = \frac{\rho_1}{\rho_2} = \frac{(\gamma - 1)M_1^2 \sin^2 \beta + 2}{(\gamma + 1)M_1^2 \sin^2 \beta}$$

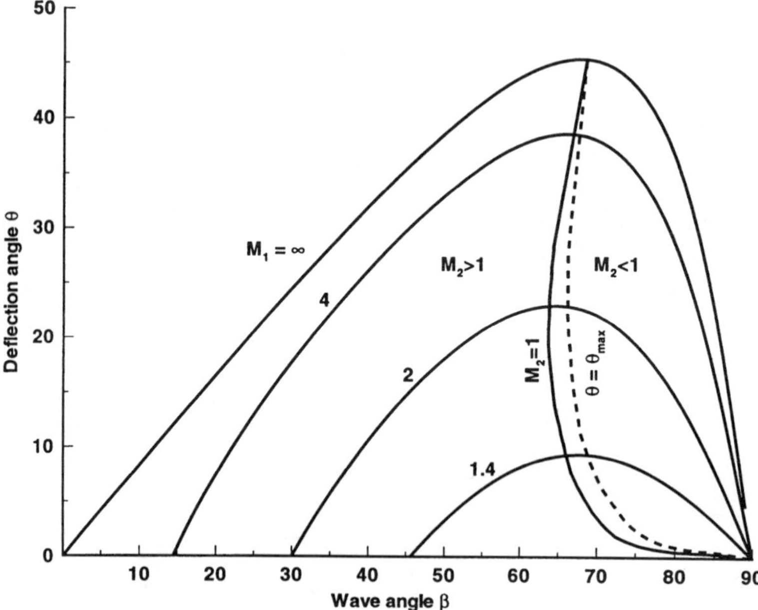

FIGURE 7.13 Oblique shock characteristics.

This can be simplified to

$$\tan\theta = 2\cot\beta \frac{M_1^2 \sin^2 \beta - 1}{M_1^2(\gamma + \cos 2\beta) + 2} \tag{7.32}$$

Dennard and Spencer (1964) have tabulated oblique shock properties as a function of M_1. Let us now make some observations from the preceding relations.

From the normal shock relations, $M_1 \sin \beta \geqslant 1$. This defines a minimum wave angle for a given Mach number. The maximum wave angle, of course, corresponds to the normal shock or $\beta = \pi/2$. Therefore, the wave angle β has the following range

$$\sin^{-1}\frac{1}{M} \le \beta \le \frac{\pi}{2} \tag{7.33}$$

Equation 7.32 becomes zero at the two limits of β. Figure 7.13 shows a plot of θ against β for various values of M_1. For each value of M_1, there is a maximum value of θ. For $\theta < \theta_{max}$, there are two possible solutions having different values of β. The larger value of β gives the stronger shock in which the flow becomes subsonic. A locus of solutions for which $M_2 = 1$ is also shown in the figure. It is seen from the figure that with weak shock solution, the flow remains supersonic except for a small range of θ slightly smaller than θ_{max}.

Let us now consider the limiting case of θ going to zero for the weak shock solution. As θ decreases to zero, β decreases to the limiting value μ, given by

$$M_1^2 \sin^2 \mu - 1 = 0$$

$$\mu = \sin^{-1} \frac{1}{M_1} \tag{7.34}$$

For this angle, the oblique shock relations show no jump in flow quantities across the wave or, in other words, there is no disturbance generated in the flow. This angle μ is called the *Mach angle* and the lines at inclination μ are called *Mach lines.*

Thin-Airfoil Theory

For a small deflection angle Δθ, it can be shown that the change in pressure in a flow at Mach M_1 is given approximately by

$$\frac{\Delta p}{p_1} \approx \frac{\gamma M_1^2}{\sqrt{M_1^2 - 1}} \Delta\theta \tag{7.35}$$

This expression holds for both compression and expansion. If Δp is measured with respect to the freestream pressure, p_1, and all deflections to the freestream direction, we can write Equation (7.35) as

$$\frac{p - p_1}{p_1} = \frac{\gamma M_1^2}{\sqrt{M_1^2 - 1}} \theta \tag{7.36}$$

where θ is positive for a compression and negative for expansion. Let us define a pressure coefficient C_p, as

$$C_p = \frac{p - p_1}{q_1}$$

where q_1 is the dynamic pressure and is equal to $\gamma p_1 M_1^2 / 2$. Equation (7.36) then gives

$$C_p = \frac{2\theta}{\sqrt{M_1^2 - 1}} \tag{7.37}$$

Equation (7.37) states that the pressure coefficient is proportional to the local flow deflection. This relation can be used to develop supersonic thin-airfoil theory. As an example, for a flat plate at angle of attack α_0 (shown in Figure 7.14), the pressure coefficients on the upper and lower surfaces are

$$C_p = \mp \frac{2\alpha_0}{\sqrt{M_1^2 - 1}}$$

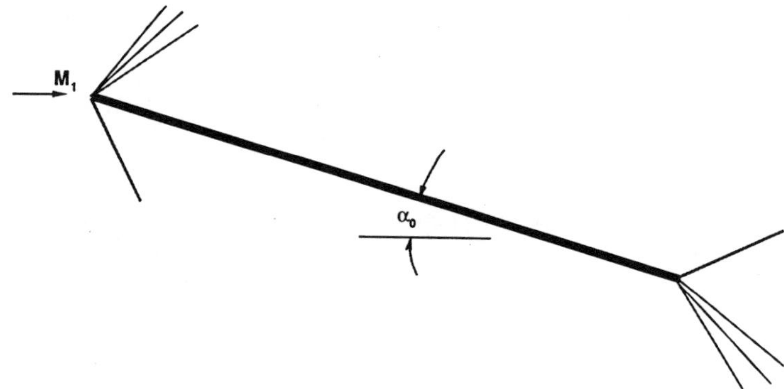

FIGURE 7.14 Lifting flat plate.

The lift and drag coefficients can be written as

$$C_L = \frac{(p_L - p_U)c\cos\alpha_0}{q_1 c} = \left(C_{p_L} - C_{p_U}\right)\cos\alpha_0$$

$$C_D = \frac{(p_L - p_U)c\sin\alpha_0}{q_1 c} = \left(C_{p_L} - C_{p_U}\right)\sin\alpha_0$$

where c is the chord length of the plate. Since α_0 is small, we can write

$$C_L = \frac{4\alpha_0}{\sqrt{M_1^2 - 1}}, \qquad C_D = \frac{4\alpha_0^2}{\sqrt{M_1^2 - 1}} \tag{7.38}$$

A similar type of expression can be obtained for an arbitrary thin airfoil that has thickness, camber, and angle of attack. Figure 7.15 shows such an airfoil. The pressure coefficients on the upper and lower surfaces can be written as

$$C_{p_U} = \frac{2}{\sqrt{M_1^2 - 1}}\frac{dy_U}{dx}, \qquad C_{p_L} = \frac{2}{\sqrt{M_1^2 - 1}}\left(-\frac{dy_L}{dx}\right) \tag{7.39}$$

For the thin airfoil, the profile may be resolved into three separate components as shown in Figure 7.15. The local slope of the airfoil can be obtained by superimposing the local slopes of the three components as

$$\frac{dy_U}{dx} = -\left(\alpha_0 + \alpha_c(x)\right) + \frac{dh}{dx} = -\alpha(x) + \frac{dh}{dx}$$

$$\frac{dy_L}{dx} = -\left(\alpha_0 + \alpha_c(x)\right) - \frac{dh}{dx} = -\alpha(x) - \frac{dh}{dx} \tag{7.40}$$

where $\alpha = \alpha_0 + \alpha_c\,(x)$ is the local total angle of attack of the camber line. The lift and drag for the thin airfoil are given by

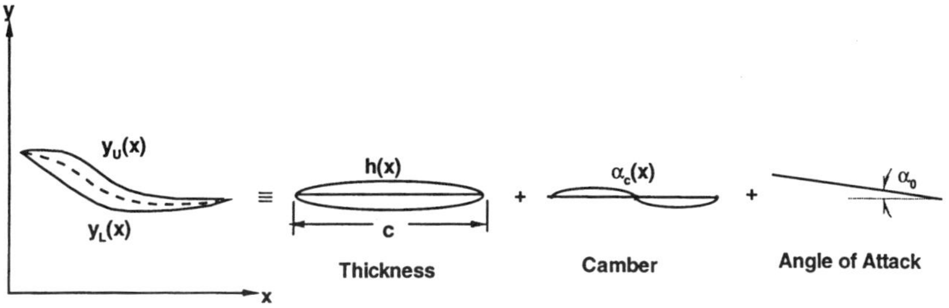

FIGURE 7.15 Arbitrary thin airfoil and its components.

$$L = q_1 \int_0^c \left(C_{P_L} - C_{P_U}\right) dx$$

$$D = q_1 \int_0^c \left[C_{P_L}\left(-\frac{dy_L}{dx}\right) + C_{P_U}\left(\frac{dy_U}{dx}\right)\right] dx$$

Let us define an average value of $\alpha(x)$ as

$$\bar{\alpha} = \frac{1}{c}\int_0^c \alpha(x)\, dx$$

Using Equation (7.40) and the fact that $\bar{\alpha}_0 = \alpha$ and $\bar{\alpha}_c = 0$ by definition, the lift and drag coefficients for the thin airfoil can be written as

$$C_L = \frac{4\alpha_0}{\sqrt{M_1^2 - 1}}$$

$$C_D = \frac{4}{\sqrt{M_1^2 - 1}}\left[\overline{\left(\frac{dh}{dx}\right)^2} + \overline{\alpha_C^2(x)} + \alpha_0^2\right] \tag{7.41}$$

Equations (7.41) show that the lift coefficient depends only on the mean angle of attack whereas the drag coefficient is a linear combination of the drag due to thickness, drag due to camber, and drag due to lift (or mean angle of attack).

References

Anderson, J.D. 1982. *Modern Compressible Flow,* McGraw-Hill, New York.

Dennard, J.S. and Spencer, P.B. 1964. Ideal-Gas Tables for Oblique-Shock Flow Parameters in Air at Mach Numbers from 1.05 to 12.0. NASA TN D-2221.

Liepmann, H.W. and Roshko, A. 1966. *Elements of Gas Dynamics,* John Wiley & Sons, New York.

Further Information

As mentioned in the beginning, this section discussed only one- or two-dimensional steady, inviscid compressible flows under perfect gas assumption. Even this discussion was quite brief because of space limitations. For more details on the subject as well as for compressible unsteady viscous flows, readers are referred to Anderson (1982) and Liepmann and Roshko (1966).

8

Multiphase Flow

John C. Chen

Lehigh University

8 Multiphase Flow...105
Introduction • Fundamentals • Gas–Liquid Two-Phase Flow • Gas–Solid, Liquid–Solid Two-Phase Flows

Multiphase Flow

Introduction

Classic study of fluid mechanics concentrates on the flow of a single homogeneous phase, e.g., water, air, steam. However, many industrially important processes involve simultaneous flow of multiple phases, e.g., gas bubbles in oil, wet steam, dispersed particles in gas or liquid. Examples include vapor–liquid flow in refrigeration systems, steam–water flows in boilers and condensers, vapor–liquid flows in distillation columns, and pneumatic transport of solid particulates. In spite of their importance, multiphase flows are often neglected in standard textbooks. Fundamental understanding and engineering design procedures for multiphase flows are not nearly so well developed as those for single-phase flows. An added complexity is the need to predict the relative concentrations of the different phases in the multiphase flows, a need that doesn't exist for single-phase flows.

Inadequate understanding not withstanding, a significant amount of data have been collected and combinations of theoretical models and empirical correlations are used in engineering calculations. This knowledge base is briefly summarized in this section and references are provided for additional information. While discussions are provided of solid–gas flows and solid–liquid flows, primary emphasis is placed on multiphase flow of gas–liquids since this is the most often encountered class of multiphase flows in industrial applications.

A multiphase flow occurs whenever two or more of the following phases occur simultaneously: gas/vapor, solids, single-liquid phase, multiple (immiscible) liquid phases. Every possible combination has been encountered in some industrial process, the most common being the simultaneous flow of vapor/gas and liquid (as encountered in boilers and condensers). All multiphase flow problems have features which are characteristically different from those found in single-phase problems. First, the

0-8493-0055-X/00/$0.00+$.50

relative concentration of different phases is usually a dependent parameter of great importance in multiphase flows, while it is a parameter of no consequence in single-phase flows. Second, the spatial distribution of the various phases in the flow channel strongly affects the flow behavior, again a parameter that is of no concern in single-phase flows. Finally, since the density of various phases can differ by orders of magnitude, the influence of gravitational body force on multiphase flows is of much greater importance than in the case of single-phase flows. In any given flow situation, the possibility exists for the various phases to assume different velocities, leading to the phenomena of slip between phases and consequent interfacial momentum transfer. Of course, the complexity of laminar/turbulent characteristics occurs in multiphase flows as in single-phase flows, with the added complexity of interactions between phases altering the laminar/turbulent flow structures. These complexities increase exponentially with the number of phases encountered in the multiphase problem. Fortunately, a large number of applications occur with just two phase flows, or can be treated as pseudo-two-phase flows.

Two types of analysis are used to deal with two-phase flows. The simpler approach utilizes homogeneous models which assume that the separate phases flow with the same identical local velocity at all points in the fluid. The second approach recognizes the possibility that the two phases can flow at different velocities throughout the fluid, thereby requiring separate conservation equations for mass and momentum for each phase. Brief descriptions of both classes of models are given below.

Fundamentals

Consider n phases in concurrent flow through a duct with cross-sectional area A_c. Fundamental quantities that characterize this flow are

$\dot{m}_i$ = mass flow rate of ith phase

u_i = velocity of ith phase

α_i = volume fraction of ith phase in channel

Basic relationships between these and related parameters are

$$G_i = \text{mass flux of } i\text{th phase} = \frac{\dot{m}_i}{A_c} \tag{8.1}$$

$$v_i = \text{superficial velocity of } i\text{th phase} = \frac{G_i}{\rho_i} \tag{8.2}$$

$$u_i = \text{actual velocity of } i\text{th phase} = \frac{v_i}{\alpha_i} \tag{8.3}$$

$$x_i = \text{flow quality of } i\text{th phase} = \frac{\dot{m}_i}{\sum_i^n m_i} = \frac{G_i}{\sum_{i=1}^n G_i} \tag{8.4}$$

α_i = volume fraction of ith phase

$$= \frac{\left(\frac{x_i}{\rho_i u_i}\right)}{\sum_{i=1}^{n}\left(\frac{x_i}{\rho_i u_i}\right)} \tag{8.5}$$

In most engineering calculations, the above parameters are defined as average quantities across the entire flow area, A_c. It should be noted, however, that details of the multiphase flow could involve local variations across the flow area. In the latter situation, G_i, v_i, and α_i are often defined on a local basis, varying with transverse position across the flow area.

Pressure drop along the flow channel is associated with gravitational body force, acceleration forces, and frictional shear at the channel wall. The total pressure gradient along the flow axis can be represented as

$$\frac{dP}{dz} = \left(\frac{dP}{dz}\right)_g + \left(\frac{dP}{dz}\right)_a + \left(\frac{dP}{dz}\right)_f \tag{8.6}$$

where

$$\left(\frac{dP}{dz}\right)_g = -g\cos\theta \cdot \sum_{i=1}^{n}\alpha_i\rho_i \tag{8.7}$$

θ = angle of channel from vertical

and

$$\left(\frac{dP}{dz}\right)_a = -\sum_{i=1}^{n} G_i \frac{du_i}{dz} \tag{8.8}$$

$$\left(\frac{dP}{dz}\right)_f = -\frac{\rho u^2}{2D} f \tag{8.9}$$

ρ = density of multiphase mixture

$$= \sum_{i=1}^{n}\rho_i\alpha_i \tag{8.10}$$

u = an average mixture velocity

$$= \frac{1}{\rho}\sum_{i=1}^{n} G_i \tag{8.11}$$

f = equivalent Darcy friction factor for the multiphase flow

In applications, the usual requirement is to determine pressure gradient (dP/dz) and the volume fractions (α_i). The latter quantities are of particular importance since the volume fraction of individual phases affects all three components of the pressure gradient, as indicated in Equations (8.7) to (8.11). Correlations of various types have been developed for prediction of the volume fractions, all but the simplest of which utilize empirical parameters and functions.

The simplest flow model is known as the homogeneous equilibrium model (HEM), wherein all phases are assumed to be in neutral equilibrium. One consequence of this assumption is that individual phase velocities are equal for all phases everywhere in the flow system:

$$u_i = u \quad \text{for all } i \tag{8.12}$$

This assumption permits direct calculation of the volume fractions from known mass qualities:

$$\alpha_i = \frac{x_i}{\rho_i \sum_{i=1}^{n} \left(\frac{x_i}{\rho_i} \right)} \tag{8.13}$$

The uniform velocity for all phases is the same as mixture velocity:

$$u = \frac{1}{\rho} \sum_{i=1}^{n} G_i \tag{8.14}$$

where

$$\frac{1}{\rho} = \sum_{i=1}^{n} \left(\frac{x_i}{\rho_i} \right) \tag{8.15}$$

This homogeneous model permits direct evaluation of all three components of axial pressure gradient, if flow qualities (x_i) are known:

$$\left(\frac{dP}{dz} \right)_g = -\frac{g \cos\theta}{\sum_{i=1}^{n} \left(\frac{x_i}{\rho_i} \right)} \tag{8.16}$$

$$\left(\frac{dP}{dz} \right)_a = -\left(\sum_{i=1}^{n} G_i \right) \cdot \frac{du}{dz} \tag{8.17}$$

$$\left(\frac{dP}{dz} \right)_f = -\frac{\rho u^2}{2D_f} \cdot f \tag{8.18}$$

where u and ρ are given by Equations (8.14) and (8.15).

Predicting the coefficient of friction (f to clear) remains a problem, even in the homogeneous model. For cases of fully turbulent flows, experience has shown that a value of 0.02 may be used as a first-order approximation for (f to clear). More-accurate estimates require empirical correlations, specific to particular classes of multiphase flows and subcategories of flow regimes.

The following parts of this section consider the more common situations of two-phase flows and describe improved design methodologies specific to individual situations.

Gas–Liquid Two-Phase Flow

The most common case of multiphase flow is two-phase flow of gas and liquid, as encountered in steam generators and refrigeration systems. A great deal has been learned about such flows, including delineation of flow patterns in different flow regimes, methods for estimating volume fractions (gas void fractions), and two-phase pressure drops.

Flow Regimes

A special feature of multiphase flows is their ability to assume different spatial distributions of the phases. These different flow patterns have been classified in flow regimes, which are themselves altered by the direction of flow relative to gravitational acceleration. Figures 8.1 and 8.2 (Delhaye, 1981) show the flow patterns commonly observed for co-current flow of gas and liquid in vertical and horizontal channels, respectively. For a constant liquid flow rate, the gas phase tends to be distributed as small bubbles at low gas flow rates. Increasing gas flow rate causes agglomeration of bubbles into larger slugs and plugs. Further increasing gas flow rate causes separation of the phases into annular patterns wherein liquid concentrates at the channel wall and gas flows in the central core for vertical ducts. For horizontal ducts, gravitational force tends torain the liquid annulus toward the bottom of the channel, resulting in stratified and stratified wavy flows. This downward segregation of the liquid phase can be overcome by kinetic forces at high flow rates, causing stratified flows to revert to annular flows. At high gas flow rates, more of the liquid tends to be entrained as dispersed drops; in the limit one obtains completely dispersed mist flow.

Flow pattern maps are utilized to predict flow regimes for specific applications. The first generally successful flow map was that of Baker (1954) for horizontal flow, reproduced here in Figure 8.3. For vertical flows, the map of Hewitt and Roberts (1969), duplicated in Figure 8.4, provides a simple method for determining flow regimes. Parameters used for the axial coordinates of these flow maps are defined as follows:

$$\lambda = \left(\frac{\rho_g \rho_\ell}{\rho_a \rho_w}\right)^{1/2} \tag{8.19}$$

$$\psi = \left(\frac{\sigma_w}{\sigma}\right)\left[\left(\frac{\mu_\ell}{\mu_w}\right)\left(\frac{\rho_w}{\rho_\ell}\right)^2\right]^{1/3} \tag{8.20}$$

$$j = \text{volumetric flux, } \frac{G}{\rho} \tag{8.21}$$

Void Fractions

In applications of gas–liquid flows, the volume fraction of gas (α_g) is commonly called "void fraction" and is of particular interest. The simplest method to estimate void fraction is by the HEM. From Equation (8.13), the void fraction can be estimated as

$$\alpha_g = \frac{x_g}{x_g + (1 - x_g)\dfrac{\rho_g}{\rho_\ell}} \tag{8.22}$$

where α_g, x_g, ρ_g, ρ_ℓ are cross-sectional averaged quantities.

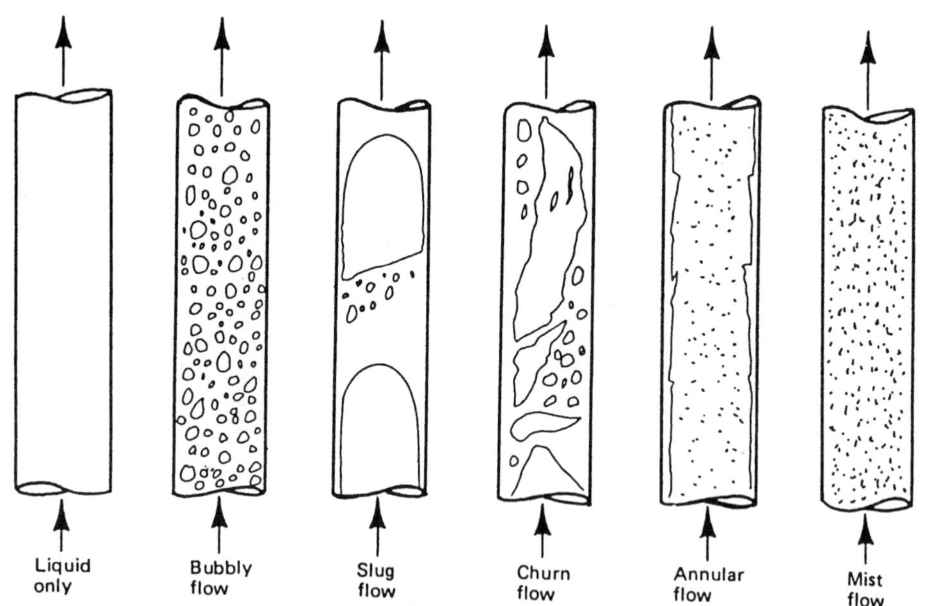

FIGURE 8.1 Flow patterns in gas–liquid vertical flow. (From Lahey, R.T., Jr. and Moody, F.I. 1977. *The Thermal Hydraulics of a Boiling Water Nuclear Reactor,* The American Nuclear Society, LaGrange, IL. With permission.)

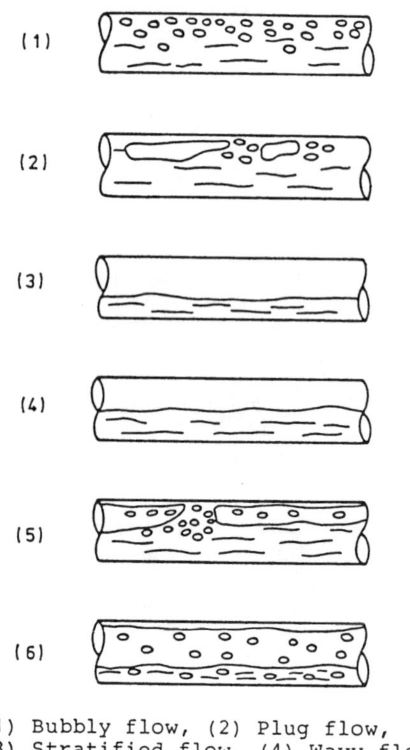

FIGURE 8.2 Flow patterns in gas–liquid horizontal flow.

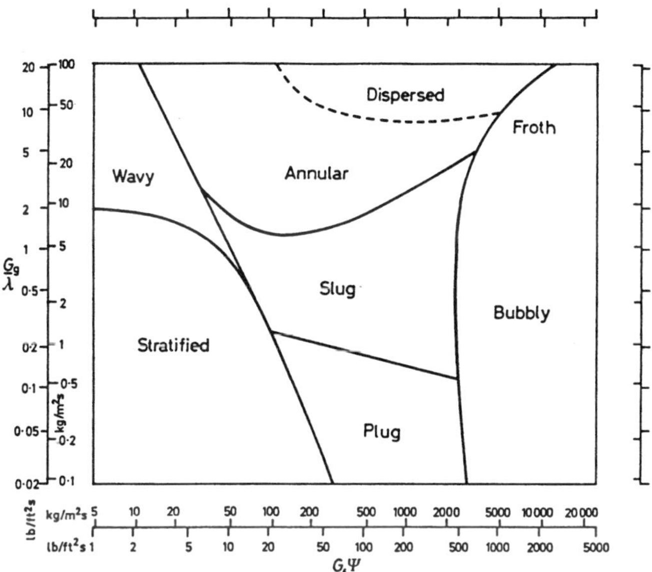

FIGURE 8.3 Flow pattern map for horizontal flow (Baker, 1954). (From Collier, J.G. 1972. *Convective Boiling and Condensation,* McGraw-Hill, London. With permission.)

In most instances, the homogenous model tends to overestimate the void fraction. Improved estimates are obtained by using separated-phase models which account for the possibility of slip between gas and liquid velocities. A classic separated-phase model is thatof Lockhart and Martinelli (1949). The top portion of Figure 8.5 reproduces the Lockhart–Martinelli correlation for void fraction (shown as α) as a function of the parameter X which is defined as

$$X = \left[\left(\frac{dP}{dz} \right)_{f\ell} \div \left(\frac{dP}{dz} \right)_{fg} \right]^{1/2} \tag{8.23}$$

where

$\left(\frac{dP}{dz} \right)_{f\ell}$ = frictional pressure gradient of liquid phase flowing alone in channel

$\left(\frac{dP}{dz} \right)_{fg}$ = frictional pressure gradient of gas phase flowing alone in channel

Often, flow rates are sufficiently high such that each phase if flowing alone in the channel would be turbulent. In this situation the parameter X can be shown to be

$$X_{tt} = \left(\frac{1 - x_g}{x_g} \right)^{0.9} \left(\frac{\rho_g}{\rho_\ell} \right)^{0.5} \left(\frac{\mu_\ell}{\mu_g} \right)^{0.1} \tag{8.24}$$

Another type of separated-phase model is the drift-flux formulation of Wallis (1969). This approach focuses attention on relative slip between phases and results in slightly different expressions depending

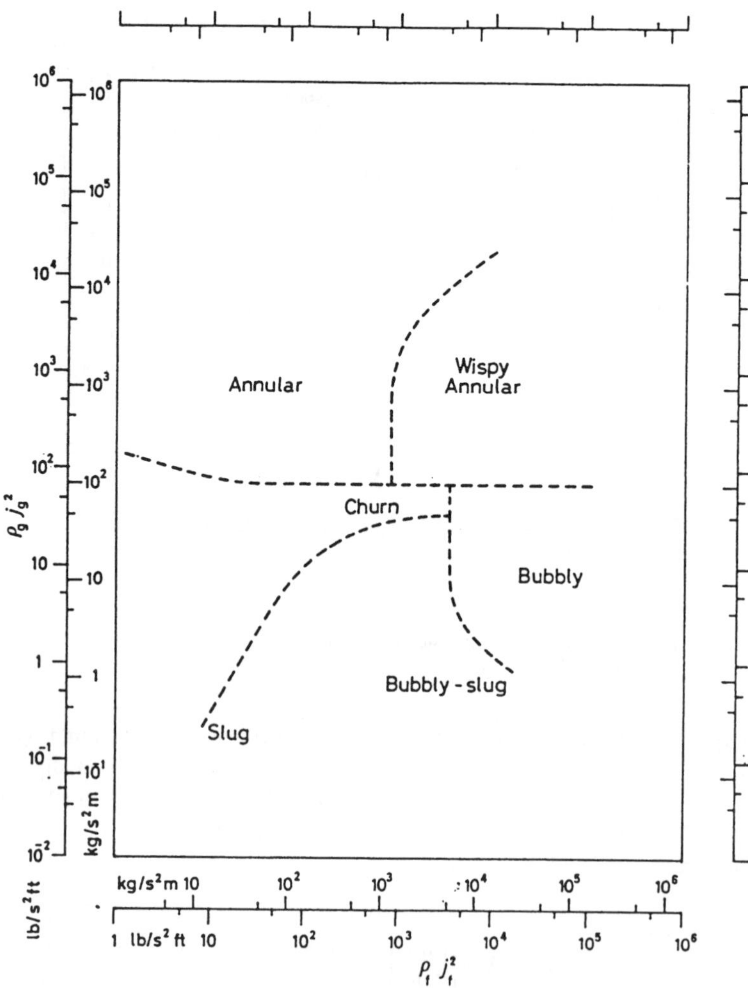

FIGURE 8.4 Flow pattern map for vertical flow (Hewitt and Roberts, 1969). (From Collier, J.G. 1972. *Convective Boiling and Condensation,* McGraw-Hill, London. With permission.)

on the flow regime. For co-current upflow in two of the more common regimes, the drift-flux model gives the following relationships between void fraction and flow quality:

Bubbly flow or churn-turbulent flow:

$$\alpha_g = \frac{x_g}{\left(\frac{u_o \rho_g}{G}\right) + C_o\left[x_g + \left(1 - x_g\right)\frac{\rho_g}{\rho_\ell}\right]} \tag{8.25}$$

Dispersed drop (mist) flow:

$$x_g = \frac{1 - \left(1 - \alpha_g\right)\left(\frac{u_o \rho_\ell}{G}\alpha_g^2 + 1\right)}{1 - \left(1 - \alpha_g\right)\left(1 - \frac{\rho_\ell}{\rho_g}\right)} \tag{8.26}$$

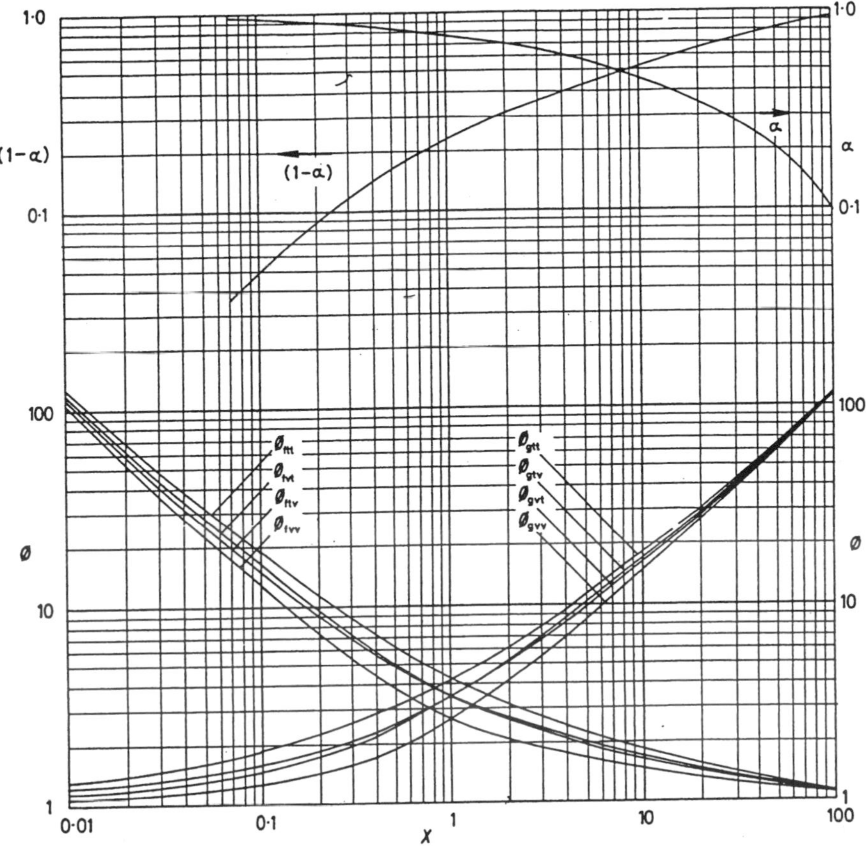

FIGURE 8.5 Correlations for void fraction and frictional pressure drop (Lockhart and Martinelli, 1949). (From Collier, J.G. 1972. *Convective Boiling and Condensation,* McGraw-Hill, London. With permission.)

where u_o= terminal rise velocity of bubble, in bubbly flow, or terminal fall velocity of drop in churn-turbulent flow

C_o = an empirical distribution coefficient $\simeq$ 1.2

Pressure Drop

Equations 8.16 through 8.18 permit calculation of two-phase pressure drop by the homogeneous model, if the friction coefficient (*f*) is known. One useful method for estimating (*f*) is to treat the entire two-phase flow as if it were all liquid, except flowing at the two-phase mixture velocity. By this approach the frictional component of the two-phase pressure drop becomes

$$\left(\frac{dP}{dz}\right)_f = \left[1 + x_g\left(\frac{\rho_\ell}{\rho_g} - 1\right)\right] \cdot \left(\frac{dP}{dz}\right)_{f\ell G} \tag{8.27}$$

where $(dP/dz)_{f\ell G}$ = frictional pressure gradient if entire flow (of total mass flux G) flowed as liquid in the channel.

The equivalent frictional pressure drop for the entire flow as liquid, $(dP/dz)_{f\ell g}$, can be calculated by standard procedures for single-phase flow. In using Equations (8.16) through (8.18), the void fraction would be calculated with the equivalent homogeneous expression Equation (8.13).

A more accurate method to calculate two-phase pressure drop is by the separated-phases model of Lockhart and Martinelli (1949). The bottom half of Figure 8.5 shows empirical curves for the Lockhart–Martinelli frictional multiplier, ϕ:

$$\phi_i = \left[\left(\frac{dP}{dz}\right)_f \div \left(\frac{dP}{dz}\right)_{fi}\right]^{1/2} \tag{8.28}$$

where (i) denotes either the fluid liquid phase (f) or gas phase (g). The single-phase frictional gradient is based on the ith phase flowing alone in the channel, in either viscous laminar (v) or turbulent (t) modes. The most common case is where each phase flowing alone would be turbulent, whence one could use Figure 8.5 to obtain

$$\begin{aligned}\left(\frac{dP}{dz}\right)_f &= \text{frictional pressure gradient for two-phase flow}\\ &= \phi_{gtt}^2 \cdot \left(\frac{dP}{dz}\right)_{fg}\end{aligned} \tag{8.29}$$

where $(dP/dz)_{fg}$ is calculated for gas phase flowing alone and $X = X_{tt}$ as given by Equation (8.24).

The correlation of Lockhart–Martinelli has been found to be adequate for two-phase flows at low-to-moderate pressures, i.e., with reduced pressures less than 0.3. For applications at higher pressures, the revised models of Martinelli and Nelson (1948) and Thom (1964) are recommended.

Gas–Solid, Liquid–Solid Two-Phase Flows

Two-phase flows can occur with solid particles in gas or liquid. Such flows are found in handling of granular materials and heterogeneous reaction processing. Concurrent flow of solid particulates with a fluid phase can occur with various flow patterns, as summarized below.

Flow Regimes

Consider vertical upflow of a fluid (gas or liquid) with solid particles. Figure 8.6 illustrates the major flow regimes that have been identified for such two-phase flows. At low flow rates, the fluid phase percolates between stationary particles; this is termed flow through a fixed bed. At some higher velocity a point is reached when the particles are all suspended by the upward flowing fluid, the drag force between particles and fluid counterbalancing the gravitational force on the particles. This is the point of minimum fluidization, marking the transition from fixed to fluidized beds. Increase of fluid flow rate beyond minimum fluidization causes instabilities in the two-phase mixture, and macroscopic bubbles or channels of fluid are observed in the case of gaseous fluids. In the case of liquid fluids, the two-phase mixture tends to expand, often without discrete bubbles or channels. Further increase of fluid velocity causes transition to turbulent fluidization wherein discrete regions of separated phases (fluid slugs or channels and disperse suspensions of particles) can coexist. Depending on specific operating conditions (e.g., superficial fluid velocity, particle size, particle density, etc.), net transport of solid particles with the flowing fluid can occur at any velocity equal to or greater than that associated with slug flow and turbulent flow. Further increases in fluid velocity increase the net transport of solid particles. This can occur with large-scale clusters of solid particles (as exemplified by the fast fluidization regime) or with dilute dispersions of solid particles (as often utilized in pneumatic conveying). For engineering application of fluid–solid two-phase flows, the important thresholds between flow regimes are marked by the fluid velocity for minimum fluidization, terminal slip, and saltation threshold.

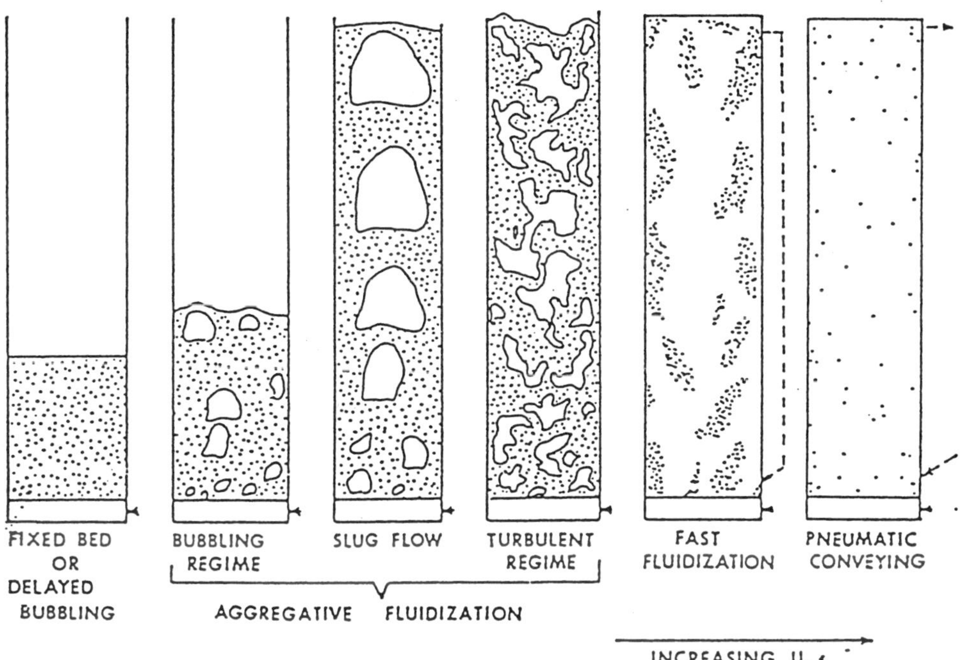

FIGURE 8.6 Flow patterns for vertical upflow of solid particles and gas or liquid. (From Chen, J.C. 1994. *Proc. Xth Int. Heat Transfer Conf.*, Brighton, U.K., 1:369–386. With permission.)

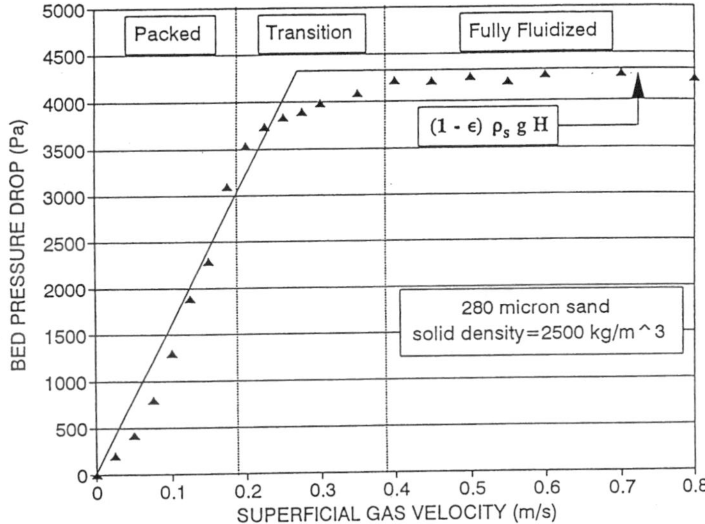

FIGURE 8.7 Transition at minimum fluidization. (From Chen, J.C. 1996. In *Annual Review of Heat Transfer*, Vol. VII, Begal House, Washington, D.C. With permission.)

Minimum Fluidization

The transition from flow through packed beds to the fluidization regime is marked by the minimum fluidization velocity of the fluid. On a plot pressure drop vs. superficial fluid velocity, the point of minimum fluidization is marked by a transition from a linearly increasing pressure drop to a relatively constant pressure drop as shown in Figure 8.7 for typical data, for two-phase flow of gas with sand particles of 280 μm mean diameter (Chen, 1996). The threshold fluid velocity at minimum fluidization is traditionally derived from the Carman–Kozeny equation,

$$U_{mf} = \frac{(\rho_s - \rho_f)(\phi\, dp)^2 g}{150\mu_f} \cdot \frac{\alpha_{mf}^2}{(1-\alpha_{mf})} \tag{8.30}$$

where ϕ= sphericity of particles (unity for spherical particles)

α_{mf} = volumetric fraction of fluid at minimum fluidization

Small, light particles have minimum fluidization voidage (α_{mf}) of the order 0.6, while larger particles such as sand have values closer to 0.4.

An alternative correlation for estimating the point of minimum fluidization is that of Wen and Yu (1966):

$$\frac{U_{mf} d_p \rho_f}{\mu_f} = (33.7 + 0.041 Ga)^{0.5} - 33.7 \tag{8.31}$$

where $Ga = \rho_f d_p^3 (\rho_s - \rho_f) g / \mu_f^2$.

When the fluid velocity exceeds U_{mf}, the two-phase mixture exists in the fluidized state in which the pressure gradient is essentially balanced by the gravitational force on the two-phase mixture:

$$\frac{dP}{dz} = g[\alpha_s \rho_s + \alpha_f \rho_f] \tag{8.32}$$

This fluidized state exists until the fluid velocity reaches a significant fraction of the terminal slip velocity, beyond which significant entrainment and transport of the solid particles occur.

Terminal Slip Velocity

For an isolated single particle the maximum velocity relative to an upflowing fluid is the terminal slip velocity. At this condition, the interfacial drag of the fluid on the particle exactly balances the gravitational body force on the particle:

$$U_t = (U_f - U_s)_t = \left[\frac{4 d_p (\rho_s - \rho_f)}{3\rho_f} \cdot \frac{1}{C_D}\right]^{1/2} \tag{8.33}$$

where C_D = coefficient of drag on the particle.

The coefficient of drag on the particle (C_D) depends on the particle Reynolds number:

$$\mathrm{Re}_p = \frac{\rho_f d_p (U_f - U_s)}{\mu_f} \tag{8.34}$$

The following expressions may be used to estimate C_D as appropriate:

$$C_D = \frac{32}{\mathrm{Re}_p}, \qquad \mathrm{Re}_p \le 1$$
$$C_D = \frac{18.5}{\mathrm{Re}_p^{0.67}}, \qquad 1 \le \mathrm{Re}_p \le 10^3 \tag{8.35}$$

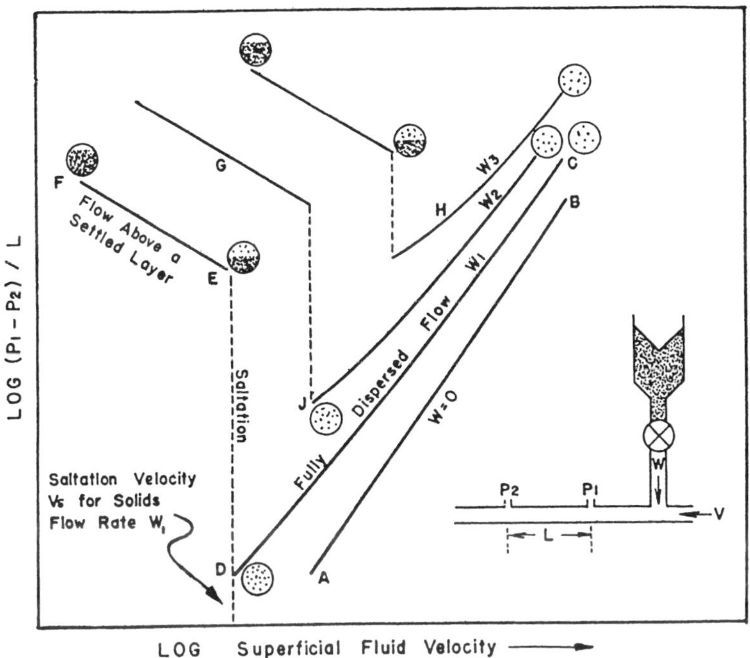

FIGURE 8.8 Flow characteristics in horizontal pneumatic conveying. (From Zeng, F.A. and Othmer, D.F. 1960. *Fluidization and Fluid-Particle Systems,* Reinhold, New York. With permission.)

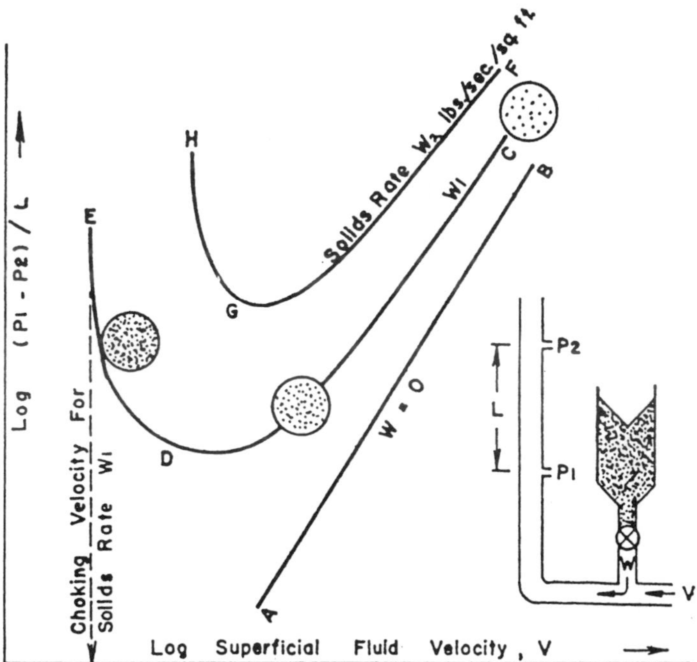

FIGURE 8.9 Flow characteristics in vertical pneumatic conveying. (From Zeng, F.A. and Othmer, D.F. 1960. *Fluidization and Fluid-Particle Systems,* Reinhold, New York. With permission.)

Pneumatic Conveying

A desirable mode of pneumatic conveying is two-phase flow with solid particles dispersed in the concurrent flowing fluid. Such dispersed flows can be obtained if the fluid velocity is sufficiently high. For both horizontal and vertical flows, there are minimum fluid velocities below which saltation of the solid particles due to gravitational force occurs, leading to settling of the solid particles in horizontal channels and choking of the particles in vertical channels. Figures 8.8 and 8.9 for Zenz and Othmer (1960) show these different regimes of pneumatic conveying for horizontal and vertical transport, respectively. Figure 8.8 shows that for a given rate of solids flow (*W*) there is a minimum superficial fluid velo city below which solid particles tend to settle into a dense layer at the bottom of the horizontal channels. Above this saltation threshold, fully dispersed two-phase flow is obtained. In the case of vertical transport illustrated in Figure 8.9, there is a minimum fluid velocity below which solid particles tend to detrain from the two-phase suspension. This choking limit varies not only with particle properties but also with the actual rate of particle flow. Well-designed transport systems must operate with superficial fluid velocities greater than these limiting saltation and choking velocities.

Zenz and Othmer (1960) recommend the empirical correlations represented in Figure 8.10 estimating limiting superficial fluid velocities at incipient saltation or choking, for liquid or gas transport of uniformly sized particles. Note that these correlations are applicable for either horizontal or vertical concurrent flow. Figure 8.10 is duplicated from the original source and is based on parameters in engineering units, as noted in the figure. To operate successfully in dispersed pneumatic conveying of solid particles, the superficial fluid velocity must exceed that determined from the empirical correlations of Figure 8.10.

Nomenclature

A_c cross-sectional flow area of channel
C_o Wallis' distribution coefficient
d_p diameter of solid particles
f_D Darcy friction factor
G mass flow flux, kg/m^2 · sec
j volumetric flow flux, m/sec
$\dot{m}$ mass flow rate, kg/sec
P pressure, N/m^2
u velocity in axial flow direction, m/sec
v superficial velocity in axial flow direction, m/sec
x mass flow quality
z axial coordinate

Greek Letters

α volume fraction
λ parameter in Baker flow map
ϕ sphericity of solid particles
ϕ_i frictional multiphase for pressure drag, Equation (8.28)
ψ parameter in Baker flow map
σ surface tension
θ angle from vertical

Subscripts

a air
f fluid phase
g gas phase
l liquid phase
mf minimum fluidization
p particle
s solid phase
t terminal slip
w water

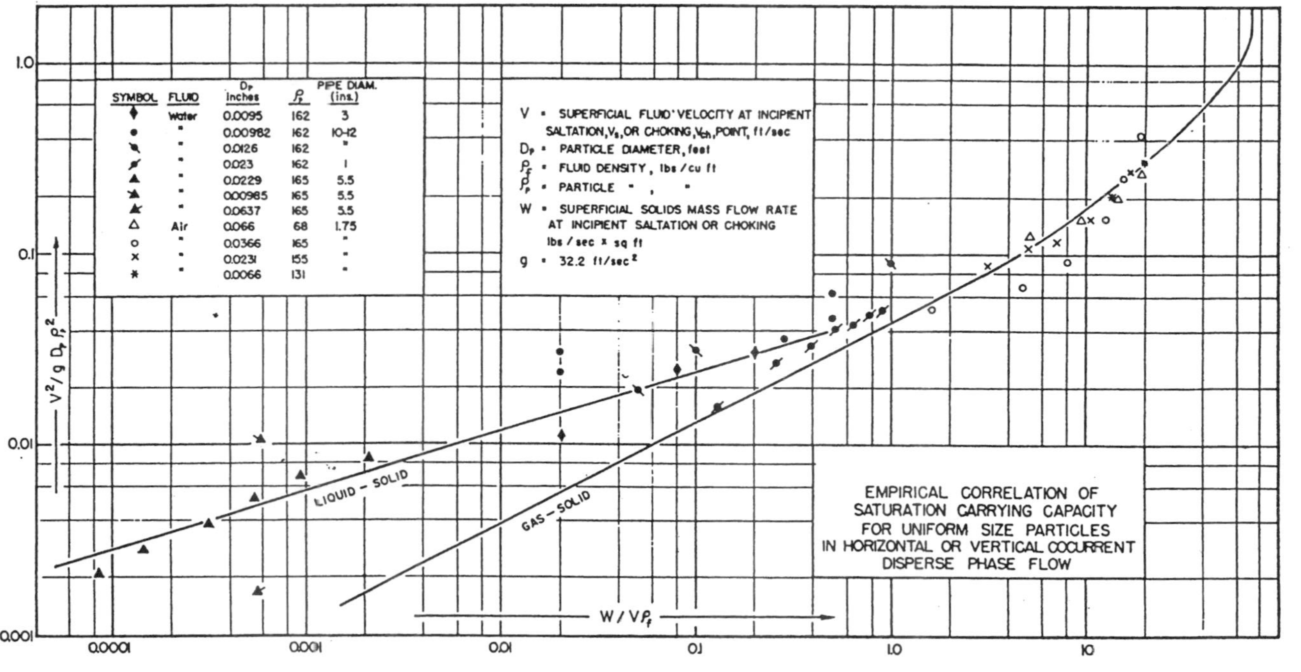

FIGURE 8.10 Correlations for limiting velocities in pneumatic conveying. (From Zeng, F.A. and Othmer, D.F. 1960. *Fluidization and Fluid-Particle Systems,* Reinhold, New York. With permission.)

References

Baker, O. 1954. Design of pipelines for simultaneous flow of oil and gas, *Oil Gas J.*.

Chen, J.C. 1994. Two-phase flow with and without phase changes: suspension flows. Keynote lecture, *Proc. Xth Int. Heat Transfer Conf.,* Brighton, U.K., 1:369–386.

Chen, J.C. 1996. Heat transfer to immersed surfaces in bubbling fluidized beds, in *Annual Review of Heat Transfer*, Vol. VII, Bengel House, Washington, D.C.

Collier, J.G. 1972. *Convective Boiling and Condensation,* McGraw-Hill, London.

Delhaye, J.M. 1981. Two-phase flow patterns, in *Two-Phase Flow and Heat Transfer,* A.E. Bergles, J.G. Collier, J.M. Delhaye, G.F. Newitt, and F. Mayinger, Eds., Hemisphere Publishing, McGraw-Hill, New York.

Hewitt, G.F. and Roberts, D.N. 1969. Studies of Two-Phase Flow Patterns by Simultaneous X-Ray and Flash Photography, Report AERE-M 2159.

Lahey, R.T., Jr. and Moody, F.I. 1977. *The Thermal Hydraulics of a Boiling Water Nuclear Reactor,* The American Nuclear Society, La Grange, IL.

Lockhart, R.W. and Martinelli, R.C. 1949. Proposed correlation of data for isothermal two-phase two-component flow in pipes, *Chem. Eng. Progr.,* 45:39.

Martinelli, R.C. and Nelson, D.B. 1984. Prediction of pressure drop during forced-circulation boiling of water, *Trans. ASME,* 70:695–702.

Thom, J.R.S. 1964. Prediction of pressure drop during forced circulation boiling of water, *Int. J. Heat Mass Transfer,* 7:709–724.

Wallis, G.B. 1969. *One-Dimensional Two-Phase Flow,* McGraw-Hill, New York.

Wen, C.Y. and Yu, Y.H. 1966. A generalized method of predicting the minimum fluidization velocity, *AIChE J.,* 12:610–612.

Zenz, F.A. and Othmer, D.F. 1960. *Fluidization and Fluid-Particle Systems,* Reinhold, New York.

9

Non-Newtonian Flows

Thomas F. Irvine, Jr.
State University of New York, Stony Brook

9

Massimo Capobianchi
State University of New York, Stony Brook

Non-Newtonian Flows 121
Introduction • Classification of Non-Newtonian Fluids • Apparent Viscosity • Constitutive Equations • Rheological Property Measurements • Fully Developed Laminar Pressure Drops for Time-Independent Non-Newtonian Fluids • Fully Developed Turbulent Flow Pressure Drops • Viscoelastic Fluids

Non-Newtonian Flows

Introduction

An important class of fluids exists which differ from Newtonian fluids in that the relationship between the shear stress and the flow field is more complicated. Such fluids are called non-Newtonian or rheological fluids. Examples include various suspensions such as coal–water or coal–oil slurries, food products, inks, glues, soaps, polymer solutions, etc.

An interesting characteristic of rheological fluids is their large "apparent viscosities". This results in laminar flow situations in many applications, and consequently the engineering literature is concentrated on laminar rather than turbulent flows. It should also be mentioned that knowledge of non-Newtonian fluid mechanics and heat transfer is still in an early stage and many aspects of the field remain to be clarified.

In the following sections, we will discuss the definition and classification of non-Newtonian fluids, the special problems of thermophysical properties, and the prediction of pressure drops in both laminar and turbulent flow in ducts of various cross-sectional shapes for different classes of non-Newtonian fluids.

0-8493-0055-X/00/$0.00+$.50

Classification of Non-Newtonian Fluids

It is useful to first define a Newtonian fluid since all other fluids are non-Newtonian. Newtonian fluids possess a property called viscosity and follow a law analogous to the Hookian relation between the stress applied to a solid and its strain. For a one-dimensional Newtonian fluid flow, the shear stress at a point is proportional to the rate of strain (called in the literature the *shear rate*) which is the velocity gradient at that point. The constant of proportionality is the dynamic viscosity, i.e.,

$$\tau_{y,x} = \mu \frac{du}{dy} = \mu\dot{\gamma} \tag{9.1}$$

where x refers to the direction of the shear stress y the direction of the velocity gradient, and $\dot{\gamma}$ is the shear rate. The important characteristic of a Newtonian fluid is that the dynamic viscosity is independent of the shear rate.

Equation (9.1) is called a constitutive equation, and if $\tau_{x,y}$ is plotted against $\dot{\gamma}$, the result is a linear relation whose slope is the dynamic viscosity. Such a graph is called a *flow curve* and is a convenient way to illustrate the viscous properties of various types of fluids.

Fluids which do not obey Equation (9.1) are called non-Newtonian. Their classifications are illustrated in Figure 9.1 where they are separated into various categories of purely viscous time-independent or time-dependent fluids and viscoelastic fluids. Viscoelastic fluids, which from their name possess both viscous and elastic properties (as well as memory), have received considerable attention because of their ability to reduce both drag and heat transfer in channel flows. They will be discussed in a later subsection

Purely viscous time-dependent fluids are those in which the shear stress in a function only of the shear rate but in a more complicated manner than that described in Equation (9.1). Figure 9.2 illustrates the characteristics of purely viscous time-independent fluids. In the figure, (a) and (b) are fluids where the shear stress depends only on the shear rate but in a nonlinear way. Fluid (a) is called pseudoplastic (or shear thinning), and fluid (b) is called dilatant (or shear thickening). Curve (c) is one which has an initial yield stress after which it acts as a Newtonian fluid, called Buckingham plastic, and curve (d), called Hershel-Buckley, also has a yield stress after which it becomes pseudoplastic. Curve (e) depicts a Newtonian fluid.

Figure 9.3 shows flow curves for two common classes of purely viscous time-dependent non-Newtonian fluids. It is seen that such fluids have a hysteresis loop or memory whose shape depends upon the time-dependent rate at which the shear stress is applied. Curve (a) illustrates a pseudoplastic time-dependent fluid and curve (b) a dilatant time-dependent fluid. They are called, respectively, thixotropic and rheopectic fluids and are complicated by the fact that their flow curves are difficult to characterize for any particular. applicationApparent Viscosity

Although non-Newtonian fluids do not have the property of viscosity, in the Newtonian fluid sense, it is convenient to define an apparent viscosity which is the ratio of the local shear stress to the shear rate at that point.

$$\mu_a = \frac{\tau}{\dot{\gamma}} \tag{9.2}$$

The apparent viscosity is not a true property for non-Newtonian fluids because its value depends upon the flow field, or shear rate. Nevertheless, it is a useful quantity and flow curves are often constructed with the apparent viscosity as the ordinate and shear rate as the abscissa. Such a flow curve will be illustrated in a later subsection.

Constitutive Equations

A constitutive equation is one that expresses the relation between the shear stress or apparent viscosity and the shear rate through the rheological properties of the fluid. For example, Equation (9.1) is the constitutive equation for a Newtonian fluid.

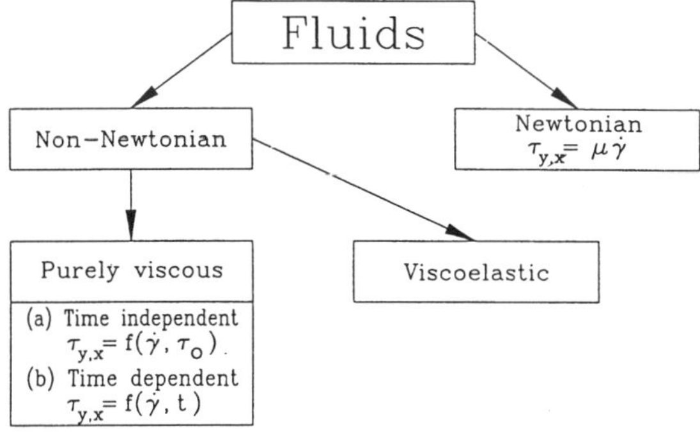

FIGURE 9.1 Classification of fluids.

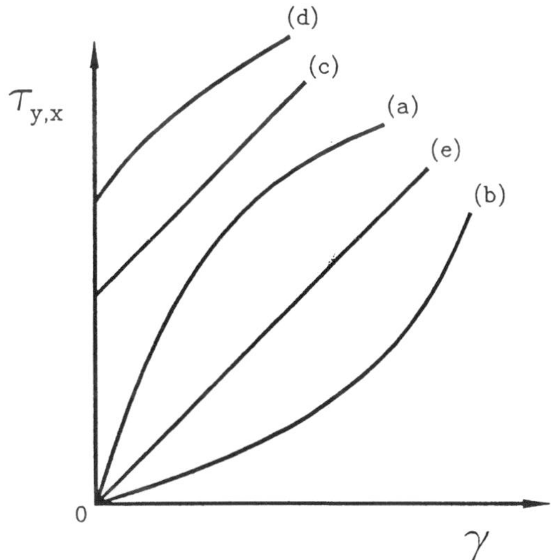

FIGURE 9.2 Flow curves of purely viscous, time-independent fluids: (a) pseudoplastic; (b) dilatant; (c) Bingham plastic; (d) Hershel–Buckley; (e) Newtonian.

Many constitutive equations have been developed for non-Newtonian fluids with some of them having as many as five rheological properties. For engineering purposes, simpler equations are normally satisfactory and two of the most popular will be considered here.

Since many of the non-Newtonian fluids in engineering applications are pseudoplastic, such fluids will be used in the following to illustrate typical flow curves and constitutive equations. Figure 9.4 is a qualitative flow curve for a typical pseudoplastic fluid plotted with logarithmic coordinates. It is seen in the figure that at low shear rates, region (a), the fluid is Newtonian with a constant apparent viscosity of μ_o (called the *zero shear rate viscosity*). At higher shear rates, region (b), the apparent viscosity begins to decrease until it becomes a straight line, region (c). This region (c) is called the power law region and is an important region in fluid mechanics and heat transfer. At higher shear rates than the power law region, there is another transition

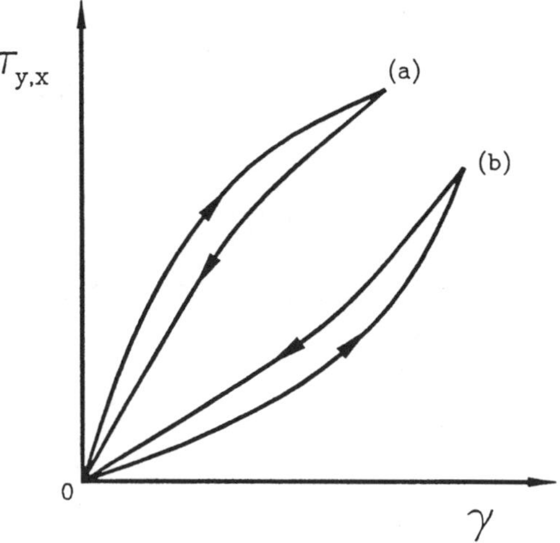

FIGURE 9.3 Flow curves for purely viscous, time-dependent fluids: (a) thixotropic; (b) rheopectic.

region (d) until again the fluid becomes Newtonian in region (e). As discussed below, regions (a), (b), and (c) are where most of the engineering applications occur.

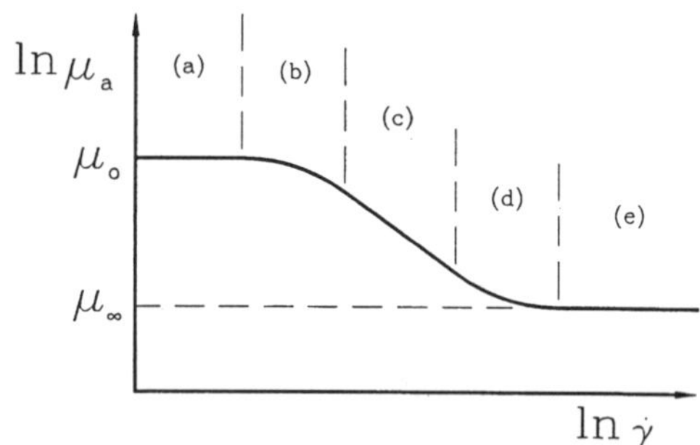

FIGURE 9.4 Illustrative flow curve for a pseudoplastic fluid (a) Newtonian region; (b) transition region I: (c) power law region; (d) transition region II; (e) high-shear-rate Newtonian region.

Power Law Constitutive Equation

Region (c) in Figure 9.4, which was defined above as the power law region, has a simple constitutive equation:

$$\tau = K\dot{\gamma}^n \tag{9.3}$$

or, from Equation (9.2):

$$\mu_a = K\dot{\gamma}^{n-1} \tag{9.4}$$

Here, K is called the fluid consistency and n the flow index. Note that if $n = 1$, the fluid becomes Newtonian and K becomes the dynamic viscosity. Because of its simplicity, the power law constitutive equation has been most often used in rheological studies, but at times it is inappropriate because it has several inherent flaws and anomalies. For example, if one considers the flow of a pseudoplastic fluid ($n < 1$) through a circular duct, because of symmetry at the center of the duct the shear rate (velocity gradient) becomes zero and thus the apparent viscosity from Equation (9.4) becomes infinite. This poses conceptual difficulties especially when performing numerical analyses on such systems. Another difficulty arises when the flow field under consideration is not operating in region (c) of Figure 9.4 but may have shear rates in region (a) and (b). In this case, the power law equation is not applicable and a more general constitutive equation is needed.

Modified Power Law Constitutive Equation

A generalization of the power law equation which extends the shear rate range to regions (a) and (b) is given by

$$\mu_a = \frac{\mu_o}{1 + \frac{\mu_o}{K}\dot{\gamma}^{1-n}} \tag{9.5}$$

Examination of Equation (9.5) reveals that at low shear rates, the second term in the denominator becomes small compared with unity and the apparent viscosity becomes a constant equal to μ_O. This represents the Newtonian region in Figure 9.4. On the other hand, as the second term in the denominator becomes large compared with unity, Equation (9.5) becomes Equation (9.4) and represents region (c), the power law region. When both denominator terms must be considered, Equation (9.5) represents region (b) in Figure 9.4.

An important advantage of the modified power law equation is that it retains the rheological properties K and n of the power law model plus the additional property μ_o. Thus, as will be shown later, in the flow and heat transfer equations, the same dimensionless groups as in the power law model will appear plus an additional dimensionless parameter which describes in which of the regions (a), (b), or (c) a particular system is operating. Also, solutions using the modified power law model will have Newtonian and power law solutions as asymptotes.

Equation (9.5) describes the flow curve for a pseudoplastic fluid ($n < 1$). For a dilatant fluid, ($n > 1$), an appropriate modified power law model is given by

$$\mu_a = \mu_o\left[1 + \frac{K}{\mu_o}\dot{\gamma}^{n-1}\right] \tag{9.6}$$

Many other constitutive equations have been proposed in the literature (Skelland, 1967; Cho and Hartnett, 1982; Irvine and Karni, 1987), but the ones discussed above are sufficient for a large number of engineering applications and agree well with the experimental determinations of rheological properties.

Rheological Property Measurements

For non-Newtonian fluids, specifying the appropriate rheological properties for a particular fluid is formidable because such fluids are usually not pure substances but various kinds of mixtures. This means that the properties are not available in handbooks or other reference materials but must be measured for each particular application. A discussion of the various instruments for measuring rheological properties is outside the scope of the present section, but a number of sources are available which describe different rheological property measurement techniques and instruments: Skelland (1967), Whorlow (1980), Irvine and Karni (1987), and Darby (1988). Figure 9.5 is an illustration of experimental flow curves measured with a falling needle viscometer and a square duct viscometer for polymer solutions of different concentrations. Also known in the figure as solid lines is the modified power law equation used to represent the experimental data. It is seen that Equation (9.5) fits the experimental data within ±2%. Table 9.1 lists the rheological properties used in the modified power law equations in Figure 9.5. It must be emphasized that a proper knowledge of these properties is vital to the prediction of fluid mechanics and heat transfer phenomena in rheological fluids.

TABLE 9.1 Rheological Properties Used in the Modified Power Law Equations in Figure 9.5 for Three Polymer Solutions of CMC-7H4

CMC	K (N · sec^n/m^2)	n	μ_o (N · sec/m^2)n
5000 wppm	2.9040	0.3896	0.21488
2500 wppm	1.0261	0.4791	0.06454
1500 wppm	0.5745	0.5204	0.03673

Source: Park, S. et al., *Proc. Third World Conf. Heat Transfer, Fluid Mechanics, and Thermodynamics,* Vol. 1, Elsevier, New York, 1993, 900–908.

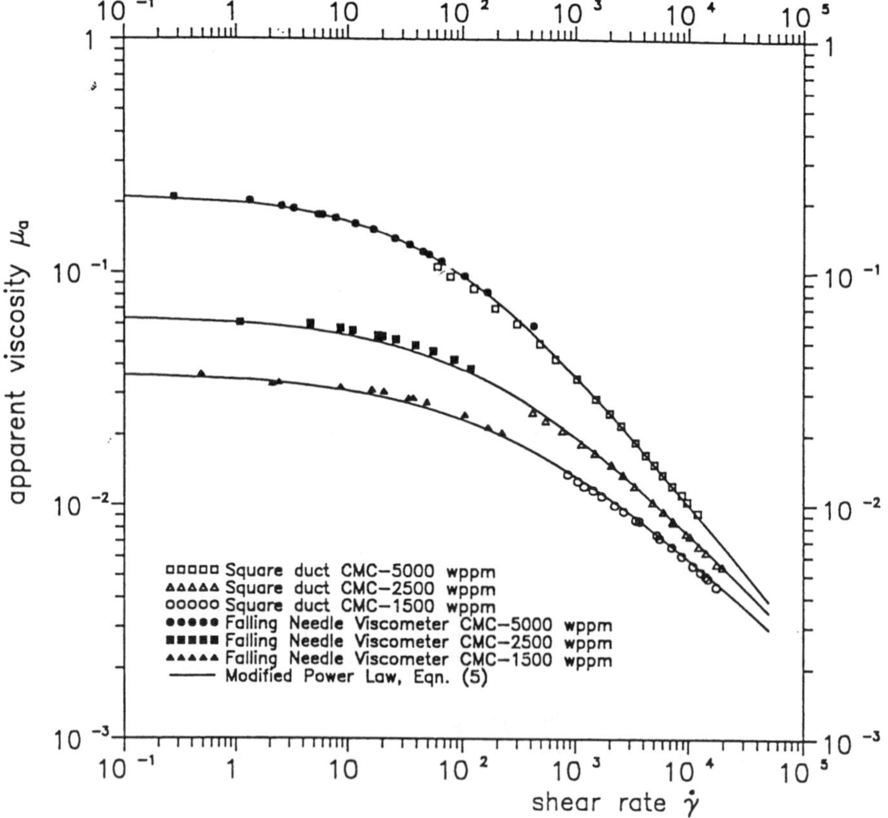

FIGURE 9.5 Experimental measurements of apparent viscosity vs. shear rate for polymer solutions (CMC-7H4) at different concentrations. (From Park, S. et al., in *Proc. Third World Conf. Heat Transfer, Fluid Mechanics, and Thermodynamics,* Vol. 1, Elsevier, New York, 1993, 900–908.

Fully Developed Laminar Pressure Drops for Time-Independent Non-Newtonian Fluids

Modified Power Law Fluids

This important subject will be considered by first discussing modified power law fluids. The reason is that such solutions include both friction factor–Reynolds number relations and a shear rate parameter. The latter allows the designer to determine the shear rate region in which his system is operating and thus the appropriate solution to be used, i.e., regions (a), (b), or (c) in Figure 9.4.

For laminar fully developed flow of a modified power law fluid in a circular duct, the product of the friction factor and a certain Reynolds number is a constant depending on the flow index, n, and the shear rate parameter, β.

$$f_D \cdot \mathrm{Re}_m = \mathrm{constant}(n, \beta) \tag{9.7}$$

where f_D is the Darcy friction factor and Re_m the modified power law Reynolds number, i.e.,

$$f_D = \frac{2\frac{\Delta p}{L} D_H}{\rho \bar{u}^2} \quad \text{(Darcy friction factor)}^*$$

$$\mathrm{Re}_m = \frac{\rho \bar{u} D_H}{\mu^*}$$

$$\mu^* = \frac{\mu_o}{1+\beta}$$

$$\beta = \frac{\mu_o}{K}\left(\frac{\bar{u}}{D_H}\right)^{1-n}$$

where β is the shear rate parameter mentioned previously which can be calculated by the designer for a certain operating duct ($\bar{u}$ and d) and a certain pseudoplastic fluid (μ_o, K, n). The solution for a circular tube has been calculated by Brewster and Irvine (1987) and the results are shown in Figure 9.6 and in Table 9.2. Referring to 9.6, we can see that when the $\log_{10} \beta$ is less than approximately –2, the duct is operating in region (a) of Figure 9.4 which is the Newtonian region and therefore classical Newtonian solutions can be used. Note that in the Newtonian region, Re_m reverts to the Newtonian Reynolds number given by

$$\mathrm{Re}_N = \frac{\rho \bar{u} D_H}{\mu_o} \tag{9.8}$$

When the value of $\log_{10} \beta$ is approximately in the range $-2 \le \log_{10} \beta \le 2$, the duct is operating in the transition region (b) of Figure 9.4 and the values of $f_D \cdot \mathrm{Re}_m$ must be obtained from Figure 9.6 or from Table 9.2.

When $\log_{10} \beta$ is greater than approximately 2, the duct is operating in the power law region (c) of Figure 9.4 and power law friction factor Reynolds number relations can be used. They are also indicated in Figure 9.6 and Table 9.2. In this region, Re_m becomes the power law Reynolds number given by

$$\mathrm{Re}_g = \frac{\rho \bar{u}^{2-n} D_H^n}{K} \tag{9.9}$$

For convenience, Brewster and Irvine (1987) have presented a correlation equation which agrees within 0.1% with the results tabulated in Table 9.2.

---

* It should be noted that the Fanning friction factor is also used in the technical literature. The Fanning friction factor is $^1/_4$ of the Darcy friction factor, and will be characterized by the symbol f_F.

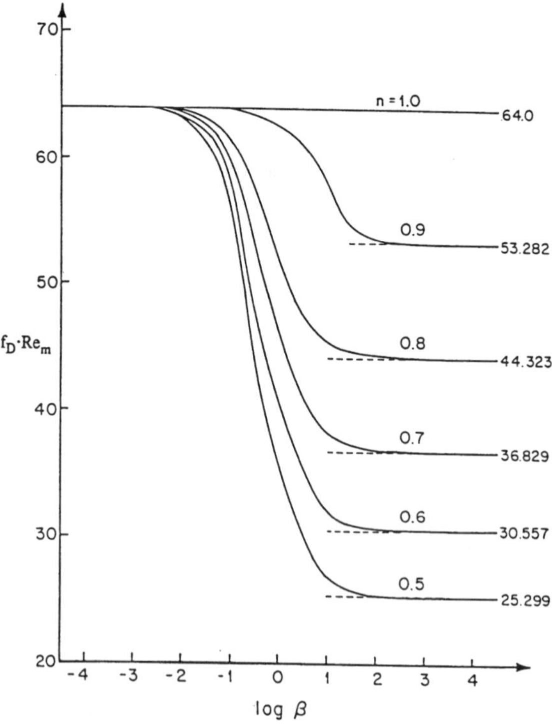

FIGURE 9.6 Product of friction factor and modified Reynolds number vs. $\log_{10} \beta$ for a circular duct. (From Brewster, R.A. and Irvine, T.F., Jr., *Wärme und Stoffübertragung,* 21, 83–86, 1987.

TABLE 9.2 Summary of Computed Values of $f_D \cdot Re_m$ for Various Values of n and β for a Circular Duct

	$f_D \cdot Re_m$ for Flow Index: n =					
β	**1.0**	**0.9**	**0.8**	**0.7**	**0.6**	**0.5**
10^{-5}	64.000	64.000	64.000	64.000	63.999	63.999
10^{-4}	64.000	63.999	63.997	63.995	63.993	63.990
10^{-3}	64.000	63.987	63.972	63.953	63.930	63.903
10^{-2}	64.000	63.873	63.720	63.537	63.318	63.055
10^{-1}	64.000	62.851	61.519	59.987	58.237	56.243
10^{0}	64.000	58.152	52.377	46.761	41.384	36.299
10^{1}	64.000	54.106	45.597	38.308	32.082	26.771
10^{2}	64.000	53.371	44.458	36.985	30.716	25.451
10^{3}	64.000	53.291	44.336	36.845	30.573	25.314
10^{4}	64.000	53.283	44.324	36.831	30.559	25.300
10^{5}	64.000	53.282	44.323	36.830	30.557	25.299
Exact solution	64.000	53.282	44.323	36.829	30.557	25.298

Source: Brewster, R.A. and Irvine, T.F., Jr., *Wärme und Stoffübertragung,* 21, 83–86, 1987. With permission.

$$f_D \cdot \mathrm{Re}_m = \frac{1+\beta}{\dfrac{1}{64} + \dfrac{\beta}{2^{3n+3}\left(\dfrac{3n+1}{4n}\right)^n}} \tag{9.10}$$

Thus, Equation (9.10) contains all of the information required to calculate the circular tube laminar fully developed pressure drop for a pseudoplastic fluid depending upon the shear rate region(s) under consideration, i.e., regions (a), (b), or (c) of Figure 9.4. Note that in scaling such non-Newtonian systems, both Re_m and β must be held constant. Modified power law solutions have been reported for two other duct shapes. Park et al. (1993) have presented the friction factor–Reynolds number relations for rectangular ducts and Capobianchi and Irvine (1992) for concentric annular ducts.

Power Law Fluids

Since the power law region of modified power law fluids ($\log_{10} \beta \geq 2$) is often encountered, the friction factor–Reynolds number relations will be discussed in detail in this subsection.

An analysis of power law fluids which is most useful has been presented by Kozicki et al. (1967). Although the method is approximate, its overall accuracy (±5%) is usually sufficient for many engineering calculations. His expression for the friction factor–Reynolds number product is given by

$$f_D \cdot \mathrm{Re}^* = 2^{6n} \tag{9.11}$$

where

$$\mathrm{Re}^* = \text{Kozicki Reynolds number,} \quad \mathrm{Re}^* = \frac{\mathrm{Re}_g}{\left[\dfrac{a+bn}{n}\right]^n 8^{n-1}} \tag{9.12}$$

and a and b are geometric constants which depend on the cross-sectional shape of the duct. For example, for a circular duct, $a = 0.25$ and $b = 0.75$. Values of a and b for other duct shapes are tabulated in Table 9.3. For additional duct shapes in both closed and open channel flows, Kozicki et al. (1967) may be consulted.

Fully Developed Turbulent Flow Pressure Drops

In a number of engineering design calculations for turbulent flow, the shear rate range falls in region (c) of Figure 9.4. Thus, power law relations are appropriate for such pressure drop calculations.

Hartnett and Kostic (1990) have investigated the various correlations which have appeared in the literature for circular tubes and have concluded that for a circular tube the relation proposed by Dodge and Metzner (1959) is the most reliable for pseudoplastic fluids. It is given by

$$\frac{1}{f_F^{1/2}} = \frac{4.0}{n^{0.75}} \cdot \log_{10}\left[\mathrm{Re}_g'\left(f_F\right)^{1-(1/2n)}\right] - \frac{0.40}{n^{1.2}} \tag{9.13}$$

where f_F is the Fanning friction factor and

TABLE 9.3 Constants *a* and *b* for Various Duct Geometrics Used in the Method Due to Kozicki et al. (1967)

Geometry	α^*	*a*	*b*
Concentric annuli, $\alpha^* = \frac{d_i}{d_o}$	0.1	0.4455	0.9510
	0.2	0.4693	0.9739
	0.3	0.4817	0.9847
	0.4	0.4890	0.9911
	0.5	0.4935	0.9946
	0.6	0.4965	0.9972
	0.7	0.4983	0.9987
	0.8	0.4992	0.9994
	0.9	0.4997	1.0000
	1.0[a]	0.5000	1.0000
Rectangular, $\alpha^* = \frac{c}{h}$	0.0	0.5000	1.0000
	0.25	0.3212	0.8482
	0.50	0.2440	0.7276
	0.75	0.2178	0.6866
	1.00	0.2121	0.8766
Elliptical, $\alpha^* = \frac{c}{h}$	0.00	0.3084	0.9253
	0.10	0.3018	0.9053
	0.20	0.2907	0.8720
	0.30	0.2796	0.8389
	0.40	0.2702	0.8107
	0.50	0.2629	0.7886
	0.60	0.2575	0.7725
	0.70	0.2538	0.7614
	0.80	0.2515	0.7546
	0.90	0.2504	0.7510
	1.00[b]	0.2500	0.7500
	2ϕ (deg)		
Isosceles triangular (2ϕ)	10	0.1547	0.6278
	20	0.1693	0.6332
	40	0.1840	0.6422
	60	0.1875	0.6462
	80	0.1849	0.6438
	90	0.1830	0.6395
	N		
Regular polygon (*N* sides)	4	0.2121	0.6771
	5	0.2245	0.6966
	6	0.2316	0.7092
	8	0.2391	0.7241

[a] Parallel plates.

[b] Circle.

Source: Irvine, T.F., Jr. and Karni, J., in *Handbook of Single Phase Convective Heat Transfer,* John Wiley and Sons, New York, 1987, pp 20-1–20-57.

$$Re'_g = Re_g \left[\frac{8^{1-n}}{\left[\frac{3n+1}{4n} \right]^n} \right] \tag{9.14}$$

Figure 9.7 is a graphical representation of Equation (9.13) which indicates the Dodge and Metzner experimental regions by solid lines, and by dashed lines where the data are extrapolated outside of their experiments.

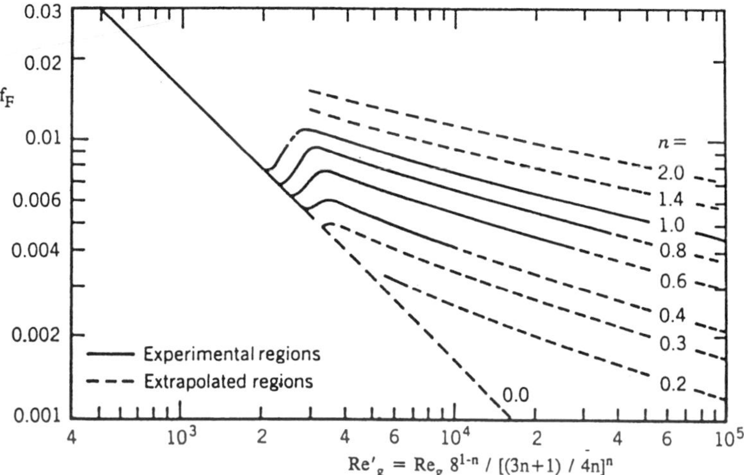

FIGURE 9.7 Dodge and Metzner relation between Fanning friction factor and Re'_g. (From Dodge, D.W. and Metzner, A.B., *AIChE J.*, 5, 189–204, 1959.)

For noncircular ducts in turbulent fully developed flow, only a limited amount of experimental data are available. Kostic and Hartnett (1984) suggest the correlation:

$$\frac{1}{f_F^{1/2}} = \frac{4}{n^{0.75}} \cdot \log_{10}\left[Re^*\left(f_F\right)^{1-(1/2n)}\right] - \frac{0.40}{n^{0.5}} \tag{9.15}$$

where f_F is again the Fanning friction factor and Re^* is the Kozicki Reynolds number:

$$Re^* = \frac{Re_g}{\left[\frac{a+bn}{n}\right]^n 8^{n-1}} \tag{9.16}$$

and a and b are geometric constants given in Table 9.3.

Viscoelastic Fluids

Fully Developed Turbulent Flow Pressure Drops

Viscoelastic fluids are of interest in engineering applications because of reductions of pressure drop and heat transfer which occur in turbulent channel flows. Such fluids can be prepared by dissolving small amounts of high-molecular-weight polymers, e.g., polyacrylamide, polyethylene oxide (Polyox), etc., in water. Concentrations as low as 5 parts per million by weight (wppm) result in significant pressure drop reductions. Figure 9.8 from Cho and Hartnett (1982) illustrates the reduction in friction factors for Polyox solutions in a small-diameter capillary tube. It is seen that at zero polymer concentration the data agree with the Blasius equation for Newtonian turbulent flow. With the addition of only 7 wppm of Polyox, there is a significant pressure drop reduction and for concentrations of 70 wppm and greater all the data fall on the Virk line which is the maximum drag-reduction asymptote. The correlations for the Blasius and Virk lines as reported by Cho and Hartnett (1982) are

$$f_F = \frac{0.079}{\mathrm{Re}^{1/4}} \quad \text{(Blasius)} \tag{9.17}$$

$$f_F = 0.20\mathrm{Re}_a^{-0.48} \quad \text{(Virk)} \tag{9.18}$$

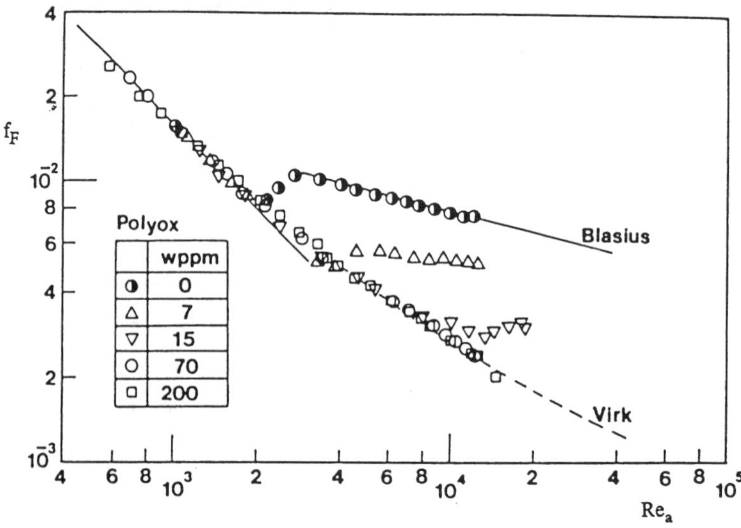

FIGURE 9.8 Reduction in friction factors for polyethylene oxide (Polyox) solutions in a small-diameter capillary tube. (From Cho, Y.I. and Harnett, J.P., *Adv. Heat Transfer,* 15, 59–141, 1982. With permission.)

At the present time, no generally accepted method exists to predict the drag reduction between the Blasius and Virk lines. Kwack and Hartnett (1983) have proposed that the amount of drag reduction between those two correlations is a function of the Weissenberg number, defined as

$$w_s = \frac{\lambda \bar{u}}{D_H} \tag{9.19}$$

where λ = characteristic time of the viscoelastic fluid. They present correlations which allow the friction factor to be estimated at several Reynolds numbers between the Blasius and Virk lines.

Fully Developed Laminar Flow Pressure Drops

The above discussion on viscoelastic fluids has only considered fully developed turbulent flows. Laminar fully developed flows can be considered as nonviscoelastic but purely viscous non-Newtonian. Therefore, the method of Kozicki et al. (1967) may be applied to such situations once the appropriate rheological properties have been determined.

Nomenclature

a = duct shape geometric constant
b = duct shape geometric constant

c = duct width (see Table 9.3) (m)
d_i = concentric annuli inner diameter (see Table 9.3) (m)
d_o = concentric annuli outer diameter (see Table 9.3) (m)
f_D = Darcy friction factor
f_F = Fanning friction factor
h = duct height (see Table 9.3) (m)
K = fluid consistency (Ns^n/m^2)
n = flow index
N = number of sides in polygon (see Table 9.3)
Re_g = generalized Reynolds number,

$$Re_g = \frac{\rho \bar{u}^{2-n} D_H^n}{K}$$

Re_m = modified power law Reynolds number,

$$Re_m = \frac{\rho \bar{u} D_H}{\mu^*}$$

Re_N = modified power law Reynolds number Newtonian asymptote,

$$Re_N = \frac{\rho \bar{u} D_H}{\mu_o}$$

Re_a = apparent Reynolds number

$$Re_a = \frac{Re_g}{\left(\frac{3n+1}{4n}\right)^{n-1} 8^{n-1}}$$

Re^* = Kozicki Reynolds number

$$Re^* = \frac{\rho \bar{u}^{2-n} D_H^n}{K\left[\frac{a+bn}{n}\right]^n 8^{n-1}}$$

Re'_g = Metzner Reynolds number

$$Re'_g = Re_g \left[\frac{8^{1-n}}{\left[\frac{3n+1}{4n}\right]^n}\right]$$

$\bar{u}$ = average streamwise velocity (m/sec)
t = time (sec)
w_s = Weissenberg number
x = direction of shear stress (m)
y = direction of velocity gradient (m)

Greek

α^* = duct aspect ratio in Table 9.3
β = shear rate parameter

$$\beta = \frac{\mu_o}{K}\left(\frac{\bar{u}}{D_H}\right)^{1-n}$$

$\dot{\gamma}$ = shear rate (L/sec)
ΔP = presure drop (N/m^2)
λ = characteristic time of viscoelastic fluid (sec)
μ_a = apparent viscosity ($N \cdot sec/m^2$)
μ_o = zero shear rate viscosity ($N \cdot sec/m^2$)
μ_∞ = high shear rate viscosity ($N \cdot sec/m^2$)
μ^* = reference viscosity

$$\mu^* = \frac{\mu_o}{1+\beta} \quad \left(\mathrm{N \cdot sec/m^2}\right)$$

τ_o = yield stress (N/m^2)
$\tau_{y,x}$ = shear stress (N/m^2)
ϕ = half apex angle (see Table 9.3) (°)

References

Brewster, A.A. and Irvine, T.F. Jr. 1987. Similtude considerations in laminar flow of power law fluids in circular ducts, *Wärme und Stoffübertagung,* 21:83–86.

Capobianchi, M. and Irvine, T.F. Jr. 1992. Predictions of pressure drop and heat transfer in concentric annular ducts with modified power law fluids, *Wärme und Stoffübertagung,* 27:209–215.

Cho, Y.I. and Hartnett, J.P. 1982. Non-Newtonian fluids in circular pipe flow, in *Adv. Heat Transfer,* 15:59–141.

Darby, R. 1988. Laminar and turbulent pipe flows of non-Newtonian fluids, in *Encyclopedia of Fluid Mechanics,* Vol. 7, Gulf Publishing, Houston, 7:20–53.

Dodge, D.W. and Metzner, A.B. 1959. Turbulent flow of non-Newtonian systems, *AIChE J.,* 5:189–204.

Harnett, J.P. and Kostic, M. 1990. Turbulent Friction Factor Correlations for Power Law Fluids in Circular and Non-Circular Channels, *Int. Comm. Heat and Mass Transfer,* 17:59–65.

Irvine, T.F. Jr. and Karni, J. 1987. Non-Newtonian fluid flow and heat transfer, in *Handbook of Single Phase Convective Heat Transfer,* pp. 20-1–20-57, John Wiley and Sons, New York.

Kostic, M. and Hartnett, J.P. 1984. Predicting turbulent friction factors of non-Newtonian fluids in non-circular ducts, *Int. Comm. Heat and Mass Transfer,* 11:345–352.

Kozicki, W., Chou, C.H., and Tiu, C. 1967. Non-Newtonian flow in ducts of arbitrary cross-sectional shape, *Can. J. Chem. Eng.,* 45:127–134.

Kwack, E.Y. and Hartnett, J.P. 1983. Empirical correlations of turbulent friction factors and heat transfer coefficients for viscoelastic fluids, *Int. Comm. Heat and Mass Transfer,* 10:451–461.

Park, S., Irvine, T.F. Jr., and Capobianchi, M. 1993. Experimental and numerical study of friction factor for a modified power law fluid in a rectangular duct, *Proc. Third World Conf. Heat Transfer, Fluid Mechanics, and Thermodynamics,* Vol. 1, Elsevier, New York, 1:900–908.

Skelland, A.H.P. 1967. *Non-Newtonian Flow and Heat Transfer,* John Wiley and Sons, New York.

Whorlow, R.W. 1980. *Rheological Techniques,* Halsted Press, New York.

Further Information

It is not possible to include all of the interesting non-Newtonian topics in a section of this scope. Other items which may be of interest and importance are listed below along with appropriate references: hydrodynamic and thermal entrance lengths, Cho and Hartnett (1982); non-Newtonian flow over external surfaces, Irvine and Karni (1987); chemical, solute, and degradation effects in viscoelastic fluids, Cho and Harnett (1982); general references, Skelland (1967), Whorlow (1980), and Darby (1988).

10

Tribology, Lubrication, and Bearing Design

Francis E. Kennedy
Dartmouth College

E. Richard Booser
Consultant, Scotia, NY

Donald F. Wilcock
Tribolock, Inc.

10 Tribology, Lubrication, and Bearing Design.................. 137
Introduction • Sliding Friction and Its Consequences • Lubricant Properties • Fluid Film Bearings • Dry and Semilubricated Bearings • Rolling Element Bearings • Lubricant Supply Methods

Tribology, Lubrication, and Bearing Design

Introduction

Tribology, the science and technology of contacting surfaces involving friction, wear, and lubrication, is extremely important in nearly all mechanical components. A major focus of the field is on friction, its consequences, especially wear and its reduction through lubrication and material surface engineering. The improper solution of tribological problems is responsible for huge economic losses in our society, including shortened component lives, excessive equipment downtime, and large expenditures of energy. It is particularly important that engineers use appropriate means to reduce friction and wear in mechanical systems through the proper selection of bearings, lubricants, and materials for all contacting surfaces. The aim of this section is to assist in that endeavor.

0-8493-0055-X/00/$0.00+$.50

Sliding Friction and its Consequences

Coefficient of Friction

If two stationary contacting bodies are held together by a normal force W and a tangential force is applied to one of them, the tangential force can be increased until it reaches a magnitude sufficient to initiate sliding. The ratio of the friction force at incipient sliding to the normal force is known as the static coefficient of friction, f_s. After sliding begins, the friction force always acts in the direction opposing motion and the ratio between that friction force and the applied normal force is the kinetic coefficient of friction, f_k.

Generally, f_k is slightly smaller than f_s and both coefficients are independent of the size or shape of the contacting surfaces. Both coefficients are very much dependent on the materials and cleanliness of the two contacting surfaces. For ordinary metallic surfaces, the friction coefficient is not very sensitive to surface roughness. For ultrasmooth or very rough surfaces, however, the friction coefficient can be larger. Typical friction coefficient values are given in Table 10.1. Generally, friction coefficients are greatest when the two surfaces are identical metals, slightly lower with dissimilar but mutually soluble metals, still lower for metal against nonmetal, and lowest for dissimilar nonmetals.

TABLE 10.1 Some Typical Friction Coefficients[a]

	Static Friction Coefficient f_s		Kinetic Friction Coefficient f_k	
Material Pair	**In Air**	**In Vacuo**	**In Air, Dry**	**Oiled**
Mild steel vs. mild steel	0.75	—	0.57	0.16
Mild steel vs. copper	0.53	0.5 (oxidized) 2.0 (clean)	0.36	0.18
Copper vs. copper	1.3	21.0	0.8	0.1
Tungsten carbide vs. copper	0.35	—	0.4	—
Tungsten carbide vs. tungsten carbide	0.2	0.4	0.15	—
Mild steel vs. polytetrafluoroethylene	0.04	—	0.05	0.04

[a] The friction coefficient values listed in this table were compiled from several of the references listed at the end of this section.

The kinetic coefficient of friction, f_k, for metallic or ceramic surfaces is relatively independent of sliding velocity at low and moderate velocities, although there is often a slight decrease in f_k at higher velocities. With polymers and soft metals there may be an increase in the friction coefficient with increasing velocity until a peak is reached, after which friction may decrease with further increases in velocity or temperature. The decrease in kinetic friction coefficient with increasing velocity, which may become especially pronounced at higher sliding velocities, can be responsible for friction-induced vibrations (stick–slip oscillations) of the sliding systems. Such vibrations are an important design consideration for clutches and braking systems, and can also be important in the accurate control and positioning of robotic mechanisms and precision manufacturing systems.

Wear

Wear is the unwanted removal of material from solid surfaces by mechanical means; it is one of the leading reasons for the failure and replacement of manufactured products. It has been estimated that the costs of wear, which include repair and replacement, along with equipment downtime, constitute up to 6% of the U.S. gross national product (Rabinowicz, 1995). Wear can be classified into four primary types: sliding wear, abrasion, erosion, and corrosive wear. Owing to its importance, wear and its control have been the subject of several handbooks (Peterson and Winer, 1980; Blau, 1992), which the interested reader may consult for further information.

Types of Wear. Sliding wear occurs to some degree whenever solid surfaces are in sliding contact. There are two predominant sliding wear mechanisms, adhesion and surface fatigue. *Adhesive wear* is caused by strong adhesive forces between the two surfaces within the real area of contact. It results in the removal of small particles from at least one of the surfaces, usually the softer one. These particles can then transfer to the other surface or mix with other material from both surfaces before being expelled as loose wear debris. Adhesive wear can be particularly severe for surfaces which have a strong affinity for each other, such as those made from identical metals. *Surface fatigue wear* occurs when repeated sliding or rolling/sliding over a wear track results in the initiation of surface or subsurface cracks, and the propagation of those cracks produces wear particles in ductile materials by a process that has been called delamination. With brittle materials, sliding wear often occurs by a *surface fracture* process.

After an initial transition or "running-in" period, sliding wear tends to reach a steady state rate which is approximated by the following Archard (or Holm/Archard) wear equation:

$$V = K * W * s/H \tag{10.1}$$

where V = volume of worn material, K = dimensionless wear coefficient, s = sliding distance, W = normal load between the surfaces, and H = hardness of the softer of the two contacting surfaces.

The dimensionless wear coefficient gives an indication of the tendency of a given material combination to wear; relative wear coefficient values are given in Figure 10.1. In general, wear coefficients are highest for identical metals sliding without lubrication, and wear is decreased by adding a lubricant and by having material pairs which are dissimilar.

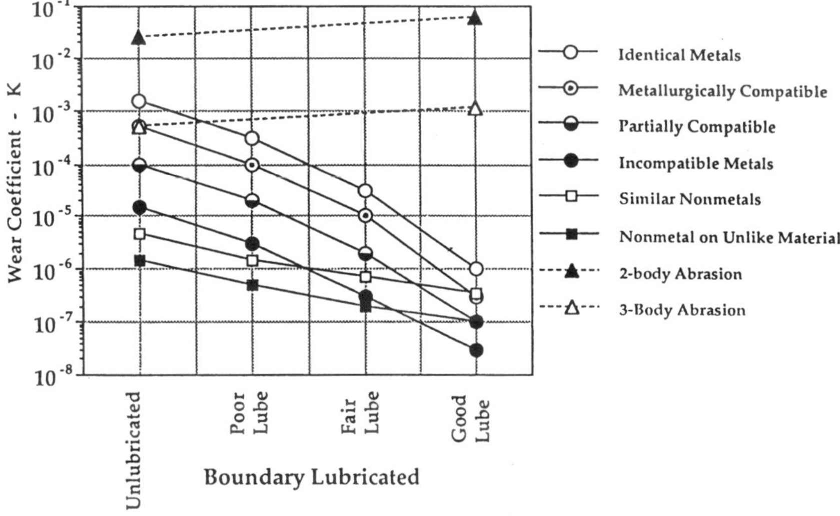

FIGURE 10.1 Typical values of wear coefficient for sliding and abrasive wear. (Modified from Rabinowicz, 1980, 1995.)

Abrasive wear occurs when a hard, rough surface slides against a softer surface (*two-body abrasion*) or when hard particles slide between softer surfaces (*three-body abrasion*). This process usually results in material removal by plowing or chip formation, especially when the abraded surface is metallic; surface fracture can occur during abrasion of brittle surfaces. In fact, abrasion mechanisms are similar to those of grinding and lapping, which could be considered as intentional abrasion. Consideration of the cutting and plowing processes shows that abrasive wear obeys the same equation (10.1) as sliding wear (Archard, 1980; Rabinowicz, 1995). Typical wear coefficients for abrasive wear are given in Figure 10.1. Since the relative size, hardness, and sharpness of the abrading particles, or surface asperities, also affect abrasive wear rates, the wear coefficients for abrasion must include recognition of those factors (Rabinowicz, 1995).

Erosion occurs when solid particles or liquid droplets impinge on a solid surface. When impingement is on a ductile metallic surface, the wear process is similar to that caused by abrasion, and is dominated by plastic deformation. Brittle surfaces, on the other hand, tend to erode by surface fracture mechanisms. The material removal rate is dependent on the angle of attack of the particles, with erosion reaching a peak at low angles (about 20°) for ductile surfaces and at high angles (90°) for brittle materials. In either case, the wear rate is proportional to the mass rate of flow of the particles and to their kinetic energy; it is inversely proportional to the hardness of the surface and the energy-absorbing potential (or toughness) of the impinged surface (Schmitt, 1980). Although erosion is usually detrimental, it can be used beneficially in such material removal processes as sandblasting and abrasive water jet machining.

Corrosive wear results from a combination of chemical and mechanical action. It involves the synergistic effects of chemical attack (corrosion) of the surface, followed by removal of the corrosion products by a wear mechanism to expose the metallic surface, and then repetition of those processes. Since many corrosion products act to protect the surfaces from further attack, the removal of those films by wear acts to accelerate the rate of material removal. Corrosive wear can become particularly damaging when it acts in a low-amplitude oscillatory contact, which may be vibration induced, in which case it is called *fretting corrosion*.

Means for Wear Reduction.

The following actions can be taken to limit sliding wear:

- Insure that the sliding surfaces are well lubricated. This can best be accomplished by a liquid lubricant (see sub-section on effect of lubrication on friction and wear), but grease, or solid lubricants such as graphite or molybdenum disulfide, can sometimes be effective when liquid lubricants cannot be used.
- Choose dissimilar materials for sliding pairs.
- Use hardened surfaces.
- Add wear-resistant coatings to the contacting surfaces (see the following subsection).
- Reduce normal loads acting on the contact.
- Reduce surface temperatures. This is particularly important for polymer surfaces.

To reduce abrasive wear:

- Use hardened surfaces.
- Add a hard surface coating.
- Reduce the roughness of hard surfaces that are in contact with softer surfaces.
- Provide for the removal of abrasive particles from contacting surfaces. This can be done by flushing surfaces with liquid and/or filtering liquid coolants and lubricants.
- Reduce the size of abrasive particles.

To reduce erosion:

- Modify the angle of impingement of solid particles or liquid droplets.
- Provide for the removal of solid particles from the stream of fluid.
- Use hardened surfaces.
- Use tough materials for surfaces.
- Add protective coating to surfaces.

Surface Engineering for Friction and Wear Reduction

Surface treatments have long been an important remedy for wear problems, and that importance has grown in recent years with the introduction of new techniques to harden surfaces or apply hard surface coatings. Available processes and characteristics for treating a steel substrate are listed in Table 10.2.

Thermal transformation hardening processes are used to harden ferrous (primarily steel) surfaces by heating the surface rapidly, transforming it to austenite, and then quenching it to form martensite. The source of heat can be one of the following: an oxyacetylene or oxypropane flame (*flame hardening*), eddy currents induced by a high-frequency electric field (*induction hardening*), a beam from a high-power laser (*laser hardening*), or a focused electron beam (*electron beam hardening*). The depth and uniformity of the hard layer depend on the rate and method of heating. These processes are characterized

TABLE 10.2 Characteristics of Surface Treatment Processes for Steel

Process	Coating or Treated Layer		Substrate Temperature (°C)
	Hardness (*HV*)	Thickness (μm)	
Surface hardening			
Flame or induction hardening	500–700	250–6000	800–1000
Laser or electron beam hardening	500–700	200–1000	950–1050
Carburizing	650–900	50–1500	800–950
Carbonitriding	650–900	25–500	800–900
Nitriding	700–1200	10–200	500–600
Boronizing	1400–1600	50–100	900–1100
Coating			
Chrome plating	850–1250	1–500	25–100
Electroless nickel	500–700	0.1–500	25–100
Hardfacing	800–2000	500–50000	1300–1400
Thermal spraying	400–2000	50–1500	<250
Physical vapor deposition	100–3000	0.05–10	100–300
Chemical vapor deposition	1000–3000	0.5–100	150–2200
Plasma-assisted chemical vapor deposition	1000–5000	0.5–10	<300
Ion implantation	750–1250	0.01–0.25	<200

by a short process time and all except electron beam hardening (which requires a moderate vacuum) can be done in air..

Thermal diffusion processes involve the diffusion of atoms into surfaces to create a hard layer. In the most widely used of these processes, *carburizing* (or case hardening), carbon diffuses into a low-carbon steel surface to produce a hard, carbon-rich case. The hardness and thickness of the case depend on the temperature, exposure time, and source of carbon (either a hydrocarbon gas, a salt bath, or a packed bed of carbon). *Carbonitriding* is a process similar to carburizing which involves the simultaneous diffusion of carbon and nitrogen atoms into carbon steel surfaces. In the *nitriding* process, nitrogen atoms diffuse into the surface of a steel which contains nitride-forming elements (such as Al, Cr, Mo, V, W, or Ti) and form fine precipitates of nitride compounds in a near-surface layer. The hardness of the surface layer depends on the types of nitrides formed. The source of nitrogen can be a hot gas (usually ammonia) or a plasma. *Nitrocarburizing* and *boronizing* are related processes in which nitrogen or boron atoms diffuse into steel surfaces and react with the iron to form a hard layer of iron carbonitride or iron boride, respectively.

Thin, hard metallic coatings can be very effective in friction and wear reduction and can be applied most effectively by *electroplating processes* (Weil and Sheppard, 1992). The most common of such coatings are *chromium*, which is plated from a chromic acid bath, and *electroless nickel*, which is deposited without electric current from a solution containing nickel ions and a reducing agent. Chromium coatings generally consist of fine-grained chromium with oxide inclusions, while electroless nickel coatings contain up to 10% of either phosphorus or boron, depending on the reducing agent used.

Thermal spray processes (Kushner and Novinski, 1992) enable a large variety of coating materials, including metals, ceramics and polymers, to be deposited rapidly on a wide range of substrates. Four different thermal spray processes are commercially available: *oxyfuel* (or flame) spraying of metallic wire or metallic or ceramic powder, *electric arc* spraying of metallic wire, *plasma arc* spraying of powder (metallic or ceramic), and *high-velocity oxyfuel* (or detonation gun) powder spray. In each thermal spray process the coating material, in either wire or powder form, is heated to a molten or plastic state, and the heated particles are propelled toward the surface to be coated where they adhere and rapidly solidify to form a coating. The hardness of the coating depends on both the sprayed material and the process parameters.

Weld hardfacing processes (Crook and Farmer, 1992) involve the application of a wear-resistant material to the surface of a part by means of a weld overlay. Weld overlay materials include ferrous alloys (such as martensitic air-hardening steel or high-chromium cast iron), nonferrous alloys (primarily cobalt- or nickel-based alloys containing hard carbide, boride, or intermetallic particles), and cemented carbides (usually tungsten carbide/cobalt cermets). In each case the surface being coated is heated to the same temperature as the molten weld layer, thus posing a limitation to the process. Weld hardfacing is best used when abrasion or sliding wear cannot be avoided (as with earthmoving or mining equipment) and the goal is to limit the wear rate.

Vapor deposition processes for wear-resistant coatings include *physical vapor deposition* (PVD), *chemical vapor deposition* (CVD), and several variants of those basic processes (Bhushan and Gupta, 1991). Each of the processes consists of three steps: (1) creation of a vapor phase of the coating material, (2) transportation of the vapor from source to substrate, and (3) condensation of the vapor phase on the substrate and growth of a thin solid film. In PVD processes the vapor is produced by either evaporation (by heating of the coating source) or sputtering (in which coating material is dislodged and ejected from the source as a result of bombardment by energetic particles). In some PVD processes the vapor becomes ionized or reacts with a gas or plasma en route to the substrate, thus modifying the structure or composition of the deposited film. In CVD processes a gas composed of a volatile component of the coating material is activated either thermally or by other means in the vicinity of the substrate, and it reacts to form a solid deposit on the surface of the hot substrate.

Both PVD and CVD methods can be used to produce a wide variety of coatings, including metals, alloys, and refractory compounds. Among the most popular vapor-deposited hard coatings for wear resistance are titanium nitride and titanium carbide. Deposition rates are relatively low compared with some other coating processes, ranging from <0.1 μm/min for some ion beam–sputtering or ion-plating processes, up to 25 μm/min or more for activated reactive evaporation or CVD processes. Most PVD processes are done in a vacuum, while CVD processes are done in a reaction chamber which may be at atmospheric pressure. *Plasma-assisted chemical vapor deposition* (PACVD) is a hybrid process in which the constituents of the vapor phase react to form a solid film when assisted by a glow discharge plasma. The advantages of PACVD over other CVD processes include lower substrate temperatures, higher deposition rates, and a wider variety of coating possibilities.

Ion implantation (Fenske, 1992) is a process in which charged particles are created in an ion source, accelerated toward the surface at high velocity, and then injected into the substrate surface. The most commonly implanted ions for surface engineering are nitrogen, carbon, boron, and titanium, although virtually any element could be implanted. The microstructure of the near-surface region is changed by the presence of the implanted ions and the result can be high near-surface hardness and wear resistance. The affected layer is very thin (<1 μm).

Effect of Lubrication on Friction and Wear

Whenever lubricated surfaces slide together at low sliding speeds or with a high applied normal load, the lubricant may not separate the two solid surfaces completely. However, the lubricant can still significantly reduce the friction coefficient by reducing the shear strength of adhesive junctions between the two surfaces. In this so-called boundary lubrication regime, the effectiveness of the lubricant can be improved if the lubricant molecules adhere well to the solid surfaces. This is best accomplished by introducing a lubricant or additive that forms a surface film through adsorption, chemisorption, or chemical reaction with the surface. The ensuing reduced shear strength of the surface film can lower the friction coefficient by as much as an order of magnitude from the dry friction value.

When a good supply of a viscous lubricant is available, the separation between the surfaces will increase as the sliding speed increases or the normal load decreases. As the separation increases, the amount of solid/solid contact between the surfaces will decrease, as will the friction coefficient and wear rate. In this "mixed friction" regime, friction is determined by the amount of plowing deformation on the softer surface by the harder surface asperities and by adhesion within the solid/solid contacts. When the surfaces become completely separated by a self-acting or externally pressurized lubricant film, the lubricating regime is hydrodynamic, wear is reduced to nearly zero, and friction reaches a low value governed by viscous shear of the lubricant. Friction coefficients in such cases can be 0.001 or lower, depending on the surface velocities and the lubricant viscosity. This is the case for most journal or thrust bearings (see subsection on fluid film bearings).

Bearings for Friction Reduction

Most mechanical systems contain moving components, such as shafts, which must be supported and held in position by stationary members. This is best done by appropriate design or selection of bearings to be used wherever the moving member is to be supported. Most bearings may be classified as either fluid film bearings, dry or semilubricated bearings, or rolling element bearings.

Fluid film bearings (see subsection below) have a conformal geometry, with a thin film of fluid separating the two surfaces. The fluid lubricant could be a liquid, such as oil, or a gas, such as air. Fluid film bearings are commonly used to support rotating cylindrical shafts, and the load on such a bearing could be either radial, in which case the bearing is called a journal bearing, or axial, for a thrust bearing. In most cases the fluid film is generated by the motion within the bearing itself, so the bearing is called self-acting or hydrodynamic. Whether or not a self-acting bearing can develop a fluid film sufficient to separate and support the two surfaces is determined by magnitude of the quantity $\mu U/W$, where μ is the (absolute) fluid viscosity, U is the relative sliding velocity, and W is the normal load. If that quantity is too small, the fluid film will be too thin and high friction will occur. This can be a problem during start-up of equipment when sliding velocities are low. That problem can be overcome by pressurizing the fluid film from an external pressure source to create a hydrostatic bearing. Whether the fluid film is externally pressurized (hydrostatic) or self-acting (hydrodynamic), separation of the solid surfaces allows wear to be essentially eliminated and friction to be very low, even when very large loads are carried by the pressurized lubricant.

Dry and semilubricated bearings (see subsection below) have conformal surfaces which are in direct contact with each other. This category includes bearings which run dry (without liquid lubrication) or those which have been impregnated with a lubricant. Dry bearings are made of a material such as a polymer or carbon-graphite which has a low friction coefficient, and they are generally used in low-load and low-speed applications. Semilubricated bearings are made of a porous material, usually metal, and are impregnated with a lubricant which resides within the pores. The lubricant, which could be oil or grease, cannot provide a complete fluid film, but usually acts as a boundary lubricant. Semilubricated bearings can carry greater loads at greater speeds than dry bearings, but not as high as either fluid film or rolling element bearings. The failure mechanism for both dry and semilubricated bearings is wear.

Rolling element bearings (see subsection below) have the advantage that rolling friction is lower than sliding friction. These bearings include rolling elements, either balls or rollers, between hardened and ground rings or plates. Their main advantage over fluid film bearings is that they have low friction both during start-up and at operating velocities, although the friction can be higher than that of fluid film bearings during steady state operation. Ball and roller bearings are most commonly lubricated by either oil or grease. In either case the lubricating film at the concentrated contacts between rolling elements and rings is very thin and the pressures in the film are very high; this is the condition known as elastohydrodynamic lubrication. Rolling element bearings fail by a number of mechanisms, often stemming from improper installation or use or from poor lubrication, but the overriding failure mechanism is rolling contact fatigue.

Each type of bearing has advantages and disadvantages, and these are summarized in Table 10.3. The Engineering Sciences Data Unit (ESDU) (1965; 1967) has developed some general guides to the

TABLE 10.3 Bearing Characteristics

	Fluid Film Bearings	Dry Bearings	Semilubricated	Rolling Element Bearings
Start-up friction coefficient	0.25	0.15	0.10	0.002
Running friction coefficient	0.001	0.10	0.05	0.001
Velocity limit	High	Low	Low	Medium
Load limit	High	Low	Low	High
Life limit	Unlimited	Wear	Wear	Fatigue
Lubrication requirements	High	None	Low/None	Low
High temperature limit	Lubricant	Material	Lubricant	Lubricant
Low temperature limit	Lubricant	None	None	Lubricant
Vacuum	Not applicable	Good	Lubricant	Lubricant
Damping capacity	High	Low	Low	Low
Noise	Low	Medium	Medium	High
Dirt/dust	Need Seals	Good	Fair	Need seals
Radial space requirement	Small	Small	Small	Large
Cost	High	Low	Low	Medium

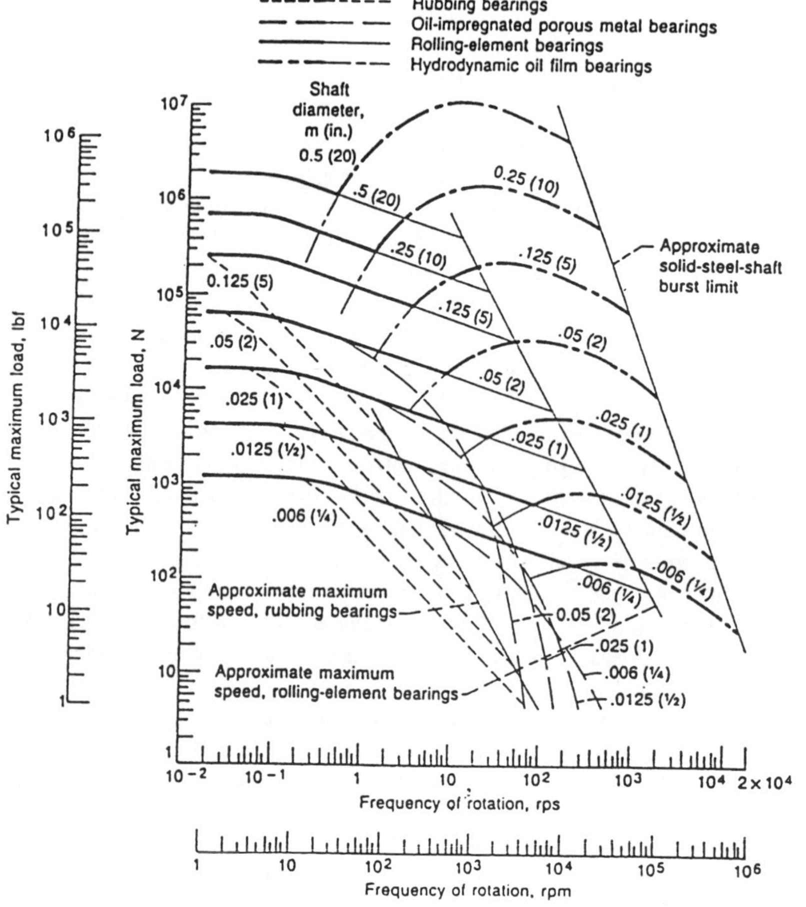

FIGURE 10.2 General guide to journal bearing–type selection. Except for rolling element bearings, curves are drawn for bearings with width/diameter = 1. A medium-viscosity mineral oil is assumed for hydrodynamic bearings. From ESDU, *General Guide to the Choice of Journal Bearing Type*, Item 67073, Institution of Mechanical Engineers, London, 1965. With permission.

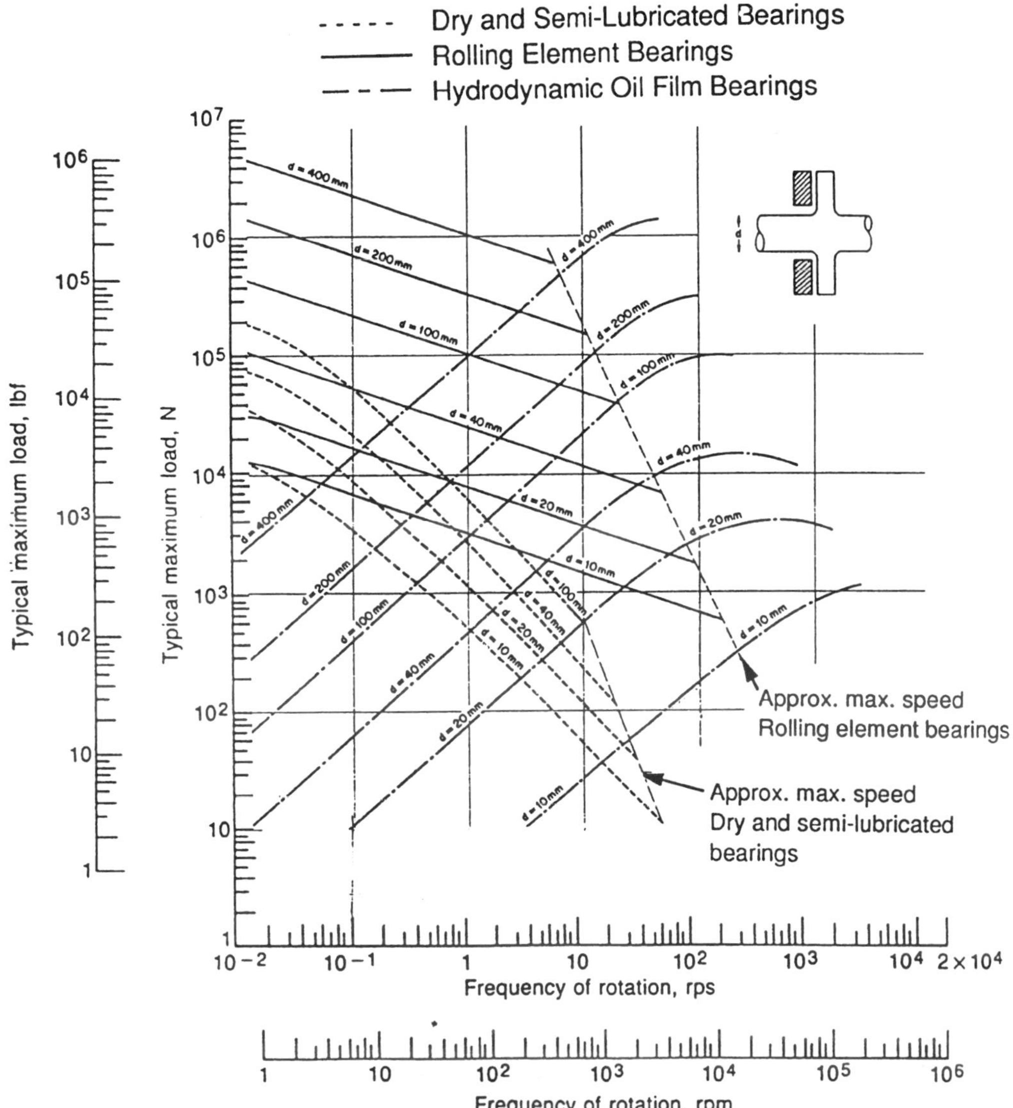

FIGURE 10.3 General guide to thrust bearing type selection. Except for rolling element bearings, curves are drawn for a ratio of inside diameter to outside diameter equal to 2 and for a nominal life of 10,000 hr. A medium-viscosity mineral oil is assumed for hydrodynamic bearings. (Based on ESDU, *General Guide to the Choice of Journal Bearing Type*, Item 67073, Institution of Mechanical Engineers, London, 1965, and Neale, M.J., *Bearings*, Butterworth-Heinemann, Oxford, 1993.)

selection of bearing type for different load and speed conditions, and those guides for journal and thrust bearing selection are given in Figures 10.2 and 10.3.

Lubricant Properties

Petroleum Oils

The vast majority of lubricants in use today are mineral oils which are obtained through the distillation of crude petroleum. Mineral oils are composed primarily of three types of hydrocarbon structures: paraffinic, aromatic, and alicyclic (naphthenic). The molecular weights of the hydrocarbons range from about 250 for low-viscosity grades, up to nearly 1000 for more-viscous lubricants.

Mineral oils by themselves do not have all of the properties required of modern lubricants. For that reason, almost all current lubricants are fortified with a chemical additive package which consists of some of the following:

Oxidation inhibitors limit oxidation of hydrocarbon molecules by interrupting the hydroperoxide chain reaction.

Rust inhibitors are surface-active additives that preferentially adsorb on iron or steel surfaces and prevent their corrosion by moisture.

Antiwear and extreme pressure agents form low shear strength films on metallic surfaces which limit friction and wear, particularly in concentrated contacts.

Friction modifiers form adsorbed or chemisorbed surface films which are effective in reducing friction of bearings during low-speed operation (boundary lubrication regime).

Detergents and dispersants reduce deposits of oil-insoluble compounds (e.g., sludge) in internal combustion engines.

Pour-point depressants lower the temperature at which petroleum oils become immobilized by crystallized wax.

Foam inhibitors are silicone polymers which enhance the separation of air bubbles from the oil.

Viscosity-index improvers are long-chain polymer molecules which reduce the effect of temperature on viscosity. They are used in multigrade lubricants.

Properties of Petroleum Oils

The lubricating oil property which is of most significance to bearing performance is viscosity. The absolute viscosity, designated as μ, could be given in SI units as pascal second ($\text{Pa} \cdot \text{sec} = \text{N} \cdot \text{sec/m}^2$) or centipoise ($1\ \text{cP} = 0.001\ \text{Pa} \cdot \text{sec}$) or in English units as $\text{lb} \cdot \text{sec/in}^2$ (or reyn). Kinematic viscosity, designated here as ν, is defined as absolute viscosity divided by density. It is given in SI units as m^2/sec or centistokes ($1\ \text{cSt} = 10^{-6}\ \text{m}^2/\text{sec}$) and in English units as $\text{in.}^2/\text{sec}$.

Viscosity data in Table 10.4 are representative of typical petroleum "turbine" and "hydraulic" oils which are widely used in industry and closely correspond to properties of most other commercially available petroleum oils. Table 10.5 gives equivalent viscosity grades for common automotive (SAE), gear (SAE and AGMA), and reciprocating aircraft engine (SAE) oils (Booser, 1995). Equivalent ISO viscosity grades are listed for the single-graded SAE automotive oils such as SAE 10W and SAE 30. For multigrade oils such as SAE 10W–30, however, the added viscosity-index improvers provide unique viscosity–temperature characteristics. Typical properties of a number of these multigrade SAE oils are included in Table 10.4.

ISO viscosity grade 32 and the equivalent SAE 10W are most widely used industrially. Lower-viscosity oils often introduce evaporation and leakage problems, along with diminished load capacity. Higher viscosity may lead to high temperature rise, unnecessary power loss, and start-up problems at low temperature. For low-speed machines, however, higher-viscosity oils ranging up to ISO 150, SAE 40 and sometimes higher are often used to obtain higher load capacity.

Oil viscosity decreases significantly with increasing temperature as shown in Fig. 10.4. While Figure 10.4 provides viscosity data suitable for most bearing calculations, oil suppliers commonly provide only the 40°C and 100°C values of kinematic viscosity in centistokes (mm^2/sec). The viscosity at other temperatures can be found by the following ASTM D341 equation relating kinematic viscosity ν in centistokes (mm^2/sec) to temperature T in degrees F:

$$\log\log(\nu + 0.7) = A - B\log(460 + T) \tag{10.2}$$

TABLE 10.4 Representative Oil Properties

	Viscosity				Density	
	Centistokes		10^{-6} reyns(lb·sec/in^2)		gm/cc	lb/in^3
	40°C	100°C	104°F	212°F	40°C	104°F
ISO Grade (Equivalent SAE)						
32 (10W)	32.0	5.36	3.98	0.64	0.857	0.0310
46 (20)	46.0	6.76	5.74	0.81	0.861	0.0311
68 (20W)	68.0	8.73	8.53	1.05	0.865	0.0313
100 (30)	100.0	11.4	12.60	1.38	0.869	0.0314
150 (40)	150.0	15.0	18.97	1.82	0.872	0.0315
220 (50)	220.0	19.4	27.91	2.36	0.875	0.0316
SAE Multigrade						
5W-30	64.2	11.0	8.15	0.99	0.860	0.0311
10W-30	69.0	11.0	8.81	1.08	0.865	0.0312
10W-40	93.5	14.3	11.9	1.45	0.865	0.0312
20W-50	165.5	18.7	21.3	2.74	0.872	0.0315

TABLE 10.5 Equivalent Viscosity Grades for Industrial Lubricants

ISO-VG Grade	Viscosity, cSt (at 40°C) Mimimum	Maximum	SAE Crankcase Oil Grades[a]	SAE Aircraft Oil Grades[a]	SAE Gear Lube Grades[a]	AGMA Gear Lube Grades Regular	EP
2	1.98	2.42	—	—	—	—	—
3	2.88	3.52	—	—	—	—	—
5	4.14	5.06	—	—	—	—	—
7	6.12	7.48	—	—	—	—	—
10	9.00	11.0	—	—	—	—	—
15	13.5	16.5	—	—	—	—	—
22	19.8	24.2	5W	—	—	—	—
32	28.8	35.2	10W	—	—	—	—
46	41.4	50.6	15W	—	75W	1	
68	61.2	74.8	20W	—	—	2	2 EP
100	90.0	110	30	65	80W-90	3	3 EP
150	135	165	40	80	—	4	4 EP
220	198	242	50	100	90	5	5 EP
320	288	352	60	120	—	6	6 EP
460	414	506	—	—	85W-140	7 comp	7 EP
680	612	748	—	—	—	8 comp	8 EP
1000	900	1100	—	—	—	8A comp	8A EP
1500	1350	1650	—	—	250	—	—

[a] Comparisons are nominal since SAE grades are not specified at 40°C viscosity; VI of lubes could change some of the comparisons.

where A and B are constants for any particular (petroleum or synthetic) oil. For ISO VG-32 oil in Table 10.4, for example, $A = 10.54805$ and $B = 3.76834$ based on the 104 and 212°F (40 and 100°C) viscosities. This gives a viscosity of 10.78 cSt at a bearing operating temperature of 160°F.

Conversion of kinematic to absolute viscosity requires use of density, which also varies with temperature. A coefficient of expansion of 0.0004/°F is typical for most petroleum oils, so the variation of density with temperature is in accordance with the following relation:

$$\rho_T = \rho_{104}\left[1 - 0.0004(T - 104)\right] \qquad (10.3)$$

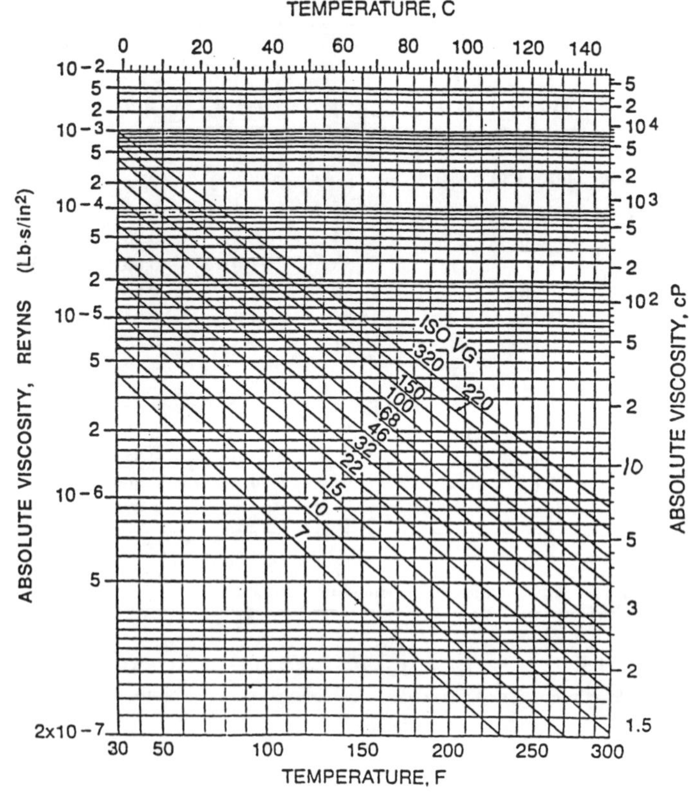

FIGURE 10.4 Viscosity–temperature chart for industrial petroleum oils (Ramondi and Szeri, 1984).

For the ISO-VG-32 oil with its density of 0.857 g/cc at 100°F (40°C), this equation gives 0.838 g/cc at 160°F. Multiplying 10.78 cSt by 0.838 then gives an absolute viscosity of 9.03 cP. Multiplying by the conversion factor of 1.45×10^{-7} gives 1.31×10^{-6} lb·sec/in.^{2} (reyns).

Heat capacity C_p of petroleum oils used in bearing calculations is given by (Klaus and Tewksburg, 1984)

$$C_p = 3860 + 4.5(T) \quad \text{in.} \cdot \text{lb}/(\text{lb} \cdot {}^\circ\text{F}) \tag{10.4}$$

The viscosity of petroleum oils is also affected significantly by pressure, and that increase can become important in concentrated contacts such as rolling element bearings where elastohydrodynamic lubrication occurs. The following relationship can be used for many oils to find the viscosity at elevated pressure:

$$\mu_p = \mu_o e^{\alpha p} \tag{10.5}$$

where μ_o is the viscosity at atmospheric pressure, μ_p is the viscosity at pressure p, and α is the pressure–viscosity exponent. The pressure–viscosity exponent can be obtained from lubricant suppliers.

Synthetic Oils

Synthetic oils of widely varying characteristics are finding increasing use for applications at extreme temperatures and for their unique physical and chemical characteristics. Table 10.6 gives a few representative examples of synthetic oils which are commercially available. Cost of the synthetics, ranging up to many times that of equivalent petroleum oils, is a common deterrent to their use where petroleum products give satisfactory performance

TABLE 10.6 Properties of Representative Synthetic Oils

Type	Viscosity, cSt at			Pour Point, °C	Flash Point, °C	Typical Uses
	100°C	40°C	-54°C			
Synthetic hydrocarbons						
Mobil 1, 5W-30[a]	11	58	—	-54	221	Auto engines
SHC 824[a]	6.0	32	—	-54	249	Gas turbines
SHC 629[a]	19	141	—	-54	238	Gears
Organic esters						
MIL-L-7808	3.2	13	12,700	-62	232	Jet engines
MIL-L-23699	5.0	24	65,000	-56	260	Jet engines
Synesstic 68[b]	7.5	65	—	-34	266	Air compressors, hydraulics
Polyglycols						
LB-300-X[c]	11	60	—	-40	254	Rubber seals
50-HB-2000[c]	70	398	—	-32	226	Water solubility
Phosphates						
Fyrquel 150[d]	4.3	29	—	-24	236	Fire-resistant fluids for die casting, air compressors and hydraulic systems
Fyrquel 220[d]	5.0	44	—	-18	236	
Silicones						
SF-96 (50)	16	37	460	-54	316	Hydraulic and damping fluids
SF-95 (1000)	270	650	7,000	-48	316	Hydraulic and damping fluids
F-50	16	49	2,500	-74	288	Aircraft and missiles
Fluorochemicals						
Halocarbon 27[e]	3.7	30	—	-18	None	Oxygen compressors, liquid-oxygen systems
Krytox 103[f]	5.2	30	—	-45	None	

[a] Mobil Oil Corp.
[b] Exxon Corp.
[c] Union Carbide Chemials Co.
[d] Akzo Chemicals
[e] Halocarbon Products Corp.
[f] DuPont Co.

Greases

Grease is essentially a suspension of oil in a thickening agent, along with appropriate additives. The oil generally makes up between 75 and 90% of the weight of a grease, and it is held in place by the gel structure of the thickener to carry out its lubricating function. The simplicity of the lubricant supply system, ease of sealing, and corrosion protection make grease the first choice for many ball-and-roller bearings, small gear drives, and slow-speed sliding applications (Booser, 1995). Consistencies of greases vary from soap-thickened oils that are fluid at room temperature to hard brick-types that may be cut with a knife. Greases of NLGI Grade 2 stiffness (ASTM D217) are most common. Softer greases down to grade 000 provide easier feeding to multiple-row roller bearings and gear mechanisms. Stiffer Grade 3 is used in some prepacked ball bearings to avoid mechanical churning as the seals hold the grease in close proximity with the rotating balls.

Petroleum oils are used in most greases; the oils generally are in the SAE 30 range, with a viscosity of about 100 to 130 cSt at 40°C. Lower-viscosity oil grades are used for some high-speed applications and for temperatures below about –20°C. Higher-viscosity oils are used in greases for high loads and low speeds. Synthetic oils are used only when their higher cost is justified by the need for special properties, such as capability for operation below –20°C or above 125 to 150°C.

The most common gelling agents are the fatty acid metallic soaps of lithium, calcium, sodium, or aluminum in concentrations of 8 to 25%. Of these, the most popular is lithium 12-hydroxystearate;

greases based on this lithium thickener are suitable for use at temperatures up to 110°C, where some lithium soaps undergo a phase change. Greases based on calcium or aluminum soaps generally have an upper temperature limit of 65 to 80°C, but this limit can be significantly raised to the 120 to 125°C range through new complex soap formulations. Calcium-complex soaps with improved high-temperature stability, for instance, are prepared by reacting both a high-molecular-weight fatty acid (e.g., stearic acid) and a low-molecular-weight fatty acid (acetic acid) with calcium hydroxide dispersed in mineral oil.

Inorganic thickeners, such as fine particles of bentonite clay, are inexpensively applied by simple mixing with oil to provide nonmelting greases for use up to about 140°C. Polyurea nonmelting organic powders are used in premium petroleum greases for applications up to about 150 to 170°C.

TABLE 10.7 Properties of Selected Solid Lubricants

	Acceptable Usage Temperature, °C				Average Friction Coefficient, f		
	Minimum		Maximum				
Material	In Air	In N_2 or Vacuum	In Air	In N_2 or Vacuum	In Air	In N_2 or Vacuum	Remarks
Molybdenum disulfide, MoS_2	–240	–240	370	820	0.10–0.25	0.05–0.10	Low f, carries high load, good overall lubricant, can promote metal corrosion
Graphite	–240	—	540	Unstable in vacuum	0.10–0.30	0.02–0.45	Low f and high load capacity in air, high f and wear in vacuum, conducts electricity
PTFE	–70	–70	290	290	0.02–0.15	0.02–0.15	Lowest f of solid lubricants, load capacity moderate and decreases at elevated temperature
Calcium flouride–barium fluoride eutectic, CaF_2–BaF_2	430	430	820	820	0.10–0.25 above 540°C 0.25–0.40 below 540°C	Same as in air	Can be used at higher temperature than other solid lubricants, high f below 540°C

Modified from Booser, E.R., in *Encyclopedia of Chemical Technology*, 4th ed., Vol. 15, John Wiley & Sons, New York, 1995, 463–517.

Additives, such as those mentioned in the subsection on petroleum oils, are added to grease to improve oxidation resistance, rust protection, or extreme pressure properties. Because of the incompatibility of oils, thickeners, and additives, greases of different types should be mixed only with caution.

Solid Lubricants

Solid lubricants provide thin solid films on sliding or rolling/sliding surfaces to reduce friction and wear. They are particularly useful for applications involving high operating temperatures, vacuum, nuclear radiation, or other environments which limit the use of oils or greases. Solid lubricant films do not prevent moving surfaces from contacting each other, so they cannot eliminate wear and their lives are limited by wear. The properties of some of the most common solid lubricants are given in Table 10.7.

The most important inorganic solid lubricants are layer–lattice solids such as molybdenum disulfide (MoS_2) and graphite. These materials are characterized by strong covalent or ionic bonding between atoms in individual layers, but relatively weak van der Waals bonds between layers, enabling the layers to slide easily relative to one another. Graphite is a very effective lubricant film when moisture or water vapor is present, since adsorbed water vapor lubricates the sliding layers, but it has poor friction properties in vacuum or other low-humidity applications. Molybdenum disulfide does not require the presence of adsorbed water vapor, so it is widely used in vacuum or space applications.

The most commonly used organic solid lubricant is polytetrafluoroethylene (PTFE) which can be used either as a fused surface coating or as a self-lubricating material (see subsection on plastics). Its low friction is attributed to the smooth profile of the PTFE molecule. The chemical inertness of PTFE makes it attractive for chemical and food-processing applications.

New ceramic-based solid lubricants have been developed for high-temperature applications, such as heat engines or space uses. One of the most promising of these is a calcium fluoride–barium fluoride eutectic, which can be used at temperatures exceeding 800°C.

Fluid Film Bearings

Journal Bearings

A journal bearing consists of an approximately cylindrical bearing body or sleeve around a rotating cylindrical shaft. In general, journal bearings are found in motors, pumps, generators, appliances, and internal combustion engines in which a fluid lubricant is used; and in smaller mechanisms such as switches, clocks, small motors, and circuit breakers in which a solid lubricant such as graphite, grease, or certain plastics serves to reduce friction. Air (gas) bearings are designed to utilize both fluid mechanics principles when operating and solid lubricant–surfaced materials for start, stop, and emergency operations.

A hydrodynamic journal bearing maintains separation of shaft from bearing because the lubricant viscosity and the speed of the shaft create pressure in the converging portion of the fluid film which carries load. The governing equations were first developed by Reynolds (1886). Their solution has led to numerous computer solutions, including those used for this section.

Journal Bearing Design. Figure 10.5 shows schematics of frequently used types of journal bearing in which one or more lobes of cylindrical shape are positioned around the shaft, their axis being assumed parallel to the shaft axis. The features of each design and applications where it is often found are listed in Table 10.8.

Noncontact journal bearings are designed to assure a continuous supply of lubricant to the load-carrying section, and the bearing grooves in Figure 10.5 are designed for that purpose. Oil must be resupplied to the bearing because of the continuous loss of lubricant as it is forced from the bearing by the load-carrying pressures generated within it. The subsection on lubricant supply methods describes some of the many systems designed to assure this supply and to cool the lubricant at the same time.

Controlling Variables. Definitions of the variables involved in journal bearing analysis are contained in Table 10.9. Because of the large range of many variables, nondimensional quantities are often used which are independent of the dimensional unit system involved. Examples are given in the English system unless otherwise stated.

Calculating Bearing Performance. Journal bearing performance can be calculated directly from dedicated computer codes which take account of load, speed, oil type, and delivery system, as well as bearing dimensions. This subsection presents two approximate solutions: a simple thermal approach and a set of interpolation tables based on computer solutions.

Thermal Approach. It is assumed that the bearing is operating at a constant but elevated temperature. A predicted operating temperature can then be found as a function of an assumed viscosity. A solution is found when the assumed viscosity equals the lubricant viscosity at that temperature. Three equations are used for this method. For radial loads, the power dissipation is

$$H_p = j\pi^3\mu(N/60)^2 D^3 L/C \quad \text{in.}\cdot\text{lb/sec} \tag{10.6}$$

where $j = 1$ for a shaft-centered design. The lubricant flow rate is

$$Q = Q_o + qCR^2\omega/2 \quad \text{in.}^3/\text{sec} \tag{10.7}$$

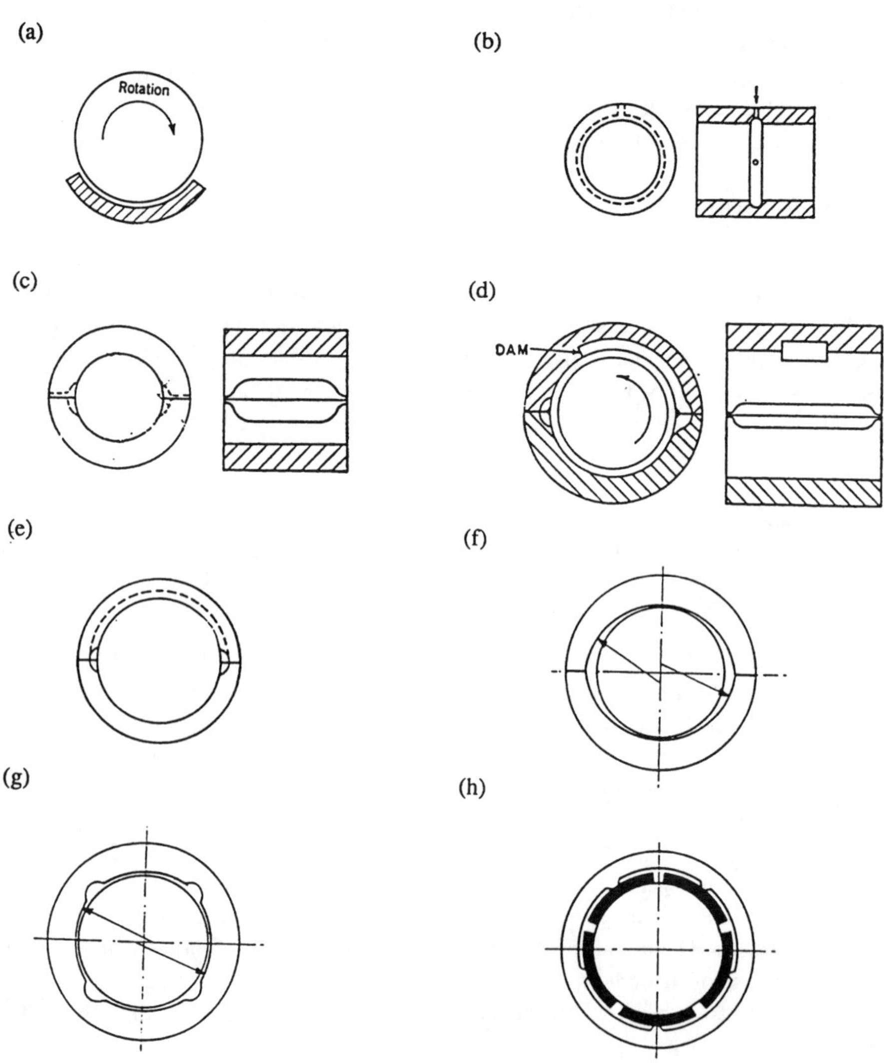

FIGURE 10.5 Types of pressure-fed journal bearings: (a) Partial arc. (b) Circumferential groove. (c) Cylindrical bearing–axial groove. (d) Pressure dam. (e) Cylindrical overshot. (f) Elliptical. (g) Multilobe. (h) Tilting pad.

TABLE 10.8 Journal Bearing Designs

Type	Typical Loading	Applications
Partial arc	Unidirectional load	Shaft guides, dampers
Circumferential groove	Variable load direction	Internal combustion engines
Axial groove types		
Cylindrical	Medium to heavy unidirectional load	General machinery
Pressure dam	Light loads, unidirectional	High-speed turbines, compressors
Overshot	Light loads, unidirectional	Steam turbines
Multilobe	Light loads, variable direction	Gearing, compressors
Preloaded	Light loads, unidirectional	Minimize vibration
Tilting pad	Moderate variable loads	Minimize vibration

TABLE 10.9 Journal Bearing Parameters

B	Bearing damping coefficient	lb/in./sec.
C	Radial clearance	in.
C_α	Adiabatic constant	—
C_p	Heat capacity	in.·lb/lb°F
D	Diameter	in.
H_p	Power loss	in.·lb/sec
K	Bearing stiffness	lb/in.
L	Bearing length	in.
N	Shaft velocity	rpm
Q	Lubricant flow rate	in.³/sec
R	Shaft radius	in.
R_e	Reynolds number	—
T_e	Entrance temperature	°F
T_f	Feed temperature	°F
T_q	Torque	in.·lb
ΔT_b	Temperature rise coefficient, bottom half	°F
ΔT_t	Temperature rise coefficient, top half	°F
U	Velocity	in./sec
W	Load	lb
e	Shaft eccentricity	in.
h	Film thickness	in.
j	Ratio: power loss/shaft-centered loss	—
p	Pressure	psi
q	Flow coefficient	—
w	Load coefficient	—
x	Coordinate in direction of load	in.
y	Coordinate orthogonal to load	in.
β	Exponential temperature coefficient of viscosity	—
ε	Shaft eccentricity, nondimensional	—
γ	Angular extent of film	—
ϕ	Attitude angle	—
ρ	Density	lb/in.³
μ	Viscosity	lb·sec/in.²
ω	Shaft velocity	rad/sec
θ	Angle in direction of rotation, from bottom dead center	—
Φ	Energy dissipation	in.·lb/sec

where q is the proportion of side flow to circulating flow, and the zero speed flow, Q_0 (in³/sec), represents other flows such as from the ends of the feed grooves which are not related to the load-carrying film itself. Q_0 can usually be neglected for rough estimation, and this approximation is useful for eccentricities as high as 0.7. Note that both q and j are functions of specific design as well as load and speed. The average operating temperature for a given viscosity is

$$T_2 = T_f + \frac{\left(H_p - \Phi\right)}{\left(\rho C_p Q\right)} \quad ^\circ\text{F} \tag{10.8}$$

where T_f is the feed temperature and Φ is the energy loss due to conduction and radiation. For diameters of 2″ or more, Φ can usually be assumed to be 0. Plotting T_2 vs. viscosity for several values of μ on a plot of the viscosity vs. T for the lubricant shows the operating temperature for the bearing as the intersection.

Flow Dynamics Solution. A more general solution for journal bearing performance is based on prediction of flow characteristics in the bearing, and of physical behavior of the bearing based on the Reynolds equation. A common two-pad journal bearing with pressurized oil feed will be used to provide specific design information. The Reynolds equation is the differential equation expressing the physical behavior of the fluid film between shaft and bearing, as written for cylindrical bearings:

$$1/R^2\left[\partial/\partial\Theta\left(h^3/\mu\right)\partial p/\partial\Theta\right]+\partial/\partial z\left(h^3/\mu\right)\partial p/\partial z=6\left(U/R\right)\partial h/\partial\Theta \tag{10.9}$$

where z is the axial coordinate and Θ is the angular coordinate.

Bearing Configuration. A cross section through a common type of two-pad cylindrical bearing is shown in Figure 10.6. Two pads having a radius of $R + C$ and an angular extent of 150°, and with load applied vertically to the lower half through a shaft, make up the bearing. Lubricant is admitted under moderate feed pressure to the two 30° grooves formed at the split as shown in Figure 10.6. The shaft rotates counterclockwise, and lubricant pressures are balanced when the shaft center is displaced down and to the right.

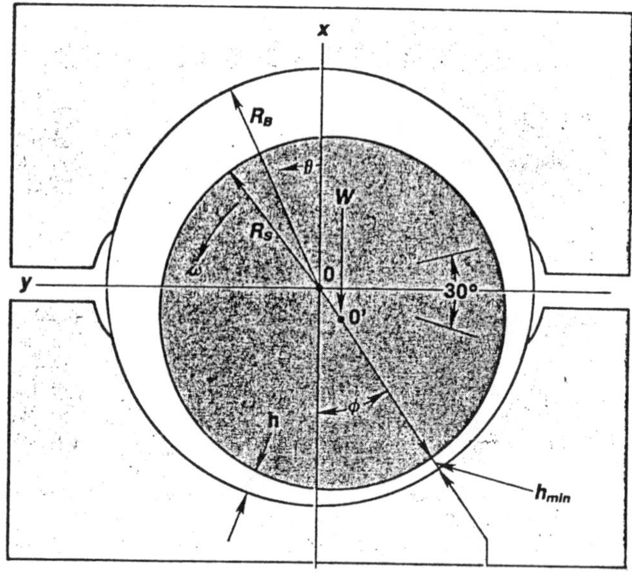

FIGURE 10.6 Geometry of split cylindrical journal bearing.

Lubricant Properties. Pressures in the lubricant film are generated as a function of the local shear rates and the local viscosity as described by the Reynolds equation. The local temperature rise is given by the local energy dissipation divided by the local flow rate:

$$\Delta T=\left[2\mu\omega R^2\Delta\theta\right]\Big/\left[h^2\rho C_p\right] \tag{10.10}$$

As an alternative to Equation 10.2, an exponential relation between viscosity and temperature is used:

$$\mu=\mu_0 e^{-\beta(T-T_0)} \tag{10.11}$$

Assuming an ISO 32 oil, viscosity μ_0 at 104°F is 98 reyns, density is 0.0310, and β is 0.0170 for the range from 104 to 212°F. The value for β may be determined from Figure 10.4. A nondimensional coefficient, C_α, the *adiabatic coefficient*, is used as an indicator of the severity of thermal heating in a design. It contains the several factors responsible for temperature rise.

$$C_\alpha=2\mu_f\beta\omega\left(R/C\right)^2/\rho C_p \tag{10.12}$$

Bearing Performance Tables. Using computer-generated solutions of the Reynolds equation for the pressure field over the bearing surface, the relevant performance properties of the bearing/lubricant system can be determined. A number of programs are commercially available for this purpose. In order to illustrate how the behavior of a journal bearing can be determined, a two-pad split cylindrical bearing with an *L*/*D* of 0.5 has been selected, and a proprietary computer code has been used to develop a group of performance quantities for such a bearing. The code accounts for the internal lubricant circulation, including mixing in the inlet groove of feed lubricant with warm lubricant from the upper half, resulting in the viscosity μ_f. The primary characteristics of the bearing are load, stiffness, and damping. Each of these factors is presented in nondimensional terms in Table 10.10, and the corresponding dimensional and other quantities are given as follows:

Film thickness, in. $$h = C(1 + e\cos\Theta) \quad (10.13a)$$

Shaft eccentricity, in. $$e = \varepsilon C \quad (10.13b)$$

Load, lb $$W = w\left[6\mu\omega(R/C)^2 DL\right] \quad (10.13c)$$

Flow, in.^{3}/sec $$Q = q\left(CR^2\,\omega/2\right) \quad (10.13d)$$

Power loss, in.·lb/sec $$H_p = j\left(2\pi\mu\omega^2 R^3\, L/C\right) \quad (10.13e)$$

Stiffness, lb/in. $$S_{xx} = K_{xx}\, 6\mu\omega R(R/C)^2 \quad (10.13f)$$

Damping, lb·sec/in. $$D_{xx} = B_{xx}\, 12\mu R(R.C)^3 \quad (10.13g)$$

The axial length/diameter ratio also influences the performance of a journal bearing. To illustrate this, Table 10.11 presents the performance of longer bearings (*L*/*D* = 0.75 and 1.00) for comparison to the more common *L*/*D* = 0.5 results in Table 10.10.

Comparing Tables 10.10 and 10.11, the use of longer bearings has several important effects on operation and performance. Comparing key variables, the effects at an eccentricity of ratio of 0.7 are as follows:

Variable	L/D = 0.5	L/D = 0.75	L/D = 1.00
Load, w	0.28	0.69	1.21
Flow, q	0.69	0.82	0.88
Attitude angle, ϕ	36.4	36.1	35.8
Power ratio, j	1.00	1.15	1.17
Stiffness, K_{xx}	1.38	3.06	5.03
Damping, B_{xx}	0.99	2.52	4.31

Effect of Turbulence. Turbulence is a mixing phenomenon that occurs in larger high-speed bearings. When this behavior occurs, the simple viscous drag behavior that has been assumed in the preceding subsections is broken up by numerous eddies which increase the drag. The Reynolds number is a nondimensional quantity that expresses this factor:

$$R_e = hU\rho/\mu \quad (10.14)$$

TABLE 10.10 Performance of L/D = 0.5 Bearing

Part 1: $C_\alpha = 0.0$

ε	0.2	0.5	0.7	0.8	0.9	0.95
ϕ	66.5	48.01	36.44	30.07	22.18	16.46
w	0.0246	0.0997	0.2784	0.5649	1.6674	4.4065
q	0.3037	0.6014	0.6927	0.6946	0.6487	0.588
j	0.7779	0.8534	1.1005	1.3905	2.008	3.084
ΔT_b	0	0	0	0	0	0
ΔT_t	0	0	0	0	0	0
K_{xx}	0.041	0.2805	1.379	4.063	22.67	—
K_{xy}	0.1465	0.3745	1.063	2.476	9.390	34.47
K_{yx}	–0.055	–0.072	0.0063	0.193	1.710	8.002
K_{yy}	0.046	0.170	0.4235	0.883	2.622	7.555
B_{xx}	0.142	0.352	0.989	2.311	8.707	32.30
B_{xy}, B_{yx}	0.023	0.094	0.236	0.522	1.547	4.706
B_{yy}	0.056	0.105	0.174	0.302	0.630	1.390

Part 2: $C_\alpha = 0.1$

ε	0.2	0.5	0.7	0.8	0.9	0.95
ϕ	69.9	50.2	38.7	32.35	24.83	19.8
w	0.022	0.087	0.233	0.451	1.184	2.621
q	0.312	0.620	0.721	0.728	0.692	0.642
j	0.686	0.723	0.863	0.997	1.253	1.545
ΔT_b	0.274	0.403	0.642	0.907	1.519	2.346
ΔT_t	0.243	0.211	0.183	0.168	0.151	0.142
K_{xx}	0.038	0.2365	1.041	2.935	13.66	50.44
K_{xy}	0.126	0.3135	0.870	1.851	3.078	18.30
K_{yx}	–0.047	–0.061	–0.021	0.139	1.068	3.961
K_{yy}	0.037	0.140	0.3585	0.669	1.784	4.327
B_{xx}	0.121	0.286	0.776	1.592	4.97	14.00
B_{xy}, B_{yx}	0.016	0.071	0.195	0.341	0.850	2.10
B_{yy}	0.047	0.086	0.156	0.216	0.394	0.757

Part 3: $C_\alpha = 0.2$

ε	0.2	0.5	0.7	0.8	0.9	0.95
ϕ	73.4	52.2	40.8	34.55	27.23	22.5
w	0.020	0.077	0.198	0.368	0.890	1.779
q	0.320	0.639	0.747	0.759	0.730	0.760
j	0.613	0.628	0.712	0.791	0.933	1.092
ΔT_b	0.520	0.7520	1.162	1.594	2.521	3.651
ΔT_t	0.472	0.415	0.363	0.333	0.301	0.284
K_{xx}	0.035	0.1925	0.830	2.156	8.86	28.6
K_{xy}	0.11	0.272	0.704	1.477	4.515	11.72
K_{yx}	–0.041	–0.062	–0.018	0.074	0.640	2.371
K_{yy}	0.029	0.125	0.2895	0.551	1.375	2.932
B_{xx}	0.104	0.242	0.596	1.21	3.90	7.830
B_{xy}, B_{yx}	0.011	0.061	0.140	0.212	0.634	1.21
B_{yy}	0.040	0.080	0.121	0.187	0.326	0.501

TABLE 10.10 (continued) Performance of *L/D* = 0.5 Bearing

Part 4 C_α = 0.4

ε	0.2	0.5	0.7	0.8	0.9	0.95
ϕ	80.2	56.0	44.5	38.4	31.3	26.7
w	0.016	0.061	0.148	0.260	0.562	1.000
q	0.331	0.6720	0.795	0.815	0.797	0.760
j	0.504	0.498	0.534	0.570	0.637	0.716
ΔT_b	0.946	1.33	1.97	2.61	3.87	5.26
ΔT_t	0.898	0.801	0.712	0.658	0.597	0.562
K_{xx}	0.029	0.137	0.538	1.295	4.56	12.6
K_{xy}	0.085	0.206	0.503	0.985	2.67	6.17
K_{yx}	−0.0315	−0.0548	0.0298	0.0233	0.321	1.136
K_{yy}	0.019	0.094	0.214	0.382	0.860	1.68
B_{xx}	0.079	0.175	0.397	0.734	1.75	3.44
B_{xy}, B_{yx}	0.0041	0.042	0.094	0.166	0.329	0.55
B_{yy}	0.030	0.064	0.092	0.131	0.120	0.276

where h is the local film thickness, U is the relative velocity of one surface with respect to the other, ρ is the fluid density, and μ is the local viscosity.

The influence of turbulence on an *L/D* = 0.5 bearing is shown in Table 10.12. Examination of Table 10.12 shows that the principal effects of operation in the turbulent regime with a Reynolds number above about 1000 are in the greater power required (j) and the maximum bearing temperature. Load capacity and flow are only moderately affected.

Example Calculation. The problem is to design a two-pad cylindrical bearing for support of a rotor having a bearing load of 8000 lb, a shaft diameter of 6 in., and a speed of 3600 rpm. Assume use of ISO VG-32 oil fed at a temperature of 120°F. Compute operating characteristics for a 3-in.-long bearing. Assume a radial clearance of 0.0044 in.

Feed viscosity, $\mu_f = 3.98 \times 10^{-6} e^{-0.00170(120-104)} = 3.03 \times 10^{-6}$ reyn
Angular velocity, $\omega = 3600 \times 2\pi/60 = 377$ rad/sec
Adiabatic coefficient: $C_\alpha = 2 \times 3.03 \times 10^{-6} \times 0.0170 \times 377 \times (3/0.0044)^2/0.0310/4320 = 0.1345$
Load coefficient (from Equation 10.13c: $w = 8000/[6 \times 3.03 \times 10^{-6} \times 377 \times 3 \times 6 \times (33/0.0044)^2] = 0.139$

The desired solution lies between Part 2 and Part 3 of Table 10.10. By using linear interpolation between the tabulated values for C_α of 0.1 and 0.2, and values of ε of 0.7 and 0.8, an approximate operating point of $C_\alpha = 0.1345$ yields the following coefficients: $\varepsilon = 0.729$, $w = 0.279$, $q = 0.733$, $j = 0.860$, and $\Delta T_b = 0.915$.

By using Equations 10.13, the dimensional operating results are:

Shaft velocity: $\omega = 3600 \times 2\pi/60 = 377$ rad/sec
Flow: $Q = 0.733 \times 0.0044 \times 3^2 \times 377 = 5.47$ in^3/sec
Power loss: $H_p = 0.860 \times 2\pi \times 3.03 \times 10^{-6} \times 377^2 \times 3^3 \times 3/0.0044 = 42.8$ in.·lb/sec
Oil temperature: $T_b = 120 + 0.915/0.0170 = 174$°F

Thrust Bearings

Types of Thrust Bearings. Oil film thrust bearings range from coin-size flat washers to sophisticated assemblies many feet in diameter. Of the six common types of thrust bearings shown in Table 10.13,

TABLE 10.11 Performance of Long Bearings

Part 1: $L/D = 0.75$, $C_\alpha = 0.0$

ε	0.2	0.5	0.7	0.8	0.9	0.95
ϕ	64.74	46.54	36.13	30.17	22.64	17.03
w	0.0705	0.2714	0.6947	1.311	3.440	8.241
q	0.392	0.738	0.825	0.811	0.737	0.6545
j	0.777	0.871	1.145	1.450	2.184	3.233
ΔT_b	0	0	0	0	0	0
ΔT_t	0	0	0	0	0	0
K_{xx}	0.121	0.706	3.065	8.506	41.5	—
K_{xy}	0.418	0.992	2.517	5.228	18.1	59.0
K_{yx}	–0.123	–0.189	0.052	0.404	3.18	16.2
K_{yy}	0.113	0.429	1.012	1.891	5.33	13.49
B_{xx}	0.423	0.982	2.52	5.16	17.7	54.4
B_{xy}, B_{yx}	0.057	0.249	0.609	1.10	3.24	7.58
B_{yy}	0.127	0.263	0.444	0.641	1.35	2.32

Part 2: $L/D = 1.00$, $C_\alpha = 0.00$

ε	0.2	0.5	0.7	0.8	0.9	0.95
ϕ	63.2	45.3	35.8	30.3	22.9	17.4
w	0.138	0.506	1.214	2.18	5.34	12.15
q	0.444	0.800	0.879	0.856	0.769	0.679
j	0.782	0.886	1.174	1.768	2.250	3.323
ΔT_b	0	0	0	0	0	0
ΔT_t	0	0	0	0	0	0
K_{xx}	0.234	1.254	5.026	13.24	60.9	—
K_{xy}	0.818	1.795	4.142	8.12	26.8	83.5
K_{yx}	–0.201	–0.313	–0.075	0.671	4.96	24.9
K_{yy}	0.198	0.732	1.64	2.95	8.04	19.5
B_{xx}	0.82	1.87	4.31	8.27	26.5	75.9
B_{xy}, B_{yx}	0.10	0.45	0.97	1.68	4.78	10.36
B_{yy}	0.21	0.46	0.70	0.98	2.02	3.24

TABLE 10.12. Influence of Turbulence ($\varepsilon = 0.7$, $C_\alpha = 0.2$, arc = 150°)

R_e	0	1000	2000	4000
ϕ	40.8	43.8	46.4	49.2
w	0.198	0.171	0.197	0.221
q	0.747	0.809	0.862	0.914
j	0.712	0.983	1.459	2.124
ΔT_b	1.162	0.585	0.918	1.404
K_{xx}	0.830	0.627	0.634	0.647
K_{xy}	0.704	0.575	0.577	0.645
K_{yx}	–0.018	–0.034	–0.047	–0.078
K_{yy}	0.289	0.257	0.282	0.330
B_{xx}	0.596	0.483	0.513	0.534
B_{xy}, B_{yx}	0.140	0.125	0.132	0.136
B_{yy}	0.121	—	—	0.104

the first five are hydrodynamic. As with journal bearings, each of these generates oil film pressure when a rotating thrust face pumps oil by shear into a zone of reduced downstream clearance. When thrust load increases, film thickness drops until a new balance is reached between inflow and outflow, raising pressure until the higher bearing load is balanced. The hydrostatic bearing uses a separate oil pump to supply the pressurized flow.

TABLE 10.13 Common Thrust Bearings and Their Range of Application

Type	O.D., in.	Unit Load, psi
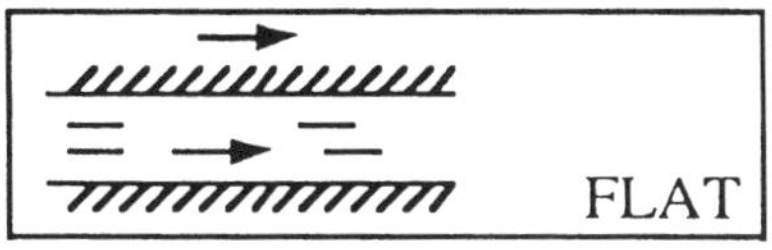 FLAT	0.5–20	20–100
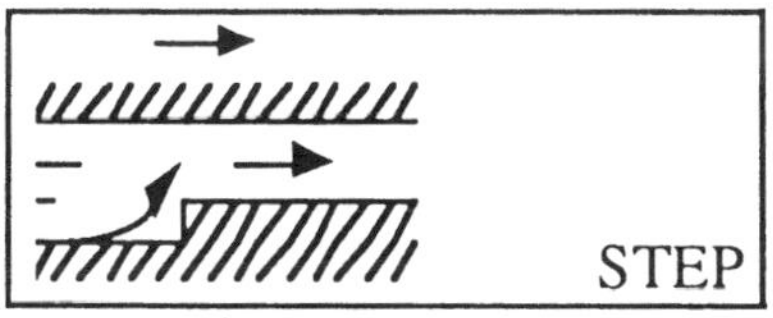 STEP	0.5–10	100–300
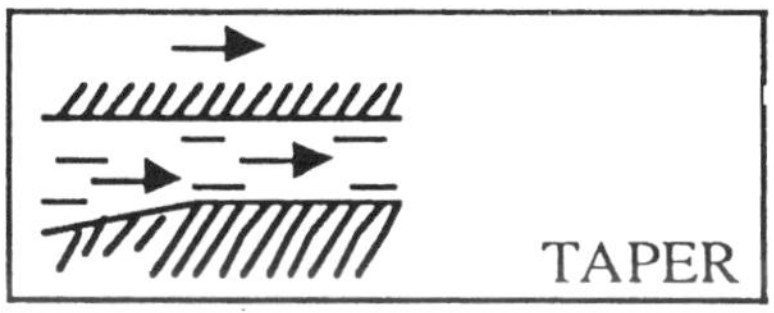 TAPER	2–35	150–300
 TILTING PAD	4–120	250–700
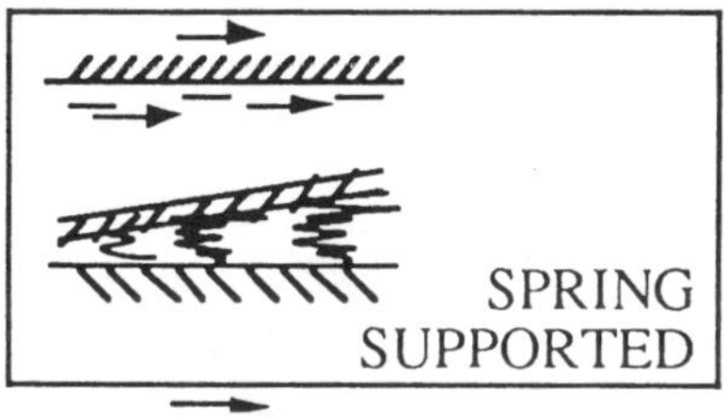 SPRING SUPPORTED	50–120	350–700
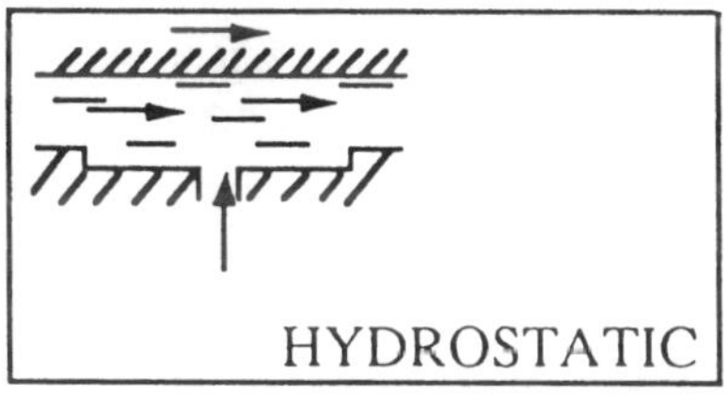 HYDROSTATIC	3–50	500–3000

Source: Booser, E.R. and Wilcock, D.F., *Machine Design*, June 20, 1991, 69–72. With permission.

Flat-land bearings, simplest to fabricate and least costly, are the first choice for simple positioning of a rotor and for light loads in electric motors, appliances, pumps, crankshafts, and other machinery. They carry less load than the other types because flat parallel planes do not directly provide the required pumping action. Instead, their action depends on thermal expansion of the oil and warping of the bearing material induced by heating from the passing oil film. The resulting slight oil wedge then gives a load rating of about 10 to 20% of that for the other types.

Step bearings also offer relatively simple design. With a coined or etched step, they lend themselves to mass production as small-size bearings and thrust washers. Step height for optimum load capacity approximately equals the minimum film thickness, often 0.001 in. or less. Circumferential length beyond the step is ideally 45% of the total bearing segment (Wilcock and Booser, 1956). Step thrust bearings are well suited for low-viscosity fluids such as water, gasoline, fuels, and solvents. Minimum film thickness in these applications is so small that features such as pivots and tapers are usually impractical. Step height must be small enough for good load capacity, yet large enough to accommodate some wear without becoming worn away. Step erosion by contaminants is sometimes a problem. *Tapered-land bearings* provide reliable, compact designs for mid- to large-size high-speed machines such as turbines, compressors, and pumps. Taper height normally should be about one to three times the minimum film thickness. For greater load capacity and to minimize wear during starting, stopping, and at low speeds, a flat land is commonly machined at the trailing edge to occupy from 10 to 30% of the circumferential length of each segment. Because operation of these bearings is sensitive to load, speed, and lubricant viscosity, they are typically designed for the rated range of operating conditions for specific machines.

Tilting-pad thrust bearings are used increasingly in turbines, compressors, pumps, and marine drives in much the same range of applications as tapered-land designs. They usually have a central supporting pivot for each of their three to ten bearing segments. Each of these thrust pad segments is free to adjust its position to form a nearly optimum oil wedge for widely varying loads, speeds, and lubricants, and with rotation in both directions. A secondary leveling linkage system is commonly introduced to support the individual pads; this provides a further advantage over tapered-land designs by accommodating some misalignment. Off-the-shelf units are available to match rotor shaft diameters from about 2 to 12 in., and custom designs range up to 120 to 170 in. in outside diameter. Recent trends to increase load capacity have led to offsetting pivots from the circumferential midpoint of a pad to about 60% beyond the leading edge, to substituting copper for steel as the backing for a tin babbitt bearing surface, and to nonflooded lubrication to minimize parasitic power loss from useless churning of oil passing through the bearing housing.

Springs or other flexible supports for thrust segments are employed for bearings ranging up to 10 ft or more in outside diameter and carrying millions of pounds of thrust. This flexible mounting avoids the high load concentration encountered by pivots in supporting large tilting-pads. Rubber backing can provide this flexible mounting for smaller thrust pads.

Hydrostatic thrust bearings are used where sufficient load support cannot be generated by oil film action within the bearing itself. This may be the case with low-viscosity fluids, or for load support on an oil film at standstill and very low speeds. The fluid is first pressurized by an external pump and then introduced into pockets in the bearing surface to float the load. A compact hydrostatic thrust bearing can sometimes be constructed with a single pocket at the end of a rotor. Larger bearings usually use three or more pressurized pockets to resist misalignment or to support off-center loads. Hydraulic flow resistance in the supply line to each pocket, or constant flow to each pocket (as with ganged gear pumps) then provides any asymmetric pressure distribution needed to support an off-center load. Bearing unit load is commonly limited to about 0.5 (0.75 with fixed flow systems) times the hydrostatic fluid supply pressure—up to 5000 psi with conventional lubricating oils.

Design Factors for Thrust Bearings. In preliminary sizing, the inside diameter d of a thrust bearing is made sufficiently larger than the shaft to allow for assembly, and to provide for any required oil flow to the thrust bearing inside diameter (see Figure 10.7). This clearance typically ranges from about $^1/_8$ in. for a 2-in. shaft to $^1/_2$ in. for a 10-in. shaft. Bearing outside diameter D is then set to provide bearing area sufficient to support total thrust load W (lb or N) with an average loading P (psi or N/m^2), using typical values from Table 10.13:

$$D = \left(\frac{4W}{\pi k_g P} + d^2 \right)^{0.5} \tag{10.15}$$

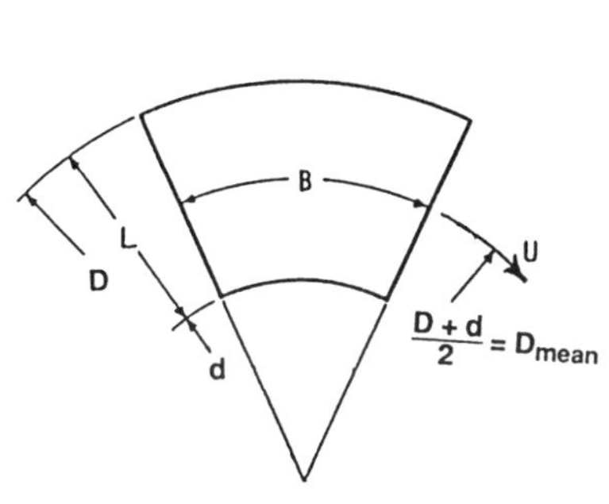

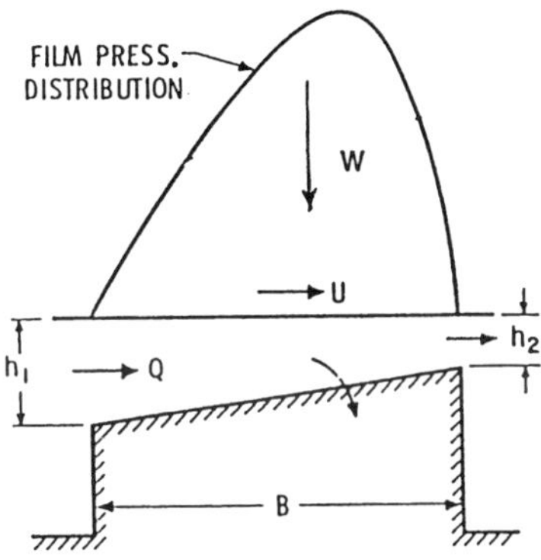

FIGURE 10.7 Sector for tapered land thrust bearing.

where k_g (typically 0.80 to 0.85) is the fraction of area between d and D not occupied by oil-distributing grooves. This bearing area is then divided by radial oil-feed groove passages, usually into "square" sectors with circumferential breadth B at their mean diameter equal to their radial length L.

While square pads usually produce optimum oil film performance, other proportions may be advantageous. With very large bearing diameters, for instance, square sectors may involve such a long circumferential dimension that the oil film overheats before reaching the next oil-feed groove. With a radially narrow thrust face, on the other hand, square sectors may become so numerous as to introduce excessive oil groove area, and their short circumferential length would interfere with hydrodynamic oil film action.

Performance Analysis. Performance analyses for sector thrust bearings using a fixed taper also hold approximately for most thrust bearing shapes (other than flat lands) with the same ratio of inlet to outlet oil film thickness (Wilcock and Booser, 1956; Neale, 1970; Fuller, 1984). Both for centrally pivoted pad thrust bearings and for spring-supported thrust bearings, use of an inlet-to-outlet film thickness ratio of two is usually appropriate in such an analysis.

While computer analyses in polar coordinates and with local oil film viscosity calculated over the whole oil film area gives more-exact solutions, the following constant viscosity approximations are made by relating a rectangular bearing sector (width B, length L) to the circular configuration of Figure 10.7. This rectangular representation allows more ready evaluation of a range of variables and gives results which are quite similar to a more accurate polar coordinate analysis.

Employing the nomenclature of Figure 10.7, the following relations give minimum film thickness h_2, frictional power loss H, and oil flow into a sector Q. The dimensionless coefficients involved are given in Table 10.14 for a range of sector proportions L/B and ratios of inlet to outlet film thicknesses h_1/h_2.

$$h_2 = K_h\left(\mu UB/P\right)^{0.5} \qquad \text{in. (m)} \qquad (10.16a)$$

$$H = K_f\,\mu U^2 BL/h_2 \qquad \text{lb}\cdot\text{in./sec (N}\cdot\text{m/sec)} \qquad (10.16b)$$

$$Q = K_q ULh_2 \qquad \text{in.}^3\text{/sec (m}^3\text{/sec)} \qquad (10.16c)$$

$$\Delta T = H/\left(Q\rho C_p\right) = \frac{K_f}{K_q K_h^2 \rho C_p} P \qquad {}^\circ\text{F (}^\circ\text{C)} \qquad (10.16d)$$

where B= circumferential breadth of sector at mean radius, in. (m)

K_h,K_f, K_q= dimensionless coefficients

L = radial length of sector, in. (m)

P = unit loading on projected area of sector, W/BL, lb/in.^{2} (N/m^2)

U = surface velocity at mean diameter of sector, in./sec (m/sec)

W = load on a sector, lb (N)

C_p = oil specific heat, in.·lb/(lb·°F) (J/(kg·°C))

h_1,h_2 = leading edge and trailing edge film thicknesses, in. (m)

ρ = oil density, lb/in.^{3} (n/m^3)

μ = oil viscosity at the operating temperature, lb·sec/in.^{2} (N·sec/m^2)

Example. The following example involves a bearing for 25,000 lb thrust load in a 1200 rpm compressor whose rotor has a 5-in.-diameter journal. ISO-32 viscosity grade oil is to be fed at 120°F. Allowing $^3/_8$ in. radial clearance along the shaft sets thrust bearing bore d = 5.75 in. Taking unit loading P = 300 psi allows a margin for uncertainty in expected thrust load, and outside diameter D is given by Equation 10.14):

$$D = \left(\frac{4 \times 25000}{\pi(0.85)(300)} + 5.75^2 \right)^{0.5} = 12.6 \text{ in.}$$

With 15% of the area used for oil feed passages, k_g = 0.85. Thrust bearing segment radial length L = (12.6 – 5.75)/2 = 3.425 in. With mean diameter (5.75 + 12.6)/2 = 9.175 in., total circumferential breadth of all pads at the mean diameter = $\pi D_m k_g$ = 24.5 in. The number of sectors (and grooves) for B = L is then 24.5/3.425 = 7.2. Using seven lands, adjusted circumferential breadth B for each sector = 24.5/7 = 3.5 in. (For simpler fabrication, six or eight sectors should also be considered.) Runner velocity at the mean diameter, U = π(9.175)(1200/60) = 576.5 in./sec.

For square pads (L/B = 1) in a pivoted-paid bearing with h_1/h_2 = 2, which represents experience with centrally pivoted pads, Table 10.14 gives the following performance coefficients:

$$K_h = 0.261, \quad K_f = 0.727, \quad K_q = 0.849$$

Temperature rise is given by Equation (10.16d) which assumes that the total frictional power loss H goes into heating the total oil flow Q passing over the pad.

$$\Delta T = \frac{K_f P}{K_q K_h^2 \rho C_p} = \frac{0.727(300)}{0.849(0.261)^2(0.0313)(4535)} = 27°\text{F}$$

Adding this ΔT to the 120°F feed temperature gives 147°F as the representative oil operating temperature with a viscosity from Figure 10.4 of 1.6 × 10^{-6} lb·sec/in.^{2}. Temperature rise to the maximum oil film temperature would be expected to be about 53°F, twice the 27°F. If this bearing were in a housing fully flooded with oil, feed temperature to each pad would become the housing oil temperature, essentially the same as oil draining from the housing.

Minimum film thickness h_2 becomes, from Equation (10.16a)

$$h_2 = 0.261\left[\left(1.6 \times 10^{-6}\right)(576.5)(3.5)/300\right]^{0.5} = 0.00086 \text{ in.}$$

With a fixed tapered land, rather than a centrally pivoted pad for which it could be assumed that h_1/h_2 = 2, several iterations might be required with different assumed values of the h_1/h_2 ratio in order to

TABLE 10.14 Thrust Bearing Performance Characteristics

L/B	0.25	0.5	0.75	1.0	1.5	2.0	∞
			$h_1/h_2 = 1.2$				
K_h	0.064	0.115	0.153	0.180	0.209	0.225	0.266
K_f	0.912	0.913	0.914	0.915	0.916	0.917	0.919
K_q	0.593	0.586	0.579	0.574	0.567	0.562	0.549
			$h_1/h_2 = 1.5$				
K_h	0.084	0.151	0.200	0.234	0.275	0.296	0.351
K_f	0.813	0.817	0.821	0.825	0.830	0.833	0.842
K_q	0.733	0.714	0.696	0.680	0.659	0.647	0.610
			$h_1/h_2 = 2$				
K_h	0.096	0.170	0.223	0.261	0.305	0.328	0.387
K_f	0.698	0.708	0.718	0.727	0.739	0.747	0.768
K_q	0.964	0.924	0.884	0.849	0.801	0.772	0.690
			$h_1/h_2 = 3$				
K_h	0.100	0.173	0.225	0.261	0.304	0.326	0.384
K_f	0.559	0.579	0.600	0.617	0.641	0.655	0.696
K_q	1.426	1.335	1.236	1.148	1.024	0.951	0.738
			$h_1/h_2 = 4$				
K_h	0.098	0.165	0.212	0.244	0.282	0.302	0.352
K_f	0.476	0.503	0.529	0.551	0.581	0.598	0.647
K_q	1.888	1.745	1.586	1.444	1.242	1.122	0.779
			$h_1/h_2 = 6$				
K_h	0.091	0.148	0.186	0.211	0.241	0.256	0.294
K_f	0.379	0.412	0.448	0.469	0.502	0.521	0.574
K_q	2.811	2.560	2.273	2.013	1.646	1.431	0.818
			$h_1/h_2 = 10$				
K_h	0.079	0.121	0.148	0.165	0.185	0.195	0.221
K_f	0.283	0.321	0.353	0.377	0.408	0.426	0.474
K_q	4.657	4.182	3.624	3.118	2.412	2.001	0.834

After Khonsari, M.M., in *Tribology Data Handbook*, CRC Press, Boca Raton, FL, 1997.

determine the performance coefficients in Table 10.14. The proper value of h_1/h_2 will be the one that gives the same final calculated value of h_2 from the above equation as was assumed in the preliminary selection of K_h, K_f, and K_q.

After finding the values for h_2 *and* K_f, the power loss H can be determined using Equation (10.16b). For this example the power loss would be H = 5510 lb·in./sec.

The total oil feed to the bearing should satisfy two requirements: (1) provide a full oil film over the bearing segment and (2) maintain reasonably cool operation with no more than 30 to 40°F rise in the oil temperature from feed to drain. Equation (10.16c) can be used to find the oil feed Q needed at the sector inlet to form a full film. The oil feed needed for a 40°F rise is given by the following heat balance using typical density and specific heat values for petroleum oil:

$$Q = H\big/\left(\rho C_p \Delta T\right) \tag{10.17}$$

The required oil feed will be the larger of the values determined by (10.16c) and (10.17).

The above calculations are for a single sector; power loss and oil feed would be multiplied by the number of sectors (seven) to obtain values for the total bearing. Consideration would normally follow for other pad geometries, and possibly other lubricants and oil flow patterns, in a search for the most-promising design. More-detailed calculations of film thickness, film temperatures, oil flow, and power loss could then be obtained by one of a number of computer codes available from bearing suppliers or other sources.

Oil Film Bearing Materials

Selection of the material for use in a journal or thrust bearing depends on matching its properties to the load, temperature, contamination, lubricant, and required life.

Babbitts. Of the common bearing materials, listed in Table 10.15, first consideration for rotating machines is usually a babbitt alloy containing about 85% tin or lead together with suitable alloying elements. With their low hardness, they have excellent ability to embed dirt, conform to shaft misalignment, and rate highest for compatibility with steel journals. Tin babbitts, containing about 3 to 8% copper and 5 to 8% antimony, are usually the first choice for their excellent corrosion resistance. SAE 12 (ASTM Grade 2) tin babbitt is widely used in both automotive and industrial bearings. The much lower cost of lead babbitt, however, with 9 to 16% antimony and as much as 12% tin for improved corrosion resistance, brings SAE 13, 14, and 15 grades into wide use for both general automotive and industrial applications (Booser, 1992).

TABLE 10.15 Characteristics of Oil Film Bearing Materials

Material	Brinell Hardness	Load Capacity, psi	Max Operating Temp., °F	Compatibility[a]	Conformability and Embeddability[a]	Corrosion Resistance[a]	Fatigue Strength[a]
Tin babbitt	20–30	800–1500	300	1	1	1	5
Lead babbitt	15–25	800–1200	300	1	1	3	5
Copper lead	20–30	1500–2500	350	2	2	5	4
Leaded bronze	60–65	3000–4500	450	3	4	4	3
Tin bronze	65–80	5000+	500	5	5	2	2
Aluminum alloy	45–65	4000+	300	4	3	1	2
Zinc alloy	90–125	3000	250	4	5	5	3
Silver overplated	—	5000+	300	2	4	2	1
Two-component, babbitt surfaced	—	3000+	300	2	4	2	3
Three-component, babbitt surfaced	—	4000+	300	1	2	2	1

[a] Arbitrary scale: 1 = best, 5 = worst.

To achieve the high fatigue strength needed in reciprocating engines, only a very thin layer (commonly 0.001″) of babbitt is used so that much of the reciprocating load is taken on a stronger backing material (DeHart, 1984; Kingsbury, 1992). For bimetal bushings such as those used in automobile engines, oil grooves and feed holes are formed in a continuous steel strip coated with babbitt. The strip is then cut to size and the individual segments are rolled into finished bearings.

For heavy-duty reciprocating engines, three-layer bearings are common. By using a steel strip backing, a thin overlay of SAE 19 or 190 lead babbitt is either electroplated or precision cast on an intermediate layer about 0.1 to 0.3″ thick of copper–nickel, copper–lead, leaded bronze, aluminum, or electroplated silver.

Copper Alloys. Copper–lead alloys containing 20 to 50% lead, either cast or sintered on a steel back, provide good fatigue resistance for heavy-duty main and connecting rod bearings for automotive, truck, diesel, and aircraft engines. The 10% lead–10% tin leaded bronze has been a traditional selection for bearings in steel mills, appliances, pumps, automotive piston pins, and trunions. This has been replaced in many applications by CA932 (SAE 660) containing 3% zinc for easier casting. The harder tin bronzes require reliable lubrication, good alignment, and 300 to 400 Brinell minimum shaft hardness. Cast tin bronze bushings are used at high loads and low speeds in farm machinery, earthmoving equipment, rolling mills, and in automotive engines for connecting rod bearings.

Utility of copper alloy bearings is limited to relatively low surface speeds by the tendency to form a copper transfer film on a steel shaft. Above about 1500 to 3000 ft/min, selective plucking of softer copper material from hotter load zones in the bearing may result in welded lumps forming on the cooler, stronger transfer layer on the mating steel shaft.

Zinc Alloys. Zinc alloys containing about 10 to 30% aluminum find some use for lower cost and better wear life in replacing leaded bronzes. They are used for both oscillating and rotating applications involving speeds up to 1400 ft/min and temperatures up to 250°F.

Aluminum Alloys (DeHart, 1984; Shabel et al., 1992). Although finding only minor use in general industrial applications because of their limited compatibility with steel journals, aluminum alloys containing 6.5% tin, 1% copper, and up to 4% silicon are used as solid, bimetal, and trimetal bearings in automotive engines, reciprocating compressors, and aircraft equipment. Good journal finish and shaft hardness of Rockwell B 85 or higher are required. The good fatigue and corrosion resistance of aluminum alloys have led to use of a number of unique alloys containing major additions of silicon, lead, or tin to provide better compatibility characteristics.

Dry and Semilubricated Bearings

Various plastics, porous bronze and porous iron, carbon–graphite, rubber, and wood are widely used for bearings operating dry or with sparse lubrication (Booser, 1992). Unique properties of these materials have led to their broad use in applications once employing oil film and ball and roller bearings. While these materials provide good performance under conditions of poor or nonexistent lubrication at low speeds, performance commonly improves the closer the approach to full film lubrication.

Plastics

Most commercial plastics find some use both dry and lubricated in slow-speed bearings at light loads (Jamison, 1994). The most commonly used thermoplastics for bearings are PTFE, nylon, and acetal resins. Thermosetting plastics used for bearings include phenolics, polyesters, and polyimides. Table 10.16 compares characteristics of typical plastic bearings with those of carbon–graphite, wood, and rubber which are used in similar applications.

In addition to the maximum temperature which can be tolerated, three operating limits shown in Table 10.16 are normally defined for plastic bearings: (1) maximum load at low speed, which reflects the compressive yield strength, (2) maximum speed for running under very light load, and (3) a Pv load-speed limit at intermediate speeds, which serves as a measure of the maximum tolerable surface temperature. Since wear volume in dry sliding is approximately proportional to total load and the distance of sliding, Pv also gives a measure of wear depth d in the modified form of Archard's relation (10.1), $d = k(Pv)t$, where t is the operating time and wear factor k = wear coefficient K/hardness H.

Typical values of this wear factor k are given in Table 10.17. Since k values involve substantial variability, prototype tests are highly desirable for any planned application. Added fillers can reduce the wear factor for the base polymer by a factor of 10 to 1000 and more (Blanchet and Kennedy, 1992). Common fillers include inorganic powders such as clay, glass fibers, graphite, molybdenum disulfide, and powdered metal, and also silicone fluid as an internally available lubricant.

Porous Metals

Bearings of compressed and sintered bronze, iron, and aluminum alloy powder are produced at the rate of millions per week for shaft sizes ranging from about 1.6 to 150 mm. These sleeve bearings and thrust

TABLE 10.16. Representative Limiting Conditions for Nonmetallic Bearing Materials

Material	Maximum Temperature, °C	*P*v Limit, MN/(m·sec)[a]	Maximum Pressure, *P*, MN/m[2b]	Maximum speed, v, m/sec
Thermoplastics				
Nylon	90	0.90	5	3
Filled	150	0.46	10	—
Acetal	100	0.10	5	3
Filled	—	0.28	—	—
PTFE	250	0.04	3.4	0.3
Filled	250	0.53	17	5
Fabric	—	0.88	400	0.8
Polycarbonate	105	0.03	7	5
Polyurethane	120	—	—	—
Polysulfone	160	—	—	—
Thermosetting				
Phenolics	120	0.18	41	13
Filled	160	0.53	—	—
Polyimides	260	4	—	8
Filled	260	5	—	8
Others				
Carbon–graphite	400	0.53	4.1	13
Wood	70	0.42	14	10
Rubber	65	—	0.3	20

[a] See Table 10.18.

[b] To convert MN/m^2 to psi, multiply by 145.

TABLE 10.17 Wear Factors for Plastic Bearings[a]

Material	Wear Factor *k*, m^2/N No Filler	Filled[b]
Nylon-6, 6	4.0	0.24
PTFE	400	0.14[c]
Acetal resin	1.3	4.9
Polycarbonate	50	3.6
Polyester	4.2	1.8
Poly(phenylene oxide)	60	4.6
Polysulfone	30	3.2
Polyurethane	6.8	3.6

[a] See Booser (1992).

[b] With 30 wt% glass fiber, unless otherwise noted.

[c] 15% glass fiber.

washers are used in a wide variety of small electric motors, appliances, business machines, machine tools, automotive accessories, and farm and construction equipment (Morgan, 1984; Cusano, 1994). Traditional powder metal bearings consist of 90% copper and 10% tin (Table 10.18). The common pore volume of 20 to 30% is usually impregnated with a petroleum oil of SAE 30 viscosity. To promote formation of an oil film, high porosity with its high oil content is employed for higher speeds, often with an oil wick or grease providing a supplementary lubricant supply. Lower porosity with up to 3.5% added graphite is used for lower speeds and oscillation where oil film formation is difficult.

Porous iron bearings are used for lower cost, often with some copper and graphite added for high load capacity at low speed. Iron with up to 40% of added 90–10 bronze powder provides many of the characteristics of porous bronze bearings while enjoying the lower cost of the iron. Porous aluminum containing 3 to 5% copper, tin, and lead finds limited use for providing cooler operation, better conformability, and lower weight.

Table 10.18 gives approximate operating limits for porous metal bearings. Generally, maximum *P* values for sleeve bearings range up to 50,000 psi·ft/min. *Pv* levels for thrust bearings should generally not exceed about 20% of this value.

TABLE 10.18 Operating Limits for Porous Metal Bearings

		Pressure limit, *P*, MN/m²			
Porous Metal	**Nominal Composition, wt%**	**Static**	**Dynamic**	**Speed Limit v, m/sec**	***Pv* Limit MN/(m·sec)**
Bronze	Cu 90, Sn 10	59	28	6.1	1.8[a]
Iron		52	25	2.0	1.3
Iron–copper	Fe 90, Cu 10	140	28	1.1	1.4
Iron–copper–carbon	Fe 96, Cu 3, C 0.7	340	56	0.2	2.6
Bronze-iron	Fe 60, Cu 36, Sn 4	72	17	4.1	1.2
Aluminum		28	14	6.1	1.8

Note: To convert MN/m² to psi, multiply by 145.

[a] Approximately equivalent to 50,000 psi · ft/min limit often quoted by U.S. suppliers.

Rolling Element Bearings

Types of Rolling Element Bearings

Rolling element bearings may be classified according to the type of rolling element, i.e., ball or roller, and the loading direction. Ball and roller bearings can be designed to carry either radial or thrust loads, or a combination of the two. Standard rolling element bearing configurations are shown in Figure 10.8, and the capabilities of the different types are summarized in Figure 10.9.

Ball bearings usually consist of a number of hardened and precisely ground balls interposed between two grooved and hardened rings or races. A cage or separator is used to keep the balls equally spaced around the groove. The most common *radial ball bearing* is a deep groove, or Conrad, type which is designed to carry large radial loads, with a small thrust load capability. The radial capacity of a deep groove bearing can be increased by inserting more balls in the bearing, by means of either a face-located filling notch (which decreases the thrust capacity) or a split inner or outer ring (which requires a means to hold the ring halves axially together). The thrust capability of a radial ball bearing can be increased by inducing angular contact between ball and rings. A single-row angular contact bearing can carry thrust load in only one direction, with the thrust load capacity being dependent on the contact angle (angle between the line of action of the force and the plane perpendicular to the shaft axis). Duplex angular contact bearings consist of two angular contact bearings mounted together so they can carry thrust loads in either direction with little axial play, or they can be mounted in tandem to double the axial and radial load-carrying capacity. Self-aligning ball bearings are designed to accommodate more shaft misalignment than is possible with other radial ball bearings.

Thrust ball bearings are used primarily in machinery with a vertically oriented shaft which requires a stiff axial support. Many such bearings have a 90° contact angle and, as a result, can carry essentially no radial load; they also have limited high-speed capability. If thrust loads in both directions are to be carried, a second row of balls must be added.

Roller bearings can be made with cylindrical, tapered, or spherical rollers. As with ball bearings, the rollers are contained between two rings, with a cage or separator used to keep the rollers separated. The cage can be guided by either the rollers or one of the rings. Since roller bearings operate with line contacts, as opposed to the point (or elliptical) contacts that occur in ball bearings, roller bearings are stiffer (less radial displacement per unit load) and have a greater load-carrying capacity than a ball bearing of similar size. Roller bearings are more expensive than ball bearings of comparable size.

Radial cylindrical roller bearings are designed to carry primarily radial loads. Cylindrical roller bearings have a high radial load capacity and low friction, so they are suitable for high-speed operation.

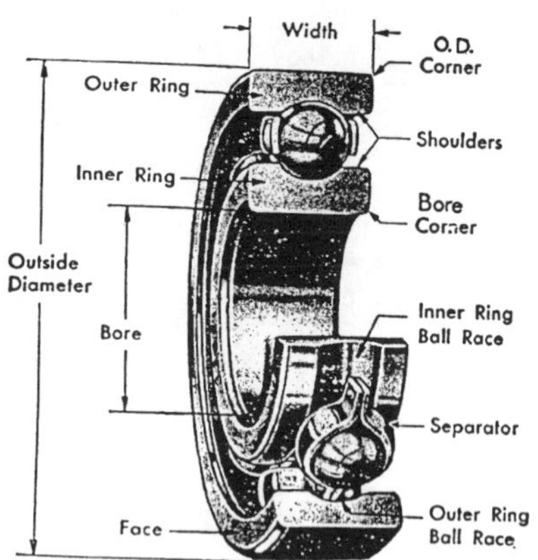

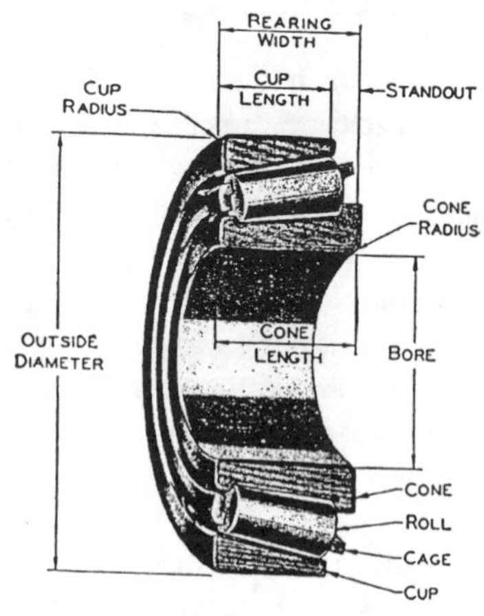

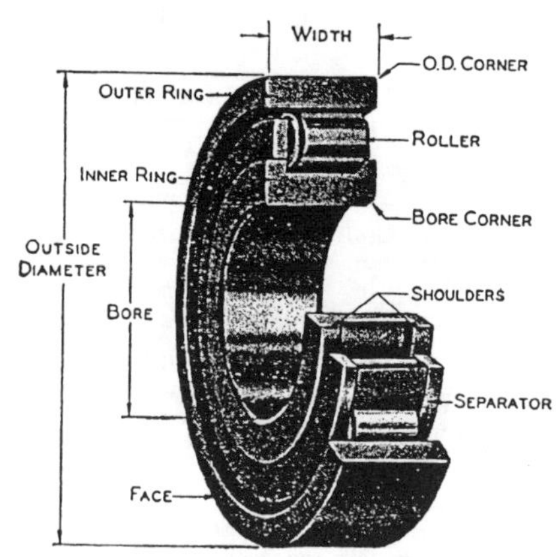

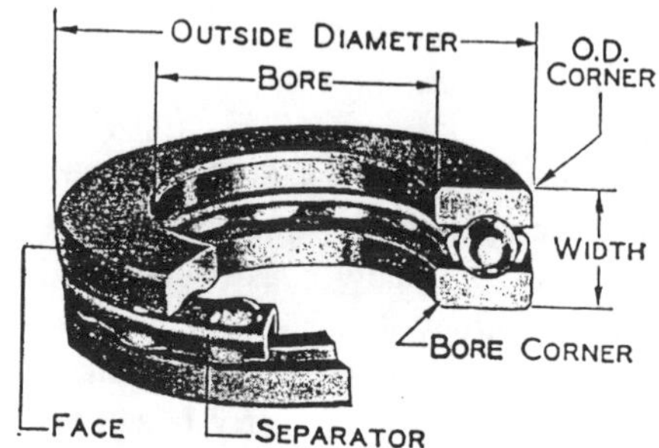

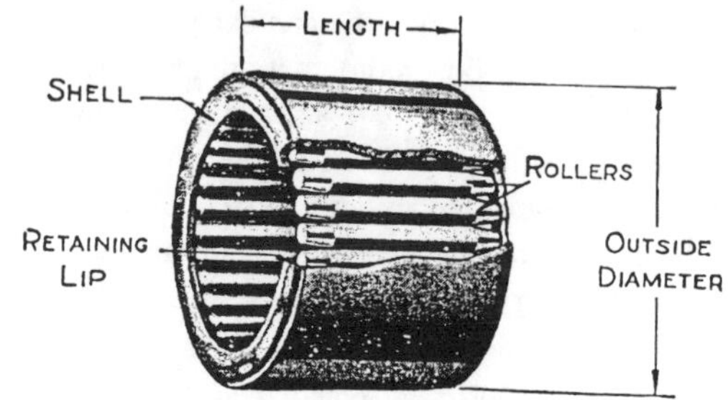

FIGURE 10.8 Major types of rolling element bearings.

Their thrust load capacity is limited to that which can be carried by contact (sliding) between the ends of the rollers and the flange or shoulder on the ring which contains them. The rollers in many cylindrical roller bearings are actually slightly crowned to relieve stress concentrations which would otherwise occur at the ends of the rollers and to compensate for misalignment of the bearing. In order to increase the load-carrying capacity of roller bearings, a second row of rollers is added instead of using longer rollers. This is because long rollers (i.e., length/diameter > 1.7) tend to skew in the roller path, thus limiting their high-speed capability and sometimes shortening their life. Needle bearings

CHARACTERISTICS OF STANDARD ROLLING ELEMENT BEARING CONFIGURATIONS

	TYPE	SIZE RANGE IN INCHES		AVERAGE RELATIVE RATINGS				DIMENSIONS	
				Capacity		Limiting Speed	Permissible Misalignment	Metric	Inch
		Bore	O.D.	Radial	Thrust				
BALL BEARINGS	CONRAD TYPE	.1181 to 41.7323	.3750 to 55.1181	Good	Fair ←→	Conrad is basis for comparison 1.00	± 0° 8′	X	X
	MAXIMUM TYPE	.6693 to 4.3307	1.5748 to 8.4646	Excellent	Poor ←→	1.00	± 0° 3′	X	
	ANGULAR CONTACT 15°/40° (15° / 40°)	.3937 to 7.4803	1.0236 to 15.7480	Good	Good (15°) Excellent (40°) ←	1.00 / 0.70	± 0° 2′	X	
	ANGULAR CONTACT 35°	.3937 to 4.3307	1.1811 to 9.4488	Excellent	Good ←	0.70	0°	X	
	SELF-ALIGNING	.1969 to 4.7244	.7480 to 9.4488	Fair	Fair ←→	1.00	± 4°	X	
CYLINDRICAL ROLLER BEARINGS	SEPARABLE INNER RING NON-LOCATING	.4724 to 19.6850	1.2598 to 28.3465	Excellent	0	1.00	± 0° 4′	X	
	SEPARABLE INNER RING ONE DIR. LOCATING	.4724 to 12.5984	1.2598 to 22.8346	Excellent	Fair ←	1.00	± 0° 4′	X	
	SELF-CONTAINED TWO DIR. LOCATING	.4724 to 3.9370	1.4567 to 8.4646	Excellent	Fair ←→	1.00	± 0° 4′	X	
TAPERED ROLLER BEARINGS	SEPARABLE	.6205 to 6.0000	1.5700 to 10.0000	Good	Good →	0.60	± 0° 2′	X	X
SPHERICAL ROLLER BEARINGS	SELF-ALIGNING	.9843 to 12.5984	2.0472 to 22.8346	Good	Fair ←→	0.50	± 4°	X	
	SELF-ALIGNING	.9843 to 35.4331	2.0472 to 46.4567	Excellent	Good ←→	0.75	± 1°	X	
NEEDLE BEARINGS	COMPLETE BEARINGS with or without locating rings & lubricating groove	.2362 to 14.1732	.6299 to 17.3228	Good	0	0.60	± 0° 2′	X	X
	DRAWN CUP	.1575 to 2.3622	.3150 to 2.6772	Good	0	0.30	± 0° 2′	X	X
THRUST BEARINGS	SINGLE DIRECTION BALL Grooved Race	.2540 to 46.4567	.8130 to 57.0866	Poor	Excellent →	0.30	0°	X	X
	SINGLE DIRECTION CYL. ROLLER	1.1811 to 23.6220	1.8504 to 31.4960	0	Excellent →	0.20	0°	X	
	SELF-ALIGNING SPHERICAL ROLLER	3.3622 to 14.1732	4.3307 to 22.0472	Poor	Excellent →	0.50	± 3°	X	

FIGURE 10.9 Characteristics of standard rolling element bearing configurations.

have long rollers, however, and they are useful when there are severe radial space limitations and when neither high load capacity nor high speeds are required.

Spherical roller bearings usually have an outer ring with a spherical inside diameter, within which are barrel-shaped rollers. This makes these bearings self-aligning, and also gives them a larger contact area between roller and ring than is the case for other rolling element bearings. Because of this, spherical roller bearings have a very high radial load-carrying capacity, along with some ability to carry thrust loads. They have higher friction between roller and ring, and this limits their high-speed capability.

Tapered roller bearings have tapered rollers, ideally shaped like truncated cones, contained between two mating cones of different angles, the inner cone and the outer cup. The contact angle of the bearing determines its thrust load capability; a steeper angle is chosen for more thrust capacity. If a single row of rollers is used, the bearing is separable and can carry thrust loads in only one direction. If the thrust is double acting, a second bearing can be mounted in a back-to-back configuration or a double row bearing can be selected. In tapered roller bearings there is sliding contact between the ends of the rollers and the guide flange on the inner cone, and this sliding contact requires lubrication to prevent wear and reduce friction.

Thrust roller bearings can be either cylindrical, needle, tapered, or spherical (Figure 10.9). In each case there is high load-carrying capacity, but the sliding that occurs between rollers and rings requires lubrication and cooling.

Rolling Element Bearing Materials

Ball and roller bearings require materials with excellent resistance to rolling contact fatigue and wear, as well as good dimensional stability and impact resistance. The rolling elements are subjected to cyclic contact pressures which can range from 70 to 3500 MPa (100 to 500 ksi) or more, and the bearing materials must be hard enough to resist surface fatigue under those conditions. Of the through-hardening steels which meet these requirements, the most popular is AISI 52100, which contains about 1% carbon and 1.5% chromium. In general, balls and rollers made from 52100 are hardened to about Rockwell C60. Standard bearings made from 52100 may suffer from unacceptable dimensional changes resulting from metallurgical transformations at operating temperatures above 140°C (285°F). Special stabilization heat treatments enable operation at higher temperatures, with successful operation at temperatures as high as 200°C (390°F) having been achieved in cases involving low loads. The strength and fatigue resistance of the material diminish if the bearing temperature increases above about 175°C (350°F, however, so above that temperature materials with better hot-hardness, such as M50 tool steel, are required. Carburizing steels such as AISI 8620 have been developed for tapered roller bearings and other heavily loaded types that benefit from the tougher core and case compressive residual stress developed during carburizing. For applications in oxidative or corrosive environments, a hardened martensitic stainless steel such as SAE 440C may be chosen. For the highest operating temperatures, ceramic materials may be used in bearings. The most promising of the ceramics for rolling element bearing applications is silicon nitride. Its major use so far in bearings has been in hybrid bearings with ceramic balls or rollers and metallic rings, but all-ceramic bearings have also been developed.

The temperature limits of these bearing materials are given in Table 10.19. For all bearing materials, great care must be taken in the processing and production stages to ensure that no defects or inclusions are present that could serve as an initiation site for fatigue cracks. For most high-performance metallic bearings, this requires a very clean steel production process, such as vacuum arc remelting. Heat treatment of the material is also important to produce the required dimensional stability. The production process for ceramic bearings is even more critical, because a defect in a ceramic bearing element could result in catastrophic fracture of the brittle material.

Bearing cages or retainers have as their primary purpose the separation of the rolling elements. In some cases, they also provide some solid lubrication to the bearing. Low-carbon steel is the most common cage material, but bronze (silicon iron bronze or aluminum bronze) and polymers (particularly nylon 6-6) are used in many applications.

TABLE 10.19 Temperature Limits for Rolling Element Bearing Materials

Material	Maximum Operating Temperature °C	Maximum Operating Temperature °F
AISI 52100	140–175	285–350
AISI 8620 (carburized)	150	300
440C stainless steel	170	340
M50 tool steel	315	600
Hybrid Si_3N_4–M50	425	800
All-ceramic (Si_3N_4)	650	1200

Selection of Rolling Element Bearings

It has been stated that if a rolling element bearing in service is properly lubricated, properly aligned, kept free of abrasive particles, moisture, and corrosive agents, and properly loaded, then all causes of damage will be eliminated except one, contact fatigue (Harris, 1991). The fatigue process results in a spall which may originate on or just beneath the contact surface. Studies of rolling contact fatigue life by Lundberg and Palmgren (1947; 1952) and others showed that most rolling element bearings have fatigue lives which follow a Weibull statistical distribution, wherein the dependence of strength on volume is explained by the dispersion in material strength. Most bearings today are designed according to the Lundberg–Palmgren model, which has been standardized by international (ISO, 1990) and national standards (e.g., ANSI/AFBMA, 1990), although recent work (Ioannides and Harris, 1985) has found that modern bearings have longer lives than those predicted by the standard methods.

The basic rating life of rolling element bearings is the L_{10} life, which is the number of revolutions at constant speed for which there is a 10% probability of failure (or 90% reliability). The basic dynamic load-carrying capacity, or load rating, of a bearing is the constant load C which corresponds to an L_{10} life of one million revolutions. For any other bearing load F, the L_{10} life can be determined by the following relationship:

$$L_{10} = (C/F)^n \tag{10.18}$$

where the load-life exponent $n = 3$ for ball bearings, and $n = 10/3$ for roller bearings.

The equivalent bearing load includes contributions from both radial and thrust loads, and can be calculated by the following expression:

$$F = XF_r + YF_a \tag{10.19}$$

where X is a radial load factor, Y is a thrust load factor, F_r is the radial load applied to the bearing, and F_a is the applied thrust (or axial) load.

Values for the dynamic load rating C, as well as the load factors X and Y for any bearing configuration can be found in manufacturers' catalogs, or they can be calculated according to formulas given in bearing texts by Harris (1991) or Eschmann et al. (1985). Those references also give life adjustment factors which can be used to adjust the desired L_{10} life to account for special operating conditions, special material selections, special lubrication conditions, or for a reliability different from 90%.

The bearing user will generally select a commercially available bearing by the following procedure:

1. Determine the axial and thrust loads acting at the bearing location.
2. Determine the required bearing life (L_{10}).
3. Select the most appropriate bearing type from among those given in Figure 10.9.
4. Use the X and Y values appropriate to the type of bearing and loading conditions in Equation (10.19) to find the equivalent dynamic bearing load F.
5. Determine the required dynamic load capacity C from Equation (10.18).

6. Select a bearing with a dynamic load capacity at least as large as the required value from a manufacturer's catalog.
7. Provide an appropriate mounting arrangement for the bearing. Manufacturers' catalogs can be consulted for guidance in designing the mounting and selecting appropriate fits for the bearing. The importance of the fit cannot be overemphasized, since improper fit can result in considerable reduction in bearing life.
8. Provide adequate lubrication for the bearing (see below). Seals and/or shields may be integrated into the bearing to retain or protect the lubricant in the bearing.

Rolling Bearing Lubrication

The primary load-carrying contacts between rolling elements and rings exhibit nearly pure rolling. There are many sliding contacts in rolling element bearings, however, including those where roller ends contact the internal flanges of rings, where rolling elements contact separator/cage, and where the separator contacts the guiding (piloting) ring of the bearing. All of those contacts must be lubricated to limit friction and wear, and either a grease or an oil can be used for that purpose.

Under most normal operating conditions, rolling element bearings can be grease lubricated. *Greases* coat surfaces with a thin boundary lubricant film of oil, thickener, and additive molecules, thus providing protection against sliding wear, and provide oil to lubricate the concentrated rolling contacts (see below). The selection of a grease depends on its effective temperature range, oil viscosity, consistency, and rust-inhibiting properties. For normal applications, a bearing should be filled with grease up to 30 to 50% of its free volume. Overfilling will cause overheating, particularly at high speeds. Grease will deteriorate with time and will leak out. For that reason, there should be a relubrication schedule, with grease being added at intervals which can be estimated by the following expression (Neale, 1993):

$$\text{relubrication interval (hours)} = \left(k/d^{1/2}\right)\left[\left(14\times 10^{6}/n\right) - 4d^{1/2}\right] \tag{10.20}$$

where $k = 10$ for radial ball bearings, 5 for cylindrical roller bearings, and 1 for spherical or tapered roller bearings; d = bearing bore diameter (mm); and n = speed (rpm)

Oil lubrication is required when high speed or high operating temperatures preclude the use of grease. It is necessary to choose an oil of proper viscosity and appropriate viscosity–temperature characteristics in order to insure sufficient thickness of oil film in the lubricated concentrated contacts. If viscosity is too low, the film thickness will not prevent metal/metal contact, but if the viscosity is too high, excessive friction will occur. Oil can be applied to a bearing by one of several methods (listed in order of increasing effectiveness at increasing bearing speed): *oil bath*, in which the rolling elements carry the oil through the bearing; *oil circulating system*, in which the oil is pumped from the bearing through an external filter and heat exchanger and back to the bearing; *oil mist*, in which an airstream carries oil droplets to the bearing; and *oil jets*, in which the oil is injected into the bearing through carefully positioned nozzles. The quantity and entry velocity of the oil must be carefully selected and controlled in order to dissipate the heat generated in the bearing.

The lubrication mechanism in the concentrated contacts of rolling element bearings is *elastohydrodynamic lubrication* (EHL). EHL typically occurs in lubricated, nonconforming elastic contacts, such as the elliptical contact that occurs between ball and raceway or the rectangular contact between roller and ring. These lubricated contacts have a very small area and the contact pressures are very high. Because of those high pressures, the contacting surfaces deform and the lubricant viscosity increases, thereby aiding its ability to sustain heavy loading without oil-film breakdown. A diagram of these phenomena is shown in Figure 10.10. The most important parameter of the concentrated contact, from the point of view of rolling element bearing performance, is minimum EHL film thickness, h_o. The following expression can be used to find minimum film thickness in most rolling element bearing contacts (Hamrock and Dowson, 1977):

$$h_o = 3.63 R_x U^{0.68} G^{-0.49} W_p^{-0.73}\left(1 - e^{-0.68\kappa}\right) \tag{10.21}$$

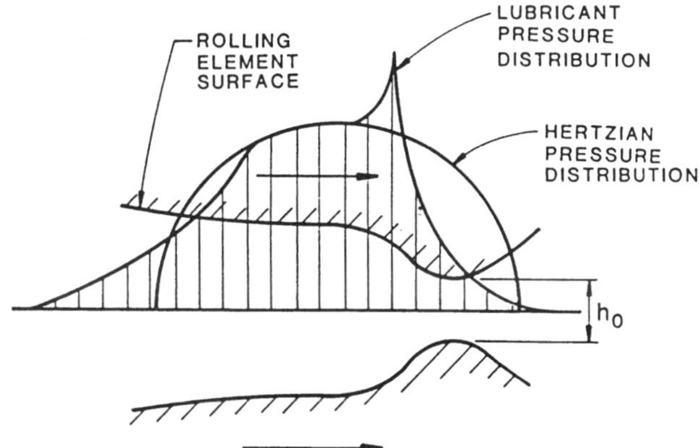

FIGURE 10.10 Typical pressure and film thickness distributions in elastohydrodynamic contact.

where $R_x = (R_{x1} R_{x2})/(R_{x1} + R_{x2})$, $R_y = (R_{y1} R_{y2})/(R_{y1} + R_{y2})$, ellipticity parameter $\kappa = R_x/R_y$, $U = \mu (u_1 + u_2)/2E'R_x$, μ = absolute viscosity, u_1 and u_2 are velocities of rolling element and ring, $E' = E/(1-\nu^2)$, E = modulus of elasticity, ν = Poisson's ratio, $G = \alpha E'$, α = pressure–viscosity exponent, $W_p = W/E' R_x^2$, and W = radial load.

The minimum film thickness must be large enough to prevent metal/metal contact within the lubricated conjunctions. The criterion for this is stated as

$$h_o \geq 1.5\left(r_{q1}^2 + r_{q2}^2\right)^{0.5} \tag{10.22}$$

where r_{q1} and r_{q2} are the rms surface roughness of the rolling element and ring, respectively. If the minimum film thickness is less than this value, complete EHL will not occur, and this could result in wear, surface fatigue, and eventual early bearing failure (i.e., well before the predicted L_{10} life).

An alternative to oil or grease lubrication for rolling element bearings operating under severe conditions is *solid lubrication*. Solid lubricants can be used effectively in high temperatures or vacuum conditions where liquid lubricants are impractical or would provide marginal performance. Solid lubricants do not prevent solid/solid contact, so wear of the lubricant coating can be expected; bearing life is governed by the depletion of the solid lubricant film.

Lubricant Supply Methods

Lubrication systems for oil film bearings can generally be grouped into three classifications: self contained devices for small machines; centralized systems, common in manufacturing plants; and circulating systems dedicated to a single piece of equipment such as a motor, turbine, or compressor. Upper speed limits for common journal bearing lubrication methods are indicated in Figure 10.11 (Wilcock and Booser, 1987). Submerging the bearing directly in an oil bath is a common alternative for vertical machines.

Self-Contained Units

Lifting of oil by *capillary action* from a small reservoir is used to feed small bearings in business machines, household appliances, electrical instruments, controls, and timer motors. In capillary tubes, the height h to which oil will rise is (Wilcock and Booser, 1987)

$$h = 2\sigma \cos\theta/(r\rho) \tag{10.23}$$

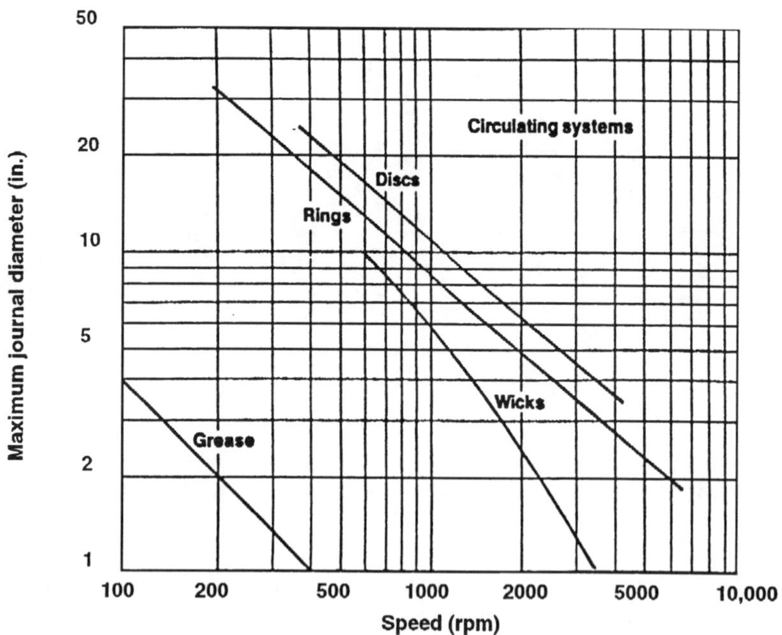

FIGURE 10.11 Upper limits for journal bearing lubrication methods. (From Wilcock, D.F. and Booser, E.R., *Mach. Des.*; April 23, 101–107, 1987. With permission.)

where σ = surface tension, lb/in., r = capillary radius (or spacing between two parallel plates), in.; ρ = oil density, lb/in.^{3}. Because oils wet most surfaces readily, the cosine of the contact angle can be taken as unity. As an example, with $\sigma = 1.7 \times 10^{-4}$ lb/in. for a petroleum oil, the rise in an 0.005-in.-radius capillary will be $h = 2(1.7 \times 10^{-4})(1)/(0.005)(0.0307) = 2.2$ in.

Wick lubrication is applied in millions of fractional horsepower motors annually. Although wicks generally are not efficient at raising oil more than about 2 in., lift may be up to 5 in. in railway journal bearings. By referring to Figure 10.12, petroleum oil delivery by a typical wick can be estimated by the following equation (Elwell, 1994):

$$Q = kAF_o\left(h_u - h\right)/(\mu L) \quad \text{in.}^3/\text{min} \tag{10.24}$$

where the constant k reflects both the capillary spacing in the wick and the surface tension of the oil; A is the wick cross-section area, in.^{2}; F_o is volume fraction of oil in the saturated wick (often about 0.75); h_u is the ultimate wicking height, about 7.5 in. for SAE Grade F-1 felt; h is oil delivery height above the reservoir surface, in.; L is wick length, in.; and μ is viscosity at the wick temperature, lb·sec/in.$^2 \times 10^6$. k is approximately 0.26 for SAE Grade F-1 felt.

Oil rings hanging over a journal, as illustrated in Figure 10.13 and Table 10.20, are used to lift oil to journal bearings in electric motors, pumps, and medium-size industrial machines (Elwell, 1994). At very low journal surface speeds below about 2 to 3 ft/sec, the ring will run synchronously with its journal. At higher speeds, increasing viscous drag on the ring in its reservoir will slow the ring surface velocity; oil ring rpm at higher speeds is often in the range of $^1/_{10}$ the journal rpm. Above about 45 ft/sec journal

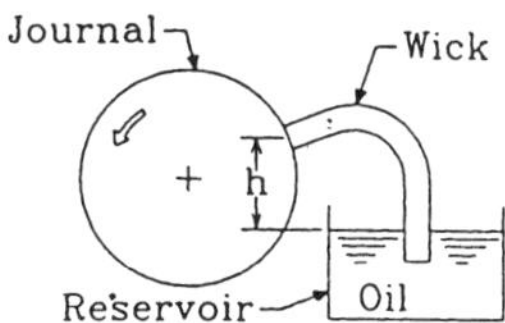

FIGURE 10.12 Wick-lubricated journal bearing. (From Elwell, R.C., in *Handbook of Lubrication and Tribology*, Vol. 3, CRC Press, Boca Raton, FL, 1994, 515–533. With permission.)

surface velocity, oil delivery drops to an unusably low level as centrifugal throw-off and windage interfere.

These self-contained systems usually supply much less oil to a bearing than needed to form a full hydrodynamic oil film (Elwell, 1994). With the starved oil supply generating an oil wedge of reduced circumferential extent, power loss will be lowered at the expense of reduced load capacity (smaller minimum film thickness).

Centralized Distribution Systems

Limitations with individual localized lubricating devices have led to widespread use of centralized systems for factory production-line equipment, construction and mining machines, and small applications. Oil or soft grease is pumped from a central reservoir in pulses or as metered continuous feed. Oil mist is piped for distances up to 300 ft for machines in petrochemical plants and steel mills. Polymer additives in the 50,000 to 150,000 molecular-weight range greatly reduce the escape of stray oil mist into the atmosphere.

Circulating Systems

Where bearing design, reliability requirements, or equipment considerations preclude use of a simpler oil feed, a circulating system is employed involving an oil reservoir, pump, cooler, and filter (Twidale and Williams, 1984). These systems become quite compact with the space limitations in aircraft, marine, or automobile engines where the reservoir may simply be the machine sump with capacity to hold only a 20- to 60-sec oil supply. Characteristics of typical oil-circulating systems for industrial use are given in Table 10.21 (Wilcock and Booser, 1987).

Dynamic Seals

Fluid seals commonly accompany bearings in a wide variety of machinery, both to control leakage of lubricating oil and process fluids and to minimize contamination. Static seals, such as O-rings and gaskets, provide sealing between surfaces which do not move relative to each other. Dynamic seals, which will be the focus of this discussion, restrain flow of fluid between surfaces in relative motion. Most dynamic seals could be classified as either contact or clearance seals. Contact seals are used when the surfaces are in sliding contact, while clearance seals imply that the surfaces are in close proximity to each other but do not contact. The major types of dynamic seals are listed in Table 10.22. Further details about the design and analysis of dynamic sealing elements can be found in the handbook article by Stair (1984) and in the book by Lebeck (1991).

As an example of a contact seal, ball bearings are often sealed to retain their lubricant and to keep out contaminants over a long lifetime. For small grease-lubricated bearings, the sealing function is often accomplished by lightly loaded contact between a rubber lip seal component and the bearing ring. For shafts ranging up to about 2 in. (5 cm) in diameter, the lip seal may be replaced by a closely fitted, but noncontacting, shield which helps contain the grease and restricts intrusion of dirt and other contaminants. For more severe sealing requirements in rolling element bearings, labyrinth clearance seals may be used (Harris, 1991).

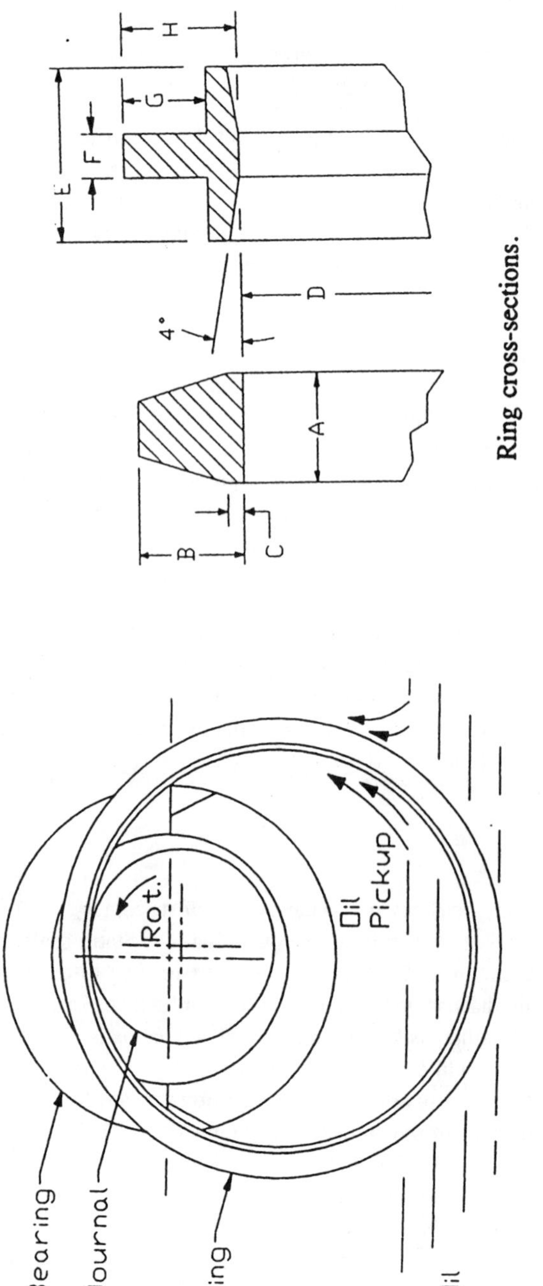

FIGURE 10.13 Oil-ring bearing elements and ring cross-sections. (From Elwell, R.C., in *Handbook of Lubrication and Tribology*, Vol. 3, CRC Press, Boca Raton, FL, 1994, 515–533. With permission.)

TABLE 10.20 Typical Oil-Ring Dimensions, mm (in.)

A	B	C	D	E	F	G	H
6 (0.24)	6 (0.24)	1 (0.04)	100 (3.94)				
7 (0.28)	7 (0.28)	1 (0.04)	135 (5.31)				
8 (0.31)	8 (0.31)	2 (0.08)	165 (6.50)				
16 (0.63)	13 (0.51)	2 (0.08)	200 (7.87)	16 (0.63)	5 (0.20)	11 (0.43)	10 (0.39)
			335 (13.2)	21 (0.83)	6 (0.24)	13 (0.51)	14 (0.55)
			585 (23.0)	25 (1.0)	7 (0.28)	14 (0.55)	20 (0.79)
			830 (32.7)	32 (1.3)	8 (0.31)	16 (0.63)	27 (1.06)

TABLE 10.21 Typical Oil Circulating Systems

Application	Duty	Oil Viscosity at 40°C (cSt)	Oil Feed (gpm)	Pump Type	Reservoir Dwell Time (min)	Type	Rating (μm)
Electrical machinery	Bearings	32–68	2	Gear	5	Strainer	50
General	Bearings	68	10	Gear	8	Dual cartridge	100
Paper mill dryer section	Bearings, gears	150–220	20	Gear	40	Dual cartridge	120
Steel mill	Bearings	150–460	170	Gear	30	Dual cartridge	150
	Gears	68–680	240	Gear	20	Dual cartridge	
Heavy duty gas turbines	Bearings, controls	32	600	Gear	5	Pleated paper	5
Steam turbine-generators	Bearings	32	1000	Centrifugal	5	Bypass 15%/hr	10

TABLE 10.22 Characteristics of Dynamic Seals

	Type of Motion					
Type of Seal	Rotating	Reciprocating	Extent of Use	Friction	Leakage	Life
Contact						
Face seals	x		H	L	L	M–H
Lip seals	x		H	L	L	L–M
Piston rings		x	H	H	L	L–M
O-Rings	x	x	M	H	L	L
Packings	x	x	H	M	M	L
Diaphragms		x	L	L	L	H
Controlled clearance						
Hydrodynamic	x		L	L	M	H
Hydrostatic	x		L	L	M	H
Floating bushing	x	x	M	M	M–H	H
Fixed geometry clearance						
Labyrinth	x		H	H	H	H
Bushing	x	x	M	H	H	M–H
Special						
Magnetic fluid	x	x	L	L	L	M
Centrifugal	x		L	M	L	H

H = High, M = Moderate, L = Low.
Modified from Stair, W.K., in *Handbook of Lubrication*, Vol. 2, CRC Press, Boca Raton, FL 1984, 581–622.

The most common seals for rotating shafts passing through fixed housings, such as pumps or gearboxes, are radial lip seals and mechanical face seals. These contact seals can be designed to handle a wide range of sealed fluids, temperatures, velocities, and pressures (Stair, 1984; Lebeck, 1991). Material selection and surface finish considerations are similar to those discussed in the subsections on sliding friction and its consequences and on dry and semilubricated bearings.

When high surface speeds are encountered, wear and frictional heating may prohibit use of rubbing contact seals. For such applications, clearance seals such as close-fitting labyrinth seals or controlled-clearance fluid film seals can be used. Fluid film seals can be either hydrostatic or hydrodynamic; both types have a pressurized film of fluid which prevents contact between the sealed surfaces and use pressure balancing to restrain leakage. The principles governing their operation are similar to those discussed in the subsection on fluid film bearings. Details of fluid film seals can be found in Shapiro (1995).

References

ANSI/AFBMA, 1990. Load Ratings and Fatigue Life for Ball Bearings, ANSI/AFBMA 9–1990, AFBMA, Washington, D.C.

Archard, J.F. 1980. Wear theory and mechanisms, in *Wear Control Handbook,* M.B. Peterson and W.O. Winer, Eds., ASME, New York.

Bhushan, B. and Gupta, B.K. 1991. *Handbook of Tribology,* McGraw-Hill, New York.

Blanchet, T.A. and Kennedy, F.E. 1992. Sliding wear mechanism of polytetrafluoroethylene (PTFE) and PTFE composites, *Wear,* 153:229–243.

Blau, P.J., Ed. 1992. *Friction, Lubrication and Wear Technology, Metals Handbook,* Vol. 18, 10th ed., ASM International, Metals Park, OH.

Booser, E.R. 1992. Bearing materials, in *Encyclopedia of Chemical Technology,* Vol. 4, pp. 1–21, John Wiley & Sons, New York.

Booser, E.R. 1995. Lubricants and lubrication, in *Encyclopedia of Chemical Technology,* 4th ed., Vol. 15, pp. 463–517, John Wiley & Sons, New York.

Booser, E.R. and Wilcock, D.F. 1987. New technique simplifies journal bearing design, *Mach. Des.,* April 23, pp. 101–107.

Booser, E.R. and Wilcock, D.F. 1991. Selecting thrust bearings, *Mach. Des.,* June 20, pp. 69–72.

Crook, P. and Farmer, H.N. 1992. Friction and wear of hardfacing alloys, in *Friction, Lubrication and Wear Technology, Metals Handbook,* Vol. 18, pp. 758–765, ASM International, Metals Park, OH.

Cusano, C. 1994. Porous metal bearings, in *Handbook of Lubrication and Tribology,* Vol. 3, pp. 491–513, CRC Press, Boca Raton, FL.

DeHart, A.O. 1984. Sliding bearing materials, in *Handbook of Lubrication,* Vol. 2, pp. 463–476, CRC Press, Boca Raton, FL.

Derner, W.J. and Pfaffenberger, E.E. 1984. Rolling element bearings, in *Handbook of Lubrication,* Vol. 2, pp. 495, CRC Press, Boca Raton, FL.

Elwell, R.C. 1994. Self-contained bearing lubrication: rings, disks, and wicks, in *Handbook of Lubrication and Tribology,* Vol. 3, pp. 515–533, CRC Press, Boca Raton, FL.

Engineering Sciences Data Unit (ESDU). 1965. *General Guide to the Choice of Journal Bearing Type,* Item 65007, Institution of Mechanical Engineers, London.

Engineering Sciences Data Unit (ESDU). 1967. *General Guide to the Choice of Thrust Bearing Type,* Item 67073, Institution of Mechanical Engineers, London.

Eschmann, P., Hasbargen, L., and Weigand, K. 1985. *Ball and Roller Bearings,* John Wiley & Sons, New York.

Fenske, G.R. 1992. Ion implantation, in *Friction, Lubrication and Wear Technology, Metals Handbook,* Vol. 18, pp. 850–860, ASM International, Metals Park, OH.

Fuller, D.D. 1984. Theory and practice of lubrication for engineers, 2nd ed., John Wiley & Sons, New York.

Hamrock, B. and Dowson, D. 1977. Isothermal elastohydrodynamic lubrication of point contacts, *ASME J. Lubr. Technol.,* 99(2): 264–276.

Harris, T.A. 1991. *Rolling Bearing Analysis,* 3rd ed., John Wiley & Sons, New York.

Ioannides, S. and Harris, T.A. 1985. A new fatigue life model for rolling bearings, *ASME J. Tribology*, 107:367–378.

ISO, 1990. Rolling Bearings Dynamic Load Ratings and Rating Life, International Standard ISO 281.

Jamison, W.E. 1994. Plastics and plastic matrix composites, in *Handbook of Lubrication and Tribology*, Vol. 3, pp. 121–147, CRC Press, Boca Raton, FL.

Khonsari, M.M. 1997. In *Tribology Data Handbook*, CRC Press, Boca Raton, FL.

Kingsbury, G.R. 1992. Friction and wear of sliding bearing materials, in *ASM Handbook*, Vol. 18 pp. 741–757, ASM International, Metals Park, OH.

Klaus, E.E. and Tewksbury, E.J. 1984. Liquid lubricants, in *Handbook of Lubrication*, Vol. 2, pp. 229–254, CRC Press, Boca Raton, FL.

Kushner, B.A. and Novinski, E.R. 1992. Thermal spray coatings, in *Friction, Lubrication and Wear Technology, Metals Handbook*, Vol. 18, pp. 829–833, ASM International, Metals Park, OH.

Lebeck, A.O. 1991. *Principles and Design of Mechanical Face Seals,* John Wiley & Sons, New York.

Lundberg, G. and Palmgren, A. 1947. Dynamic capacity of rolling bearings, *Acta Polytech. Mech. Eng. Ser.*, 1(3):196.

Lundberg, G. and Palmgren, A. 1952. Dynamic capacity of roller bearings, *Acta Polytech. Mech. Eng. Ser.*, 2(4):210.

Morgan, V.T. 1984. *Porous Metal Bearings and Their Application*, MEP-213, Mechanical Engineering Publications, Workington, U.K.

Neale, M.J. 1993. *Bearings*, Butterworth-Heinemann, Oxford.

Neale, P.B. 1970. *J. Mech. Eng. Sci.*, 12:73–84.

Peterson, M.B. and Winer, W.O., Eds., 1980. *Wear Control Handbook*, ASME, New York.

Rabinowicz, E. 1980. Wear coefficients metals, in *Wear Control Handbook*, M.B. Peterson and W.O. Winer, Eds., pp. 475–506, ASME, New York.

Rabinowicz, E. 1995. *Friction and Wear of Materials*, 2nd ed., John Wiley & Sons, New York.

Ramondi, A.A. and Szeri, A.Z. 1984. Journal and thrust bearings, in *Handbook of Lubrication,* Vol. 2, pp. 413–462, CRC Press, Boca Raton, FL.

Reynolds, O. 1886. On the theory of lubrication and its application to Mr. Beauchamp Tower's experiments, *Philos. Trans R. Soc.*, 177:157–234.

Schmitt, G.F. 1980. Liquid and solid particles impact erosion, in *Wear Control Handbook,* M.B. Peterson and W.O. Winer, Eds., pp. 231–282, ASME, New York.

Shabel, B.S. Granger, D.A., and Tuckner, W.G. 1992. Friction and wear of aluminum–silicon alloys, in *ASM Handbook*, Vol. 18, pp. 785–794, ASM International, Metals Park, OH.

Shapiro, W. 1995. Hydrodynamic and hydrostatic seals, in *Handbook of Lubrication and Tribology*, Vol. 3, pp. 445–468, CRC Press, Boca Raton, FL.

Stair, W.K. 1984. Dynamic seals, in *Handbook of Lubrication*, Vol. 2, pp. 581–622, CRC Press, Boca Raton, FL.

Twidale, A.J. and Williams, D.C.J. 1984. Circulating oil systems, in *Handbook of Lubrication*, Vol. 2, pp. 395–409, CRC Press, Boca Raton, FL.

Weil, R. and Sheppard, K. 1992. Electroplated coatings, in *Friction, Lubrication and Wear Technology, Metals Handbook*, Vol. 18, pp. 834–839, ASM International, Ohio.

Wilcock, D.F. and Booser, E.R. 1956. Bearing design and application, McGraw-Hill, New York.

Wilcock, D.F. and Booser, E.R. 1987. Lubrication techniques for journal bearings, *Machine Des.*, April 23, 101–107.

11

Pumps and Fans

Robert F. Boehm

University of Nevada

11 Pumps and Fans 181
Introduction • Pumps • Fans

Pumps and Fans

Introduction

Pumps are devices that impart a pressure increase to a liquid. Fans are used to increase the velocity of a gas, but this is also accomplished through an increase in pressure. The pressure rise found in pumps can vary tremendously, and this is a very important design parameter along with the liquid flow rate. This pressure rise can range from simply increasing the elevation of the liquid to increasing the pressure hundreds of atmospheres. Fan applications, on the other hand, generally deal with small pressure increases. In spite of this seemingly significant distinction between pumps and fans, there are many similarities in the fundamentals of certain types of these machines as well as with their application and theory of operation.

The appropriate use of pumps and fans depends upon the proper choice of device and the proper design and installation for the application. A check of sources of commercial equipment shows that many varieties of pumps and fans exist. Each of these had special characteristics that must be appreciated for achieving proper function. Preliminary design criteria for choosing between different types is given by Boehm (1987).

As is to be expected, the wise applications of pumps and fans requires knowledge of fluid flow fundamentals. Unless the fluid mechanics of a particular application are understood, the design could be less than desirable.

In this section, pump and fan types are briefly defined. In addition, typical application information is given. Also, some ideas from fluid mechanics that are especially relevant to pump and fan operation are reviewed.

Pumps

Raising of water from wells and cisterns is the earliest form of pumping (a very detailed history of early applications is given by Ewbank, 1842). Modern applications are much broader, and these find a wide variety of machines in use. Modern pumps function on one of two principles. By far the majority of

0-8493-0055-X/00/$0.00+$.50

pump installations are of the *velocity head* type. In these devices, the pressure rise is achieved by giving the fluid a movement. At the exit of the machine, this movement is translated into a pressure increase. The other major type of pump is called *positive displacement.* These devices are designed to increase the pressure of the liquid while essentially trying to compress the volume. A categorization of pump types has been given by Krutzsch (1986), and an adaptation of this is shown below.

I. Velocity head
  A. Centrifugal
    1. Axial flow (single or multistage)
    2. Radial flow (single or double suction)
    3. Mixed flow (single or double suction)
    4. Peripheral (single or multistage)
  B. Special Effect
    1. Gas lift
    2. Jet
    3. Hydraulic ram
    4. Electromagnetic
II. Positive displacement
  A. Reciprocating
    1. Piston, plunger
      a. Direct acting (simplex or duplex)
      b. Power (single or double acting, simplex, duplex, triplex, multiplex)
    2. Diaphragm (mechanically or fluid driven, simplex or multiplex)
  B. Rotary
    1. Single rotor (vane, piston, screw, flexible member, peristaltic)
    2. Multiple rotor (gear, lobe, screw, circumferential piston)

In the next subsection, some of the more common pumps are described.

Centrifugal and Other Velocity Head Pumps

Centrifugal pumps are used in more industrial applications than any other kind of pump. This is primarily because these pumps offer low initial and upkeep costs. Traditionally, pumps of this type have been limited to low-pressure-head applications, but modern pump designs have overcome this problem unless very high pressures are required. Some of the other good characteristics of these types of devices include smooth (nonpulsating) flow and the ability to tolerate nonflow conditions.

The most important parts of the centrifugal pump are the *impeller* and *volute.* An impeller can take on many forms, ranging from essentially a spinning disk to designs with elaborate vanes. The latter is usual. Impeller design tends to be somewhat unique to each manufacturer, as well as finding a variety of designs for a variety of applications. An example of an impeller is shown in Figure 11.1. This device imparts a radial velocity to the fluid that has entered the pump perpendicular to the impeller. The volute (there may be one or more) performs the function of slowing the fluid and increasing the pressure. A good discussion of centrifugal pumps is given by Lobanoff and Ross (1992).

A very important factor in the specification of a centrifugal pump is the *casing orientation* and *type.* For example, the pump can be oriented vertically or horizontally. Horizontal mounting is most common. Vertical pumps usually offer benefits related to ease of priming and reduction in required net positive suction head (see discussion below). This type also requires less floor space. Submersible and immersible pumps are always of the vertical type. Another factor in the design is the way the casing is split, and this has implications about ease of manufacture and repair. Casings that are split perpendicular to the shaft are called *radially split,* while those split parallel to the shaft axis are denoted as *axially split.* The latter can be *horizontally split* or *vertically split.* The number of *stages* in the pump greatly affects the pump-output characteristics. Several stages can be incorporated into the same casing, with an associated

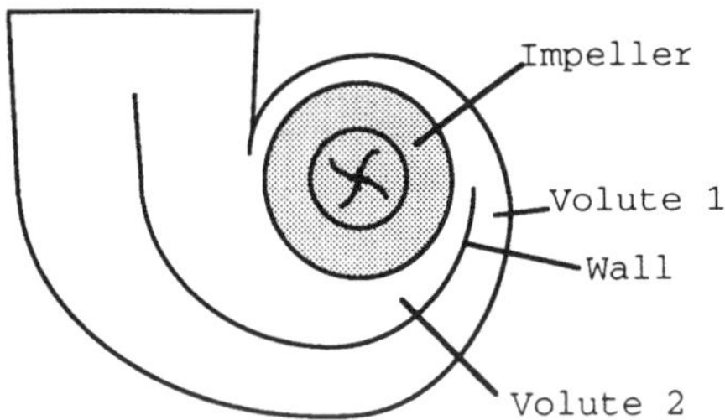

FIGURE 11.1. A schematic of a centrifugal pump is shown. The liquid enters perpendicular to the figure, and a radial velocity is imparted by clockwise spin of the impeller.

increase in pump output. Multistage pumps are often used for applications with total developed head over 50 atm.

Whether or not a pump is self-priming can be important. If a centrifugal pump is filled with air when it is turned on, the initiation of pumping action may not be sufficient to bring the fluid into the pump. Pumps can be specified with features that can minimize priming problems.

There are other types of velocity head pumps. *Jet pumps* increase pressure by imparting momentum from a high-velocity liquid stream to a low-velocity or stagnant body of liquid. The resulting flow then goes through a diffuser to achieve an overall pressure increase. *Gas lifts* accomplish a pumping action by a drag on gas bubbles that rise through a liquid.

Positive-Displacement Pumps

Positive-displacement pumps demonstrate high discharge pressures and low flow rates. Usually, this is accomplished by some type of pulsating device. A piston pump is a classic example of positive-displacement machines. Rotary pumps are one type of positive-displacement device that do not impart pulsations to the existing flow (a full description of these types of pumps is given by Turton, 1994). Several techniques are available for dealing with pulsating flows, including use of double-acting pumps (usually of the reciprocating type) and installation of pulsation dampeners.

Positive-displacement pumps usually require special seals to contain the fluid. Costs are higher both initially and for maintenance compared with most pumps that operate on the velocity head basis. Positive-displacement pumps demonstrate an efficiency that is nearly independent of flow rate, in contrast to the velocity head type (see Figure 11.2 and the discussion related to it below).

Reciprocating pumps offer very high efficiencies, reaching 90% in larger sizes. These types of pumps are more appropriate for pumping abrasive liquids (e.g., slurries) than are centrifugal pumps.

A characteristic of positive displacement pumps which may be valuable is that the output flow is proportional to pump speed. This allows this type of pump to be used for metering applications. Also a positive aspect of these pumps is that they are self-priming, except at initial start-up.

Very high head pressures (often damaging to the pump) can be developed in positive-displacement pumps if the downstream flow is blocked. For this reason, a pressure-relief-valve bypass must always be used with positive-displacement pumps.

Pump/Flow Considerations

Performance characteristics of the pump must be considered in system design. Simple diagrams of pump applications are shown in Figure 11.2. First, consider the left-hand figure. This represents a flow circuit, and the pressure drops related to the piping, fittings, valves, and any other flow devices found in the circuit must be estimated using the laws of fluid mechanics. Usually, these resistances (pressure drops) are found to vary approximately with the square of the liquid flow rate. Typical characteristics are shown in Figure 11.3. Most pumps demonstrate a flow vs. pressure rise variation that is a positive value at zero flow and decreases to zero at some larger flow. Positive-displacement pumps, as shown on the right-hand side of Figure 11.3, are an exception to this in that these devices usually cannot tolerate a zero.

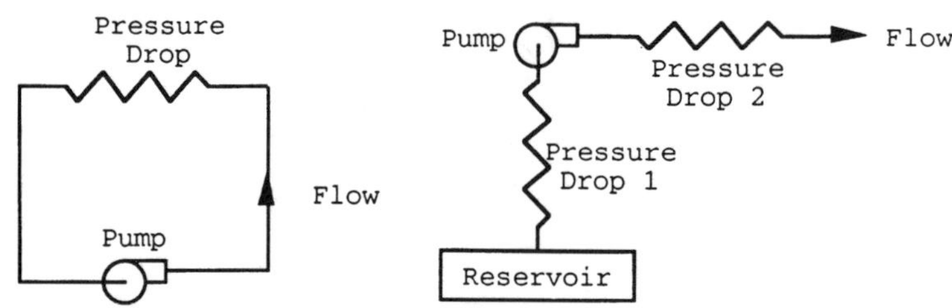

FIGURE 11.2. Typical pump applications, either in circuits or once-through arrangements, can be represented as combined fluid resistances as shown. The resistances are determined from fluid mechanics analyses.

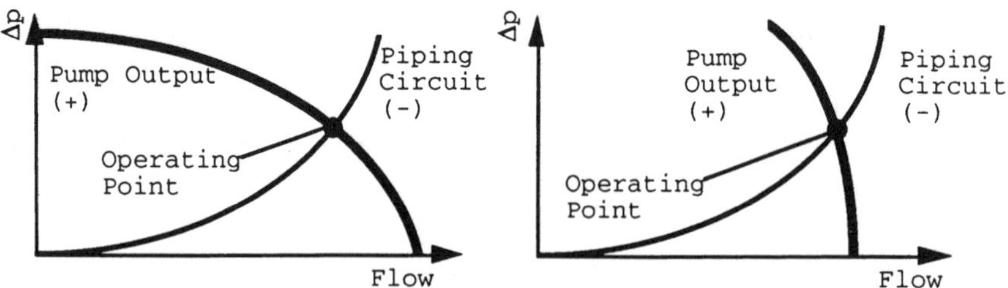

FIGURE 11.3. An overlay of the pump flow vs. head curve with the circuit piping characteristics gives the operating state of the circuit. A typical velocity head pump characteristic is shown on the left, while a positive-displacement pump curve is shown on the right.

flow. An important aspect to note is that a closed system can presumably be pressurized. A contrasting situation and its implications are discussed below.

The piping diagram show on the right-hand side of Figure 11.2 is a once-through system, another frequently encountered installation. However, the leg of piping through "pressure drop 1" shown there can have some very important implications related to *net positive suction head*, often denoted as **NPSH**. In simple terms, NPSH indicates the difference between the local pressure and the thermodynamic saturation pressure at the fluid temperature. If NPSH = 0, the liquid can vaporize, and this can result in a variety of outcomes from noisy pump operation to outright failure of components. This condition is called **cavitation**. Cavitation, if it occurs, will first take place at the lowest pressure point within the piping arrangement. Often this point is located at, or inside, the inlet to the pump. Most manufacturers specify how much NPSH is required for satisfactory operation of their pumps. Hence, the actual NPSH (denoted as **NPSHA**) experienced by the pump must be larger than the manufacturer's required NPSH (called **NPSHR**). If a design indicates insufficient NPSH, changes should be made in the system, possibly including alternative piping layout, including elevation and/or size, or use of a pump with smaller NPSH requirements.

The manufacturer should be consulted for a map of operational information for a given pump. A typical form is shown in Figure 11.4. This information will allow the designer to select a pump that satisfied the circuit operational requirements while meeting the necessary NPSH and most-efficient-operation criteria

Several options are available to the designer for combining pumps in systems. Consider a comparison of the net effect between operating pumps in series or operating the same two pumps in parallel. Examples of this for pumps with characteristics such as centrifugal units are shown in Figure 11.5. It is clear that one way to achieve high pumping pressures with centrifugal pumps is to place a number of units in series. This is a related effect to what is found in *multistage* designs.

Fans

As noted earlier, fans are devices that cause air to move. This definition is broad and can include a flapping palm branch, but the discussion here deals only with devices that impart air movement due to

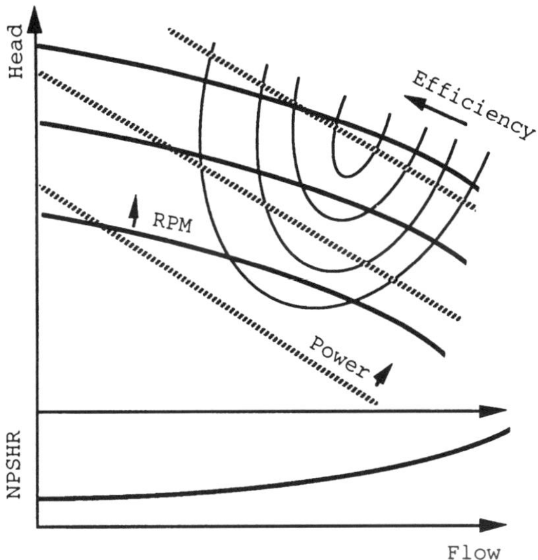

FIGURE 11.4. A full range of performance information should be available from the pump manufacturer, and this may include the parameters shown.

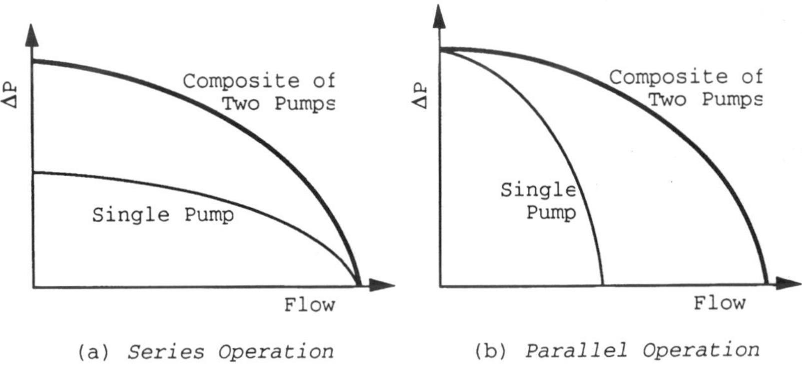

FIGURE 11.5. Series (a) and parallel (b) operation of centrifugal pumps are possible. The resultant characteristics for two identical pumps are shown.

rotation of an impeller inside a fixed casing. In spite of this limiting definition, a large variety of commercial designs are included.

Fans find application in many engineering systems. Along with the chillers and boilers, they are the heart of heating, ventilating, and air conditioning (HVAC) systems. When large physical dimensions of a unit are not a design concern (usually the case), centrifugal fans are favored over axial flow units for HVAC applications. Many types of fans are found in *power plants.* Very large fans are used to furnish air to the boiler, as well as to draw or force air through cooling towers and pollution-control equipment. *Electronic cooling* finds applications for small units. Even automobiles have several fans in them. Because of the great engineering importance of fans, several organizations publish rating and testing criteria (see, for example, ASME, 1990).

Generally fans are classified according to how the air flows through the impeller. These flows may be *axial* (essentially a propeller in a duct), *radial* (conceptually much like the centrifugal pumps discussed earlier), *mixed*, and *cross*. While there are many other fan designations, all industrial units fit one of these classifications. Mixed-flow fans are so named because both axial and radial flow

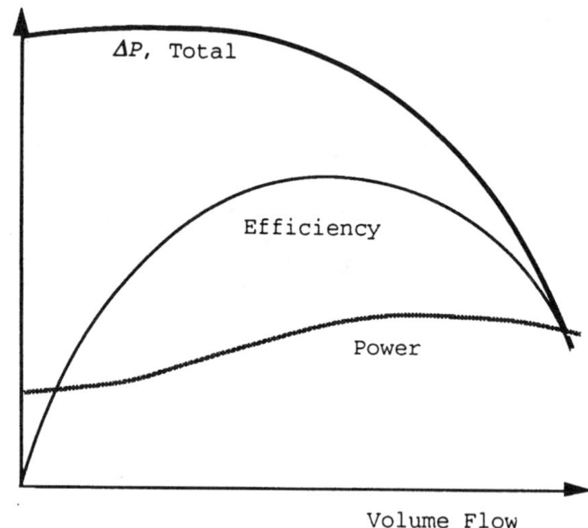

FIGURE 11.6. Shown are characteristics of a centrifugal fan. The drawbacks to operating away from optimal conditions are obvious from the efficiency variation.

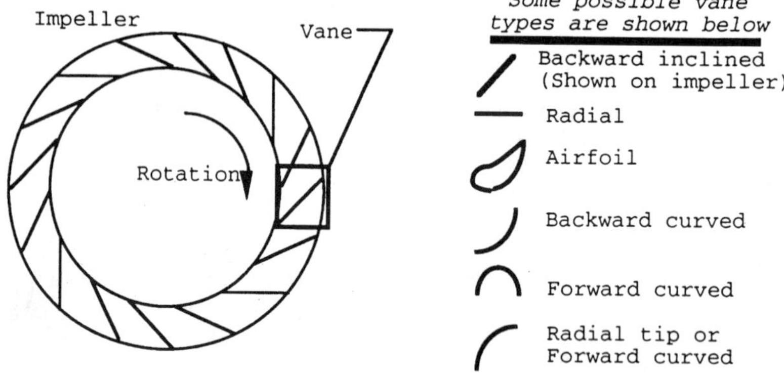

FIGURE 11.7. A variety of vane types that might be used on a centrifugal fan are shown.

occur on the vanes. Casings for these devices are essentially like those for axial-flow machines, but the inlet has a radial-flow component. On cross-flow impellers, the gas traverses the blading twice.

Characteristics of fans are shown in Figure 11.6. Since velocities can be high in fans, often both the total and the static pressure increases are considered. While both are not shown on this figure, the curves have similar variations. Of course the total ΔP will be greater than will the static value, the difference being the velocity head. This difference increases as the volume flow increases. At zero flow (the shutoff point), the static and total pressure difference values are the same. Efficiency variation shows a sharp optimum value at the design point. For this reason, it is critical that fan designs be carefully tuned to the required conditions.

A variety of vane type are found on fans, and the type of these is also used for fan classification. Axial fans usually have vanes of airfoil shape or vanes of uniform thickness. Some vane types that might be found on a centrifugal (radial-flow) fan are shown in Figure 11.7.

One aspect that is an issue in choosing fans for a particular application is fan efficiency. Typical efficiency comparisons of the effect of blade type on a centrifugal fan are shown in Figure 11.8. Since velocities can be high, the value of aerodynamic design is clear. Weighing against this are cost and other factors.

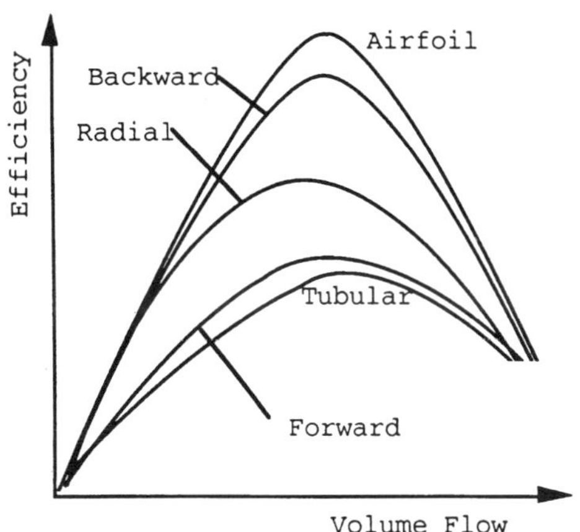

FIGURE 11.8. Efficiency variation with volume flow of centrifugal fans for a variety of vane types is shown.

An additional aspect that may be important in the choice of fans is noise generation. This may be most critical in HVAC applications. It is difficult to describe noise characteristics in brief terms because of the frequency-dependent nature of these phenomena. However, a comparison of specific sound power level (usually denoted by K_w) shows backward-curved centrifugal fans with aerodynamic blades perform best among the designs. Details of noise characteristics are given elsewhere (ASHRAE, 1991).

While each type of fan has some specific qualities for certain applications, most installations use centrifugal (radial-flow) fans. A primary exception is for very-high-flow, low-pressure-rise situations where axial (propeller) fans are used.

Similarities exist between fans and pumps because the fluid density essentially does not vary through either type of machine. Of course, in pumps this is because a liquid can be assumed to be incompressible. In fans, a gas (typically air) is moved with little pressure change. As a result, the gas density can be taken to be constant. Since most fans operate near atmospheric pressure, the ideal gas assumptions can be used in determining gas properties.

Flow control in fan applications, where needed, is a very important design concern. Methods for accomplishing this involve use of dampers (either on the inlet or on the outlet of the fan), variable pitch vanes, or variable speed control. Dampers are the least expensive to install, but also the most inefficient in terms of energy use. Modern solid state controls for providing a variable frequency power to the drive motor is becoming the preferred control method, when a combination of initial and operating costs is considered.

Defining Terms

Cavitation: Local liquid conditions allow vapor voids to form (boiling).

NPSH: Net positive suction head is the difference between the local absolute pressure of a liquid and the thermodynamic saturation pressure of the liquid based upon the temperature of the liquid. Applies to the inlet of a pump.

NPSHA: Actual net positive suction head is the NPSH at the given state of operation of a pump.

NPSHR: Required net positive suction head is the amount of NPSH required by a specific pump for a given application.

References

ASHRAE, 1991. *ASHRAE Handbook 1991, HVAC Applications,* American Society of Heating, Refrigerating, and Air Conditioning Engineers, Atlanta, Chapter 42.

ASME, 1990. *ASME Performance Test Codes, Code on Fans*, ASME PTC 11-1984 (reaffirmed 1990), American Society of Mechanical Engineers, New York.

Boehm, R.F. 1987. *Design Analysis of Thermal Systems*, John Wiley and Sons, New York, 17–26.

Ewbank, T. 1842. *A Description and Historical Account of Hydraulic and Other Machines for Raising Water*, 2nd ed., Greeley and McElrath, New York.

Krutzsch, W.C. 1986. Introduction: classification and selection of pumps, in *Pump Handbook*, 2nd ed., I. Karassik et al., Eds., McGraw-Hill, New York, Chapter 1.

Lobanoff, V. and Ross, R. 1992. *Centrifugal Pumps: Design & Application*, 2nd ed., Gulf Publishing Company, Houston.

Turton, R.K. 1994. *Rotodynamic Pump Design*, Cambridge University Press, Cambridge, England.

Further Information

Dickson, C. 1988. *Pumping Manual*, 8th ed., Trade & Technical Press, Morden, England.

Dufour, J. and Nelson, W. 1993. *Centrifugal Pump Sourcebook*, McGraw-Hill, New York.

Fans. 1992. In *1992 ASHRAE Handbook, HVAC Systems and Equipment*, American Society of Heating, Refrigerating, and Air Conditioning Engineers, Atlanta, GA, Chapter 18.

Garay, P.N. 1990. *Pump Application Book*, Fairmont Press, Liburn, GA.

Krivchencko, G.I. 1994. *Hydraulic Machines, Turbines and Pumps*, 2nd ed., Lewis Publishers, Boca Raton, FL.

Stepanoff, A.J. 1993. *Centrifugal and Axial Flow Pumps: Theory, Design, and Application* (Reprint Edition), Krieger Publishing Company, Malabar, FL.

12

Liquid Atomization and Spraying

Rolf D. Reitz
University of Wisconsin

12 Liquid Atomization and Spraying 189
Spray Characterization • Atomizer Design Considerations • Atomizer Types

Liquid Atomization and Spraying

Sprays are involved in many practical applications, including in the process industries (e.g., spray drying, spray cooling, powdered metals); in treatment applications (e.g., humidification, gas scrubbing); in coating applications (e.g., surface treatment, spray painting, and crop spraying); in spray combustion (e.g., burners, furnaces, rockets, gas turbines, diesel and port fuel injected engines); and in medicinal and printing applications. To be able to describe sprays it is necessary to obtain a detailed understanding of spray processes.

In the simplest case, the liquid to be sprayed is injected at a high velocity through a small orifice. Atomization is the process whereby the injected liquid is broken up into droplets. Atomization has a strong influence on spray vaporization rates because it increases the total surface area of the injected liquid greatly. Fast vaporization may be desirable in certain applications, but undesirable in others, where the liquid is required to impinge on a target. The trajectories of the spray drops are governed by the injected momentum of the drop, drag forces, and interactions between the drops and the surrounding gas. Control of these and other spray processes can lead to significant improvements in performance and in quality of product, and to reduction of emission of pollutants.

Spray Characterization

Practical atomizers generate sprays with a distribution of drop sizes, with average sizes in the diameter range from a few microns (1 μm = 10^{-6} m) to as large as 0.5 mm. It is important to quantify the details of the distribution depending on the application. For example, the smaller drops in a spray vaporize fast, and this is helpful to control ignition processes in some combustion systems. On the other hand, the large drops carry most of the mass and momentum of the injected liquid and these drops are able to penetrate into the high-pressure gases in engine combustion chambers. Typical average drop sizes for

0-8493-0055-X/00/$0.00+$.50

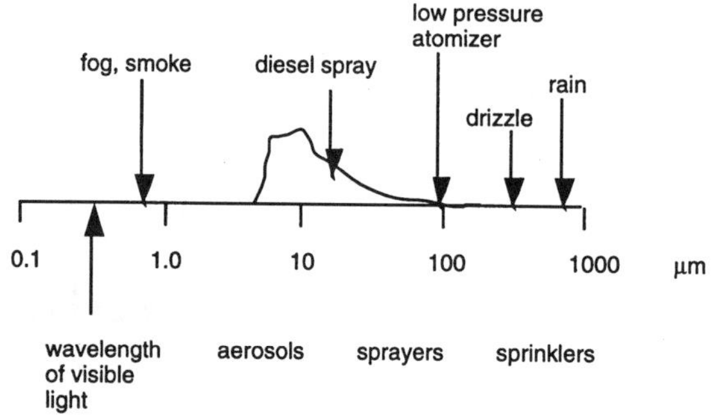

FIGURE 12.1 Typical average spray drop sizes for various classes of sprays. A representative size distribution is depicted for the diesel spray.

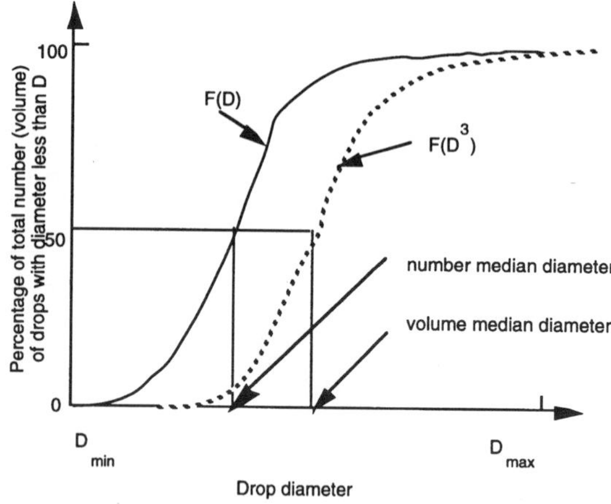

FIGURE 12.2 Cumulative spray drop number and volume distributions.

broad classes of sprays are shown schematically in Figure 12.1. It should be noted that the terminology used to describe sprays in Figure 12.1 is qualitative and is not universally agreed upon.

Methods for characterizing the size distribution of spray drops are discussed in References 1 and 2. A probability distribution function, $F(D)$, is introduced that represents the fraction of drops per unit diameter range about the diameter, D, as shown in Figure 12.2. The spray drop sizes span a range from a minimum diameter, $D_{\min}$, to a maximum diameter, $D_{\max}$. It is also convenient to introduce a mean or average drop diameter instead of having to specify the complete drop size distribution. The number median drop diameter (NMD) represents that drop whose diameter is such that 50% of the drops in the spray have sizes less than this size. Spray drop size distribution data can also be represented as a volume (or mass) distribution function, $F(D^3)$; this gives more weight to the large drops in the distribution. In this case, a volume median diameter (VMD) or a mass median diameter (MMD) can also be defined, as indicated in Figure 12.2.

Various other mean diameters are also in common use. These are summarized using the standard notation of Mugele and Evans[2] as

$$\left(D_{jk}\right)^{j-k} = \frac{\int_{D_{\min}}^{D_{\max}} D^j f(D)\, dD}{\int_{D_{\min}}^{D_{\max}} D^k f(D)\, dD} \tag{12.1}$$

where $f(D) = dF(D)/dD$ is the drop size probability density function (usually normalized such that $\int_{D_{min}}^{D_{max}} f(D)dD = 1$). Commonly used mean diameters are D_{10} (i.e., $j = 1$, $k = 0$, sometimes called the length mean diameter[3] and D_{32} (i.e., $j = 3$, $k = 2$, called the Sauter mean diameter or SMD). The Sauter mean diameter has a useful physical interpretation in combustion applications since drop vaporization rates are proportional to the surface area of the drop. It represents the size of that drop that has the same volume-to-surface area ratio as that of the entire spray.

Several distribution functions have been found to fit experimental data reasonably well. Among these are the Nukiyama–Tanasawa and the Rosin–Rammler distributions which have the general form[3] $f(D) = aD^p \exp(\{-bD\}^q)$, where the constants a, p, b, and q characterize the size distribution. The higher the parameter, q, the more uniform the distribution, and typically $1.5 < q < 4$. Other distributions have been proposed which consist of logarithmic transformations of the normal distribution, such as $f(D) = a \exp(-y^2/2)$, where $y = \delta \ln(\eta D/(D_{max} - D))$, and a, δ, and η are constants. In this case, the smaller δ, the more uniform the size distribution. It should be noted that there is no theoretical justification for any of these size distributions. Spray drop size distributions can be measured nonintrusively by using optical laser diffraction and phase/Doppler instruments. A discussion of these techniques and their accuracy is reviewed by Chigier.[4]

Atomizer Design Considerations

Atomization is generally achieved by forcing a liquid or a liquid–gas mixture through a small hole or slit under pressure to create thin liquid sheets or jets moving at a high relative velocity with respect to the surrounding ambient gas. Desirable characteristics of atomizers include the ability to atomize the liquid over a wide range of flow rates, low power requirements, and low susceptibility to blockage or fouling. In addition, atomizers should produce consistent sprays with uniform flow patterns in operation.

Atomizers can be broadly characterized as those producing hollow cone or solid cone sprays, as depicted in Figure 12.3. In solid cone (or full cone) sprays the spray liquid is concentrated along the spray axis, Figure 12.3(a). These sprays are useful in applications requiring high spray penetration, such as in diesel engines. In hollow cone sprays the axis region is relatively free of drops, giving wide spray dispersal, Figure 12.3(b). These sprays are often used in furnaces, gas turbines, and spray-coating applications.

Many different atomizer designs are found in applications. Common atomizer types include pressure, rotary, twin-fluid (air-assist, air-blast, effervescent), flashing, electrostatic, vibratory, and ultrasonic atomizers, as discussed next.

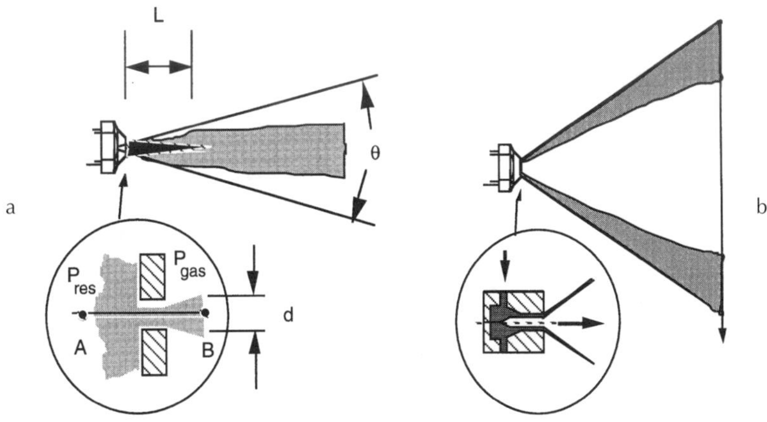

FIGURE 12.3 Schematic diagram of (a) solid cone and (b) hollow cone pressure atomizer sprays.

Atomizer Types

In *pressure atomizers* atomization is achieved by means of a pressure difference, $\Delta P = P_{res} - P_{gas}$, between the liquid in the supply reservoir pressure, P_{res}, and the ambient medium pressure, P_{gas}, across a nozzle. The simplest design is the plain orifice nozzle with exit hole diameter, d, depicted in Figure 12.3(a). The liquid emerges at the theoretical velocity $U = \sqrt{2\Delta P/\rho_{liquid}}$, the (Bernoulli) velocity along the streamline A–B in Figure 12.3(a), where ρ_{liquid} is the density of the liquid. The actual injection velocity is less than the ideal velocity by a factor called the discharge coefficient, C_D, which is between 0.6 and 0.9 for plain hole nozzles. C_D accounts for flow losses in the nozzle.

Four main jet breakup regimes have been identified, corresponding to different combinations of liquid inertia, surface tension, and aerodynamic forces acting on the jet, as shown in Figure 12.4. At low injection pressures the low-velocity liquid jet breaks up due to the unstable growth of long-wavelength waves driven by surface tension forces (Rayleigh regime). As the jet velocity is increased, the growth of disturbances on the liquid surface is enhanced because of the interaction between the liquid and the ambient gas (the first and second wind-induced breakup regimes). At high injection pressures the high-velocity jet disintegrates into drops immediately after leaving the nozzle exit (atomization regime). Criteria for the boundaries between the regimes are available.[5] Aerodynamic effects are found to become very important relative to inertial effects when the jet Weber number, $We_j > 40$, where $We_j = \rho_{gas}U^2d/\sigma$, ρ_{gas} is the gas density, and σ is the liquid surface tension.

Experiments show that the unstable growth of surface waves is aided by high relative velocities between the liquid and the gas, and also by high turbulence and other disturbances in the liquid and gas flows, and by the use of spray liquids with low viscosity and low surface tension.

Liquid breakup characteristics such as the spray drop size, the jet breakup length, and the spray angle have been related to the unstable wave growth mechanism. The wavelengths and growth rates of the

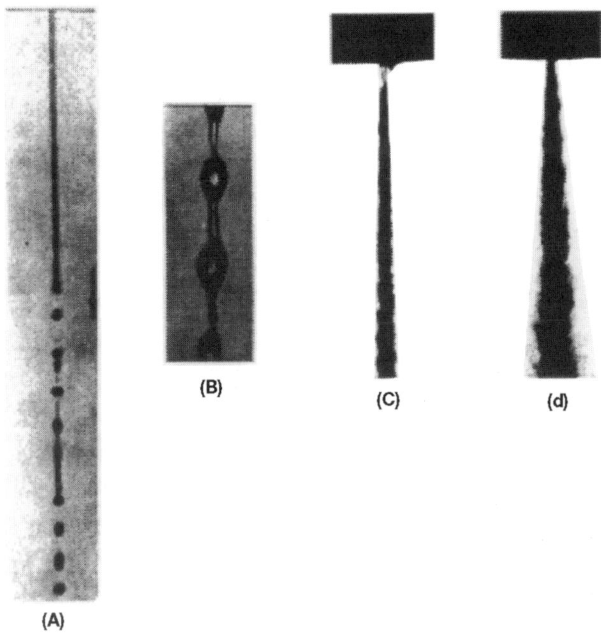

FIGURE 12.4 (a) Rayleigh breakup. Drop diameters are larger than the jet diameter. Breakup occurs many nozzle diameters downstream of nozzle. (b) First wind-induced regime. Drops with diameters of the order of jet diameter. Breakup occurs many nozzle diameters downstream of nozzle. (c) Second wind-induced regime. Drop sizes smaller than the jet diameter. Breakup starts some distance downstream of nozzle. (d) Atomization regime. Drop sizes much smaller than the jet diameter. Breakup starts at nozzle exit.

waves can be predicted using results from a linear stability analysis with[6]

$$\frac{\Lambda}{2} = 9.02\frac{\left(1+0.45Z^{0.5}\right)\left(1+0.4T^{0.7}\right)}{\left(1+0.87\,\mathrm{We}_2^{1.67}\right)^{0.6}} \tag{12.2a}$$

$$\Omega\left(\frac{\rho_1 a^3}{\sigma}\right)^{0.5} = \frac{0.34+0.38\,\mathrm{We}_2^{1.5}}{(1+Z)\left(1+1.4T^{0.6}\right)} \tag{12.2b}$$

where Λ is the wavelength, Ω is the growth rate of the most unstable surface wave, and a is the liquid jet radius. The maximum wave growth rate increases, and the corresponding wavelength decreases with increasing Weber number, $\mathrm{We}_2 = \rho_{gas}U^2a/\sigma$, where U is the relative velocity between the liquid and the gas. The liquid viscosity appears in the Ohnesorge number, $Z = \mathrm{We}_1^{1/2}/\mathrm{Re}_1$. Here, the Weber number We_1 is based on the liquid density, the Reynolds number is $\mathrm{Re}_1 = Ua/\nu_1$, ν_1 is the liquid viscosity, and the parameter $T = Z\mathrm{We}_2^{1/2}$. The wave growth rate is reduced and the wavelength is increased as the liquid viscosity increases.

The size of the drops formed from the breakup process is often assumed to be proportional to the wavelength of the unstable surface waves in modeling studies.[6] However, the drop sizes in the primary breakup region near the nozzle exist have also been found to be influenced by the length scale of the energy-containing eddies in the turbulent liquid flow.[7] There is uncertainty about atomization mechanisms since spray measurements are complicated by the high optical density of the spray in the breakup region (e.g., see Figure 12.4(d)). As the drops penetrate into the ambient gas, they interact with each other through collisions and coalescence, and the spray drop size changes dynamically within the spray as a result of secondary breakup and vaporization effects. The drop trajectories are determined by complex drop drag, breakup, and vaporization phenomena, and by interactions with the turbulent gas flow.[6]

High-pressure diesel sprays are intermittent and are required to start and stop quickly without dribble between injections. This is accomplished by means of a plunger arrangement that is actuated by a cam and spring system in mechanical "jerk" pump systems (see Figure 12.5). Modern electronic injectors include electromagnetic solenoids that permit the duration and injection pressure to be varied independently of each other and of engine speed. Experiments on diesel-type injector nozzles show that the penetration distance, S, of the tip of the spray at time, t, after the beginning of the injection is given by[8]

$$\begin{aligned} S &= 0.39Ut\left(\rho_{\text{liquid}}/\rho_{\text{gas}}\right)^{1/2} && \text{for } t < t_b \\ S &= 2.46\sqrt{U\,dt}\left(\rho_{\text{liquid}}/\rho_{\text{gas}}\right)^{1/4} && \text{for } t > t_b \end{aligned} \tag{12.3}$$

where the "breakup time" is $t_b = 40.5d(\rho_{\text{liquid}}/\rho_{\text{gas}})^{1/2}/U$. The jet breakup length (see Figure 12.3(a)), $L = Ut_b$ is independent of the injection velocity. On the other hand, for low-speed jets, or for jets injected into a low-gas-density environment, $t_b = 1.04C\ (\rho_{\text{liquid}}\, d^3/\sigma)^{1/2}$, where C is a constant typically between 12 and 16 and σ is the surface tension. In this case L increases with the injection velocity.[9] The functional form of the above jet breakup time and length correlations can be derived for an inviscid liquid in the limits of large and small Weber number, We_2 from the unstable wave growth rate in Equation (12.2) with $t_b \sim \Omega^{-1}$.

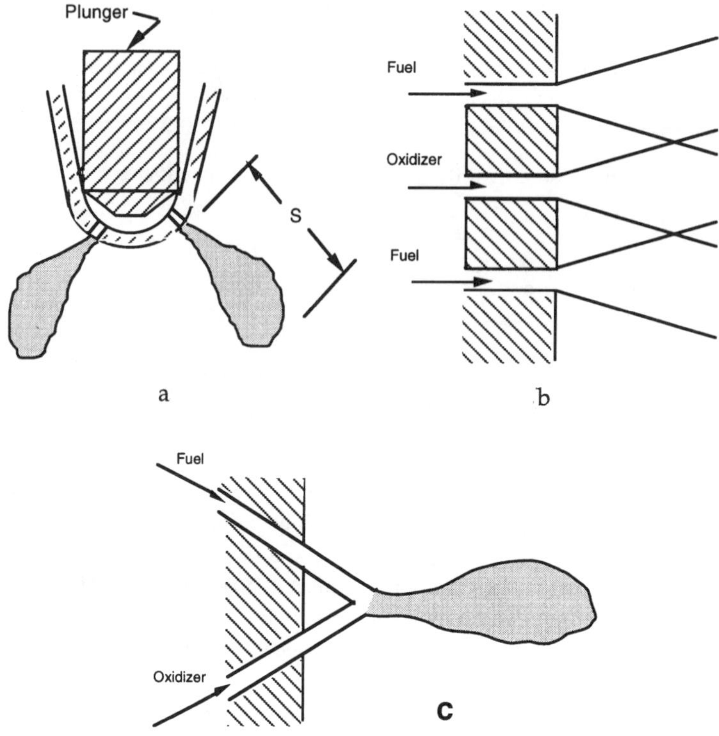

FIGURE 12.5 (a) Diesel injector multihole spray nozzle, (b) showerhead, and (c) doublet impingement nozzles.

For high-speed diesel-type jets in the atomization regime the resulting spray diverges in the form of a cone with cone angle, θ, that is usually in the range from 5 to 20°. θ increases with gas density following $\tan\theta = A(\rho_{gas}/\rho_{liquid})^{1/2}$, where A is a constant that depends on the nozzle passage length and (weakly) on the injection velocity.[9] Very high injection pressures are required to produce small drops. In diesel engines ΔP is typically as high as 200 Mpa, and drops are produced with mean diameters of the order of 10 μm (see Figure 12.1). Drop size correlations have been proposed for plain-orifice sprays, such as that presented in Table 12.1.[3] Note, however, that these correlations do not account for the fact that the spray drop size varies with time, and from place to place in the spray. Moreover, experimental correlations often do not include some parameters that are known to influence spray drop sizes, such as the nozzle passage length and its entrance geometry. Therefore, overall drop size correlations should only be used with caution.

TABLE 12.1 Representative Drop Size Correlations for Various Spray Devices (Dimensional quantities are in SI units, kg, m, s)

Device	Correlation	Notes
Plain orifice	$\mathrm{SMD} = 3.08\nu_l^{0.385}(\rho_{liquid}\sigma)^{0.737}\rho_{gas}^{0.06}\,\Delta P^{-0.54}$	Use SI units
Fan spray	$\mathrm{SMD} = 2.83 d_h\left(\sigma\mu_{liquid}^2/\rho_{gas} d_h^3\,\Delta P^2\right)^{0.25} + 0.26 d_h\left(\sigma\rho_{liquid}/\rho_{gas} d_h\,\Delta P\right)^{0.25}$	d_h = nozzle hydraulic diameter
Rotary atomizer	$\mathrm{SMD} = 0.119 Q^{0.1}\sigma^{0.5}/N d^{0.5}\rho_{liquid}^{0.4}\mu_{liquid}^{0.1}$	N = rotational speed (rev/sec), Q = volumetric flow rate, $A_{inj}\,U$
Pressure swirl	$\mathrm{SMD} = 4.52\left(\sigma\mu_{liquid}^2/\rho_{gas}\,\Delta P^2\right)^{0.25}(t\cos\theta)^{0.25} + 0.39\left(\sigma\rho_{liquid}/\rho_{gas}\,\Delta P\right)^{0.25}(t\cos\theta)^{0.75}$ $t = 0.0114 A_{inj}\rho_{liquid}^{1/2} d\cos\theta$	t = film thickness; θ = cone angle, d = discharge orifice diameter
Twin fluid/air blast	$\mathrm{SMD} = 0.48 d\left(\sigma/\rho_{gas} U^2 d\right)^{0.4}(1+1/\mathrm{ALR})^{0.4} + 0.15 d\left(\mu_{liquid}^2/\sigma\rho_{liquid} d\right)^{0.5}(1+1/\mathrm{ALR})$	ALR = air-to-liquid mass ratio
Prefilming air blast	$\mathrm{SMD} = (1+1/\mathrm{ALR})\left[0.33 d_h\left(\sigma/\rho_{gas} U^2 d_p\right)^{0.6} + 0.068 d_h\left(\mu_{liquid}^2/\sigma\rho_{liquid} d_p\right)^{0.5}\right]$	d_h = hydraulic diameter, d_p = prefilmer diameter, Figure 12.9
Ultrasonic	$\mathrm{SMD} = \left(4\pi^3\sigma/\rho_{liquid}\omega^2\right)^{1/3}$	ω = vibration frequency

Source: Lefebvre, A.H., *Atomization and Sprays*, Hemisphere Publishing, New York, 1989. With permission.

The plain orifice design is also used in twin-fluid-type liquid rocket engines in showerhead and doublet designs (Figures 12.5b and 12.5c). In the case of doublet nozzles, shown in Figure 12.6c, the impinging jets create unstable liquid sheets which break up to produce the sprays. Drop size correlations are available for liquid sheets such as those formed by discharging the liquid through a rectangular slit (see *fan spray*, Table 12.1). Thin liquid sheets or slits lead to the production of small drops. The breakup mechanism of liquid sheets is also thought to involve the unstable growth of surface waves due to surface tension and aerodynamic forces.[5]

In *rotary atomizers* centrifugal forces are used to further enhance the breakup process. In this case the liquid is supplied to the center of a spinning disk and liquid sheets or ligaments are thrown off the edges of the disk. The drop size depends on the rotational speed of the disk, as indicated in Table 12.1.

A spinning wheel or cup (turbobell) is used in some spray-painting applications. The spray shape is controlled by supplying a coflowing stream of "shaping-air."

Centrifugal forces also play a role in the breakup mechanism of *pressure swirl* atomizers (*simplex* nozzles). These atomizers give wider spray cone angle than plain orifice nozzles, and are available in hollow cone and solid cone designs. As depicted in Figure 12.3(b) the spray liquid enters a swirl chamber tangentially to create a swirling liquid sheet. The air core vortex within the swirl chamber plays an important role in determining the thickness of the liquid sheet or film at the nozzle exit. This type of nozzle produces relatively coarse sprays. A representative SMD correction is listed in Table 12.1. The spray cone angle depends on the ratio of the axial and tangential liquid velocity components at the exit of the nozzle. This type of atomizer is not well suited for use in transient applications because it tends to dribble at start-up and to shut down when the air core is not fully formed.

The basic drawback of all pressure atomizers is that the flow rate depends on the square root of ΔP. The volumetric flow rate is $Q = A_{\text{inj}}\ U$, where A_{inj} is the liquid flow area at the nozzle exit, so that a factor of 20 increase in flow rate (a typical turndown ratio from idle to full load operation of a gas turbine engine) requires a factor of 400 increase in injection pressure.

This difficulty has led to so-called wide-range atomizer designs such as those shown in Figure 12.6. The *duplex* nozzle features two sets of tangential swirl ports; the primary (or pilot) supplies fuel at low flow rates, while the secondary ports become operational at high flow rates. Another variation is the *dual-orifice* nozzle which is conceptually two simplex nozzles arranged concentrically, one supplying the primary flow and the other supplying the secondary flow. The *spill-return* nozzle is a simplex nozzle with a rear passage that returns fuel to the injection pump. In this design the flow rate is controlled by the relative spill amount, and there are no small passages to become plugged. However, the fuel is always supplied at the maximum pressure which increases the demands on the injection pump. But high swirl is always maintained in the swirl chamber and good atomization is achieved even at low flow rates.

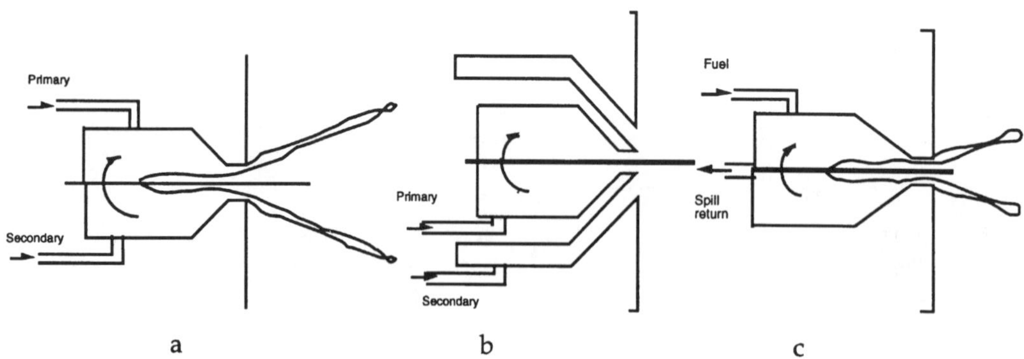

FIGURE 12.6 (a) Duplex, (b) dual orifice, and (c) spill-return-type nozzle designs.

In *twin-fluid injectors* atomization is aided by a flow of high-velocity gas through the injector passages. The high-velocity gas stream impinges on a relatively low-velocity liquid either internally (in *internal-mixing* nozzles, Figure 12.7) or externally (in *external-mixing* designs, Figure 12.8). The liquid and gas flows are typically swirled in opposite directions by means of swirl vanes to improve atomization. *Air-assist* refers to designs that use a relatively small amount of air at high (possibly sonic) velocities. *Air-blast* refers to designs that use large quantities of relatively low-velocity air which often supplies some of the air to help decrease soot formation in combustion systems[3]. (see Figure 12.9.)

In *flashing* and *effervescent* atomizers a two-phase flow is passed through the injector nozzle exit. In the former the bubbles are generated by means of a phase change which occurs as the liquid, containing a dissolved propellant gas or vapor, undergoes the pressure drop through the nozzle. This process is exploited in many household spray cans, but has the disadvantage of releasing the propellant gas required for atomization into the atmosphere. In the so-called effervescent atomizer, air bubbles are introduced

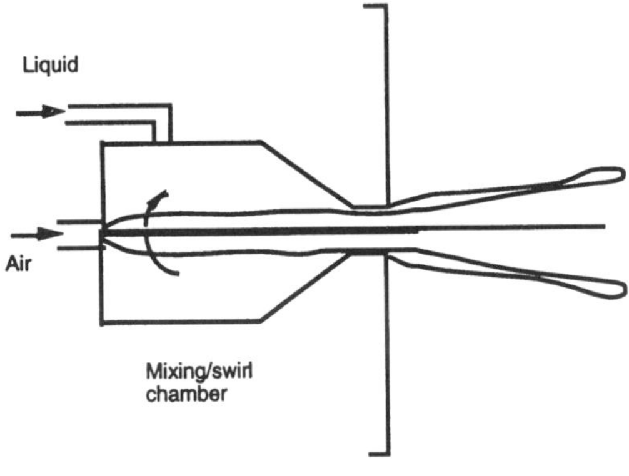

FIGURE 12.7 Internal-mixing twin-fluid injector design.

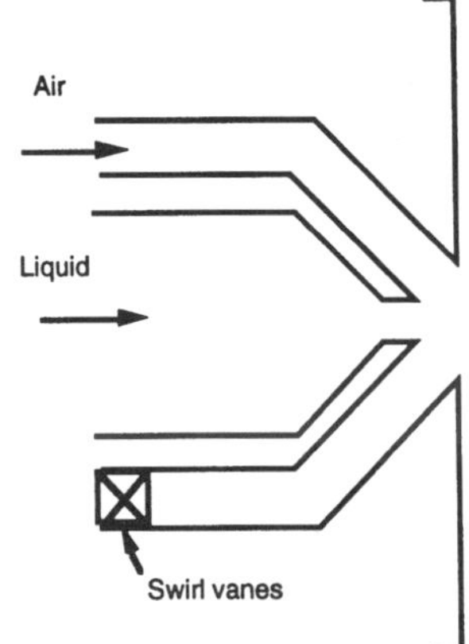

FIGURE 12.8 External-mixing twin-fluid injector design.

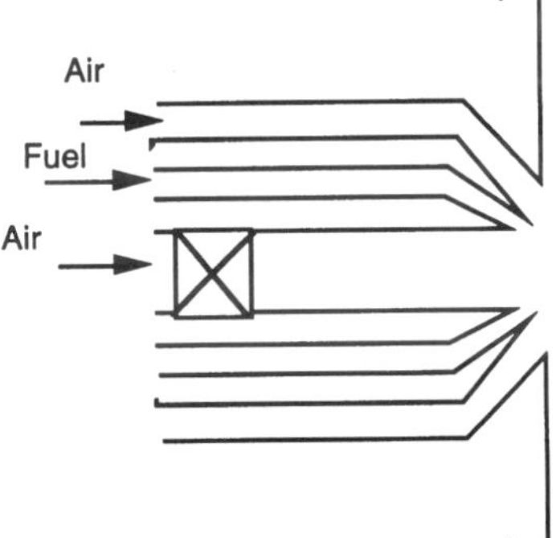

FIGURE 12.9 Prefilming air blast atomizer.

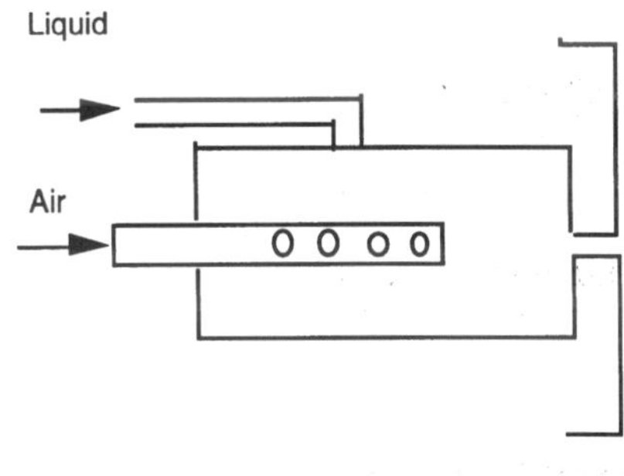

FIGURE 12.10 Internal-mixing, effervescent atomizer.

into the liquid upstream of the exit orifice, as depicted in Figure 12.10. The spray quality is found to depend weakly on the air bubble size and is independent of the nozzle exit diameter. This makes internal-mixing, air-assist atomizers very attractive for use with high-viscosity fluids and slurries where nozzle plugging would otherwise be a problem.[3]

In *electrostatic* atomizers the spray liquid is charged by applying a high-voltage drop across the nozzle. The dispersion of the spray drops is increased by exploiting electrical repulsive forces between the droplets. An electrostatic charge on the drops is also helpful in spray-coating applications, such as in automotive spray painting using electrostatic turbobell sprayers, since the charged drops are attracted to an oppositely charged target surface.

Other atomizer types include *vibratory* and *ultrasonic* atomizers (or *nebulizers*), where the drops are formed by vibrating the injector nozzle at high frequencies and at large amplitudes to produce short-wavelength disturbances to the liquid flow. Ultrasonic atomizers are used in inhalation therapy where very fine sprays (submicron sizes) are required, and an available representative drop size correlation is also listed in Table 12.1.

References

1. American Society for Testing and Materials (ASTM) Standard E799. 1988. Data Criteria and Processing for Liquid Drop Size Analysis.

2. Mugele, R. and Evans, H.D. 1951. Droplet size distributions in sprays, *Ind. Eng. Chem.*, 43, 1317–1324.
3. Lefebvre, A.H. 1989. *Atomization and Sprays*, Hemisphere Publishing, New York.
4. Chigier, N.A. 1983. Drop size and velocity instrumentation, *Prog. Energ. Combust. Sci.*, 9, 155–177.
5. Chigier, N. and Reitz, R.D. 1996. Regimes of jet breakup, in *Progress in Astronautics and Aeronautics Series*, K. Kuo, Ed., AIAA, New York, Chapter 4, pp. 109–135.
6. Reitz, R.D. 1988. Modeling atomization processes in high-pressure vaporizing sprays, *Atomisation Spray Technol.*, 3, 309–337.
7. Wu, P-K., Miranda, R.F., and Faeth, G.M. 1995. Effects of initial flow conditions on primary breakup of nonturbulent and turbulent round liquid jets, *Atomization Sprays*, 5, 175–196.
8. Hiroyasu, H. and Arai, M. 1978. Fuel spray penetration and spray angle in diesel engines, *Trans. JSAE*, 34, 3208.
9. Reitz, R.D. and Bracco, F.V. 1986. Mechanisms of breakup of round liquid jets, in *The Encyclopedia of Fluid Mechanics*, Vol. 3, N. Cheremisnoff, Ed., Gulf Publishing, Houston, TX, Chapter 10, 233–249.

Further Information

Information about recent work in the field of atomization and sprays can be obtained through participation in the Institutes for Liquid Atomization and Spraying Systems (ILASS-Americas, -Europe, -Japan, -Korea). These regional ILASS sections hold annual meetings. An international congress (ICLASS) is also held biennially. More information is available on the ILASS-Americas homepage at http://ucicl.eng.uci.edu/ilass. Affiliated with the ILASS organizations is the Institute's Journal publication *Atomization and Sprays* published by Begell House, Inc., New York.

13

Flow Measurement

Alan T. McDonald
Purdue University

Sherif A. Sherif
University of Florida

13 Flow Measurement 199
Direct Methods • Restriction Flow Meters for Flow in Ducts • Linear Flow Meters • Traversing Methods • Viscosity Measurements

Flow Measurement*

Alan T. McDonald and Sherif A. Sherif

This section deals with the measurement of mass flow rate or volume flow rate of a fluid in an enclosed pipe or duct system. Flow measurement in open channels is treated in Section 10 of this book.

The choice of a flow meter type and size depends on the required accuracy, range, cost, ease of reading or data reduction, and service life. Always select the simplest and cheapest device that gives the desired accuracy.

Direct Methods

Tanks can be used to determine the flow rate for steady liquid flows by measuring the volume or mass of liquid collected during a known time interval. If the time interval is long enough, flow rates may be determined precisely using tanks. Compressibility must be considered in gas volume measurements. It is not practical to measure the mass of gas, but a volume sample can be collected by placing an inverted

* The contributors of this chapter were asked to conform as much as possible to the nomenclature presented below, but define new terms pertaining to their particular area of specialty in the context of the chapter.

0-8493-0055-X/00/$0.00+$.50

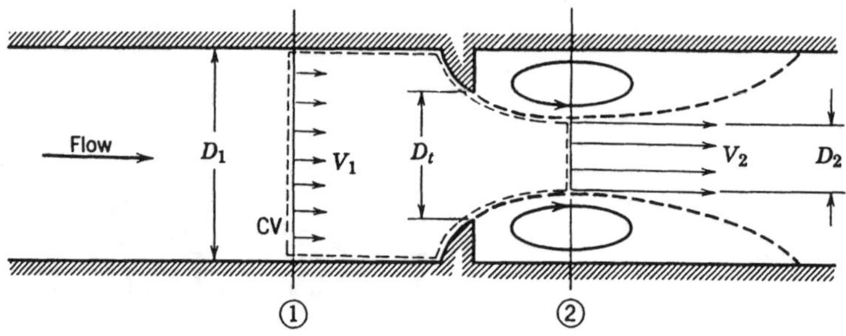

FIGURE 13.1 Internal flow through a generalized nozzle, showing control volume used for analysis.

"bell" over water and holding the pressure constant by counterweights. No calibration is required when volume measurements are set up carefully; this is a great advantage of direct methods.

Positive-displacement **flow meters** may be used in specialized applications, particularly for remote or recording uses. For example, household water and natural gas meters are calibrated to read directly in units of product. Gasoline metering pumps measure total flow and automatically compute the cost. Many positive-displacement meters are available commercially. Consult manufacturers' literature or Reference 7 for design and installation details.

Restriction Flow Meters for Flow in Ducts

Most restriction flow meters for internal flow (except the laminar flow element) are based on acceleration of a fluid stream through some form of nozzle, shown schematically in Figure 13.1. Flow separating from the sharp edge of the nozzle throat forms a recirculation zone shown by the dashed lines downstream from the nozzle. The main flow stream continues to accelerate from the nozzle throat to form a *vena contracta* at section (2) and then decelerates again to fill the duct. At the vena contracta, the flow area is a minimum, the flow streamlines are essentially straight, and the pressure is uniform across the channel section. The theoretical flow rate is

$$\dot{m}_{\text{theoretical}} = \frac{A_2}{\sqrt{1-\left(A_2/A_1\right)^2}}\sqrt{2\rho\left(p_1 - p_2\right)} \tag{13.1}$$

Equation (13.1) shows the general relationship for a **restriction flow meter**: Mass flow rate is proportional to the square root of the pressure differential across the meter taps. This relationship limits the flow rates that can be measured accurately to approximately a 4:1 range.

Several factors limit the utility of Equation (13.1) for calculating the actual mass flow rate through a meter. The actual flow area at section (2) is unknown when the vena contracta is pronounced (e.g., for orifice plates when D_t is a small fraction of D_1). The velocity profiles approach uniform flow only at large Reynolds number. Frictional effects can become important (especially downstream from the meter) when the meter contours are abrupt. Finally, the location of the pressure taps influences the differential pressure reading, $p_1 - p_2$.

The actual mass flow rate is given by

$$\dot{m}_{\text{actual}} = \frac{CA_t}{\sqrt{1-\left(A_t/A_1\right)^2}}\sqrt{2\rho\left(p_1 - p_2\right)} \tag{13.2}$$

where C is an empirical *discharge coefficient.*

If $\beta = D_t/D_1$, then $(A_t/A_1)^2 = (D_t/D_1)^4 = \beta^4$, and

$$\dot{m}_{\text{actual}} = \frac{CA_t}{\sqrt{1-\beta^4}}\sqrt{2\rho(p_1 - p_2)} \tag{13.3}$$

where $1/(1-\beta^4)^{1/2}$ is the *velocity correction factor.* Combining the discharge coefficient and velocity correction factor into a single *flow coefficient,*

$$K \equiv \frac{C}{\sqrt{1-\beta^4}} \tag{13.4}$$

yields the mass flow rate in the form:

$$\dot{m}_{\text{actual}} = KA_t\sqrt{2\rho(p_1 - p_2)} \tag{13.5}$$

Test data can be used to develop empirical equations to predict flow coefficients vs. pipe diameter and Reynolds number for standard metering systems. The accuracy of the equations (within specified ranges) is often adequate to use the meter without calibration. Otherwise, the coefficients must be measured experimentally.

For the turbulent flow regime ($\text{Re}_D > 4000$), the flow coefficient may be expressed by an equation of the form:[7]

$$K = K_\infty + \frac{1}{\sqrt{1-\beta^4}}\frac{b}{\text{Re}_{D_1}^n} \tag{13.6}$$

where subscript ∞ denotes the flow coefficient at infinite Reynolds number and constants b and n allow for scaling to finite Reynolds numbers. Correlating equations and curves of flow coefficients vs. Reynolds number are given for specific metering elements in the next three subsections following the general comparison of the characteristics of orifice plate, flow nozzle, and venturi meters in Table 13.1 (see Reference 4).

Flow meter coefficients reported in the literature have been measured with fully developed turbulent velocity distributions at the meter inlet (Section 1). When a flow meter is installed downstream from a valve, elbow, or other disturbance, a straight section of pipe must be placed in front of the meter.

TABLE 13.1.Characteristics of Orifice, Flow Nozzle, and Venturi Flow Meters

Flow Meter Type	Diagram	Head Loss	Cost
Orifice	D_1, D_t, Flow	High	Low
Flow nozzle	D_1, D_2, Flow	Intermediate	Intermediate
Venturi	D_1, D_2, Flow	Low	High

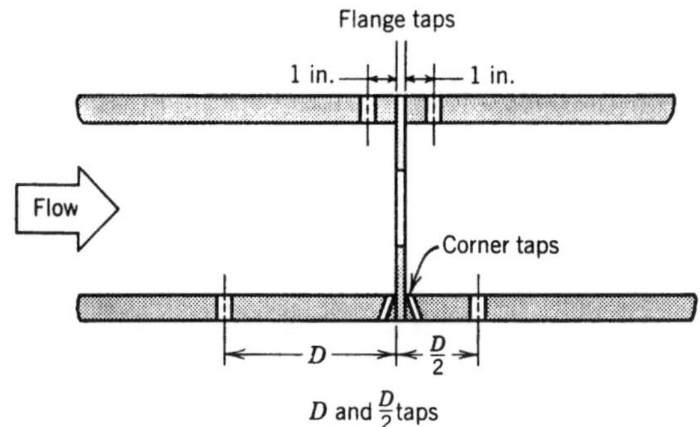

FIGURE 13.2 Orifice geometry and pressure tap locations.

Approximately 10 diameters of straight pipe upstream are required for venturi meters, and up to 40 diameters for orifice plate or flow nozzle meters. Some design data for incompressible flow are given below. The same basic methods can be extended to compressible flows.[7]

Orifice Plates

The orifice plate (Figure 13.2) may be clamped between pipe flanges. Since its geometry is simple, it is low in cost and easy to install or replace. The sharp edge of the orifice will not foul with scale or suspended matter. However, suspended matter can build up at the inlet side of a concentric orifice in a horizontal pipe; an eccentric orifice may be placed flush with the bottom of the pipe to avoid this difficulty. The primary disadvantages of the orifice are its limited capacity and the high permanent head loss caused by uncontrolled expansion downstream from the metering element.

Pressure taps for orifices may be placed in several locations as shown in Figure 13.2 (see Reference 7 for additional details). Since the location of the pressure taps influences the empirically determined flow coefficient, one must select handbook values of K consistent with the pressure tap locations.

The correlating equation recommended for a concentric orifice with corner taps is

$$C = 0.5959 + 0.0312\beta^{2.1} - 0.184\beta^{8} + \frac{91.71\beta^{2.5}}{\mathrm{Re}_{D_1}^{0.75}} \tag{13.7}$$

Equation (13.7) predicts orifice discharge coefficients within $\pm$ 0.6% for $0.2 < \beta < 0.75$ and for $10^4 < \mathrm{Re}_{D_1} < 10^7$. Some flow coefficients calculated from Equation (13.7) are presented in Figure 13.3. Flow coefficients are relatively insensitive to Reynolds number for $\mathrm{Re}_{D_1} > 10^5$ when $\beta > 0.5$.

A similar correlating equation is available for orifice plates with D and $D/2$ taps. Flange taps require a different correlation for every line size. Pipe taps, located at $2^1/_2\,D$ and $8D$, no longer are recommended.

Flow Nozzles

Flow nozzles may be used as metering elements in either plenums or ducts, as shown in Figure 13.4; the nozzle section is approximately a quarter ellipse. Design details and recommended locations for pressure taps are given in Reference 7.

The correlating equation recommended for an ASME long-radius flow nozzles[7] is

$$C = 0.9975 - \frac{6.53\beta^{0.5}}{\mathrm{Re}_{D_1}^{0.5}} \tag{13.8}$$

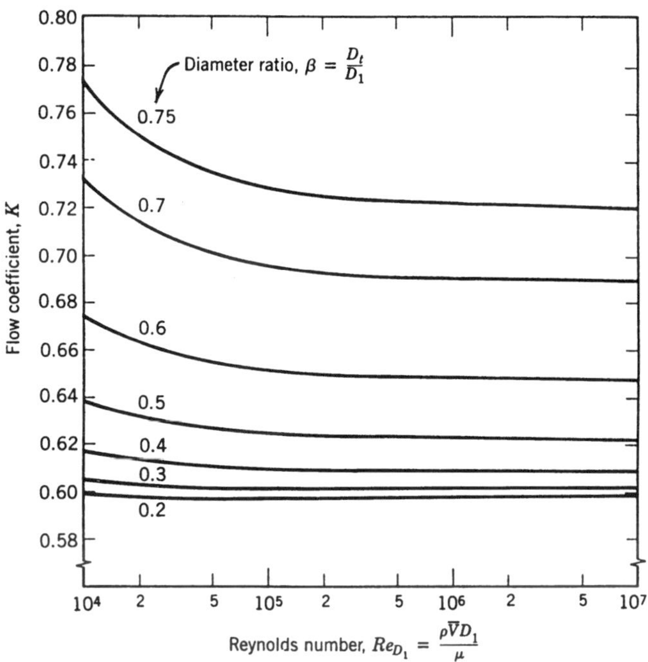

FIGURE 13.3 Flow coefficients for concentric orifices with corner taps.

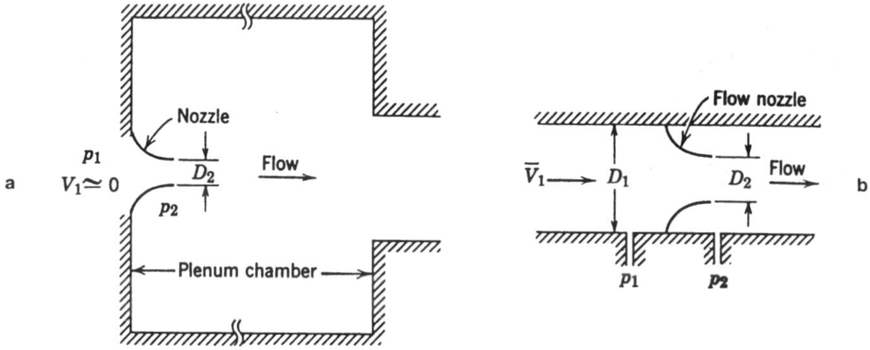

FIGURE 13.4 Typical installations of nozzle flow meters. (a) In plenum, (b) In duct.

Equation (13.8) predicts discharge coefficients for flow nozzles within ± 2.0% for $0.25 < \beta < 0.75$ for $10^4 < Re_{D_1} < 10^7$. Some flow coefficients calculated from Equation 13.8 are presented in Figure 13.5. (K can be greater than 1 when the velocity correction factor exceeds 1.) For plenum installation, nozzles may be fabricated from spun aluminum, molded fiberglass, or other inexpensive materials. Typical flow coefficients are in the range $0.95 < K < 0.99$; the larger values apply at high Reynolds numbers. Thus, the mass flow rate can be computed within approximately ± 2% using $K = 0.97$.

Venturis

Venturi meters are generally made from castings machined to close tolerances to duplicate the performance of the standard design, so they are heavy, bulky, and expensive. The conical diffuser section downstream from the throat gives excellent pressure recovery; overall head loss is low. Venturi meters are self-cleaning because of their smooth internal contours.

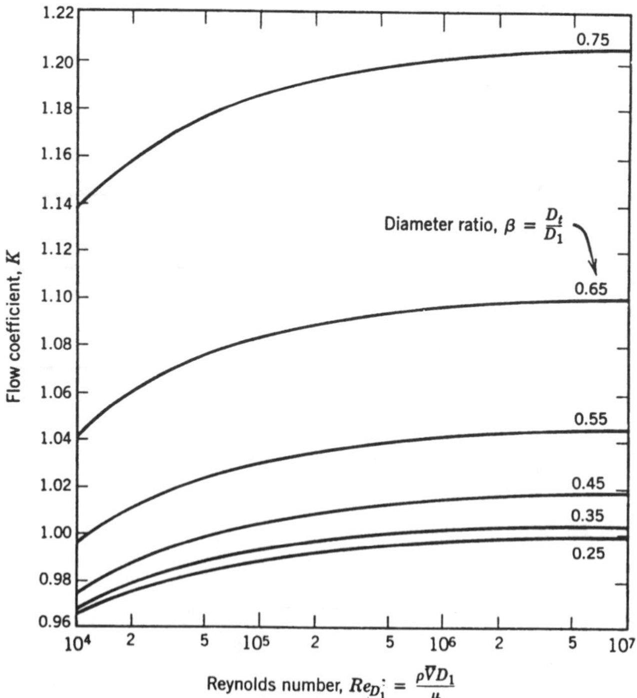

FIGURE 13.5 Flow coefficients for ASME long-radius flow nozzles.

Experimentally measured discharge coefficients for venturi meters range from 0.980 to 0.995 at high Reynolds numbers ($Re_{D_1} > 2 \times 10^5$). Thus, $C = 0.99$ can be used to calculate mass flow rate within about ±1% at high Reynolds number.[7] Consult manufacturers' literature for specific information at Reynolds numbers below 10^5.

Orifice plates, flow nozzles, and venturis all produce pressure drops proportional to flow rate squared, according to Equation 13.4. In practice, a meter must be sized to accommodate the largest flow rate expected. Because the pressure drop vs. flow rate relationship is nonlinear, a limited range of flow rate can be measured accurately. Flow meters with single throats usually are considered for flow rates over a 4:1 range.[7]

Unrecoverable head loss across a metering element may be expressed as a fraction of the differential pressure across the element. Unrecoverable head losses are shown in Figure 13.6.[7]

Laminar Flow Elements

The laminar flow element (LFE)* produces a pressure differential proportional to flow rate. The LFE contains a metering section subdivided into many passages, each of small enough diameter to assure fully developed laminar flow. Because the pressure drop in laminar duct flow is directly proportional to flow rate, the pressure drop vs. flow rate relationship is linear. The LFE may be used with reasonable accuracy over a 10:1 flow rate range. Because the relationship between pressure drop and flow rate for laminar flow depends on viscosity, which is a strong function of temperature, fluid temperature must be known in order to obtain accurate metering.

An LFE costs approximately as much as a venturi, but is much lighter and smaller. Thus, the LFE is widely used in applications where compactness and extended range are important.

* Patented and manufactured by Meriam Instrument Co., 10920 Madison Ave., Cleveland, OH.

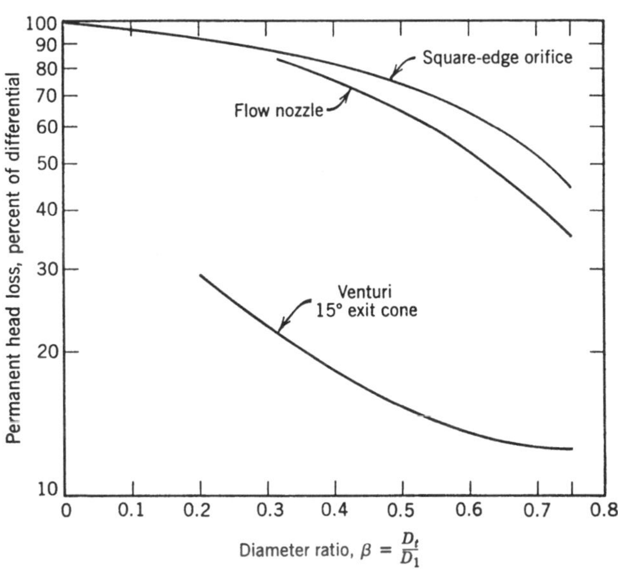

FIGURE 13.6 Permanent head loss produced by various flow metering elements.

Linear Flow Meters

Several flow meter types produce outputs proportional to flow rate. Some meters produce signals without the need to measure differential pressure. The most common linear flow meters are discussed briefly below.

Float meters indicate flow rate directly for liquids or gases. An example is shown in Figure 13.7. The float is carried upward in the tapered clear tube by the flowing fluid until drag force and float weight are in equilibrium. Float meters (often called rotameters) are available with factory calibration for a number of common fluids and flow rate ranges.x

A *turbine flow meter* is a free-running, vaned impeller mounted in a cylindrical section of tube (Figure 13.8). With proper design, the rate of rotation of the impeller may be made proportional to volume flow rate over as much as a 100:1 flow rate range. Rotational speed of the turbine element can be sensed using a magnetic or modulated carrier pickup external to the meter. This sensing method therefore requires no penetrations or seals in the duct. Thus, turbine flow meters can be used safely to measure flow rates in corrosive or toxic fluids. The electrical signal can be displayed, recorded, or integrated to provide total flow information.

Vortex shedding from a bluff obstruction may be used to meter flow. Since Strouhal number, $St = f L/V$, is approximately constant ($St \cong 0.21$), vortex shedding frequency f is proportional to flow velocity. Vortex shedding causes velocity and pressure changes. Pressure, thermal, or ultrasonic sensors may be used to detect the vortex shedding frequency, and thus to infer the fluid velocity. (The velocity profile does affect the constancy of the shedding frequency.) Vortex flow meters can be used over a 20:1 flow rate range.[7]

Electromagnetic flow meters create a magnetic field across a pipe. When a conductive fluid passes through the field, a voltage is generated at right angles to the field and velocity vectors. Electrodes placed on a pipe diameter detect the resulting signal voltage, which is proportional to the average axial velocity when the profile is axisymmetric. The minimum flow speed should be above about 0.3 m/sec, but there are no restrictions on Reynolds number. The flow rate range normally quoted is 10:1.[7]

Ultrasonic flow meters also respond to average velocity at a pipe cross section. Two principal types of ultrasonic meters measure propagation time for clean liquids, and reflection frequency shift (Doppler effect) for flows carrying particulates. The speed of an acoustic wave increases in the flow direction and

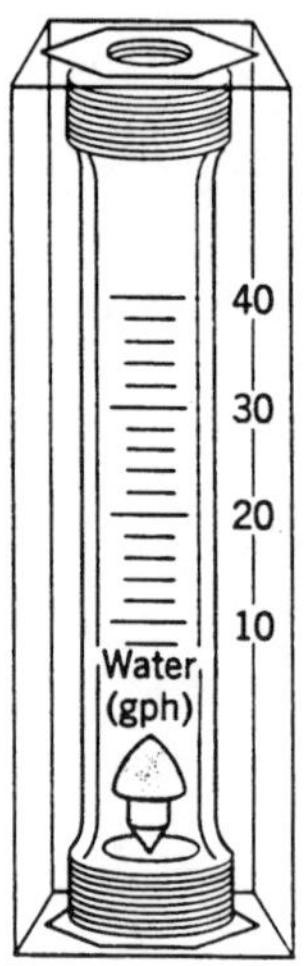

FIGURE 13.7 Float-type variable-area flow meter. (Courtesy of Dwyer Instrument Co., Michigan City, IN.)

decreases when transmitted against the flow. For clean liquids, an acoustic path inclined to the pipe axis is used to infer flow velocity. Multiple paths are used to estimate volume flow rate accurately.

Doppler effect ultrasonic flow meters depend on reflection of sonic waves (in the megahertz range) from scattering particles in the fluid. When the particles move at flow speed, the frequency shift is proportional to flow speed; for a suitably chosen path, output is proportional to volume flow rate; ultrasonic meters may require calibration in place. One or two transducers may be used; the meter may be clamped to the outside of the pipe. Flow rate range is 10:1.[7]

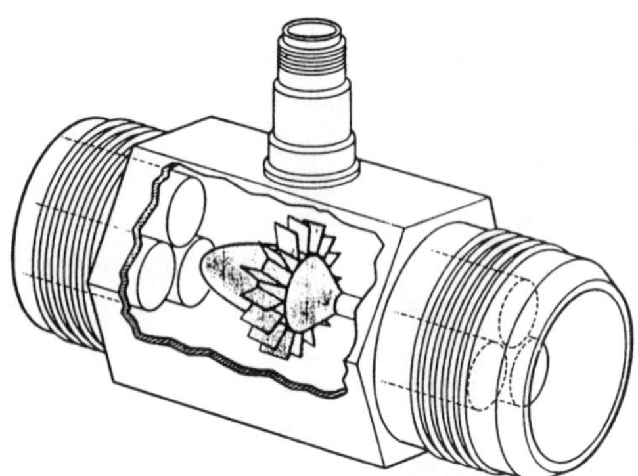

FIGURE 13.8 Turbine flow meter. (Courtesy of Potter Aeronautical Corp., Union, NJ.)

A relatively new type of true mass flow meter is based on the effect of *Coriolis acceleration* on natural frequency of a bent tube carrying fluid. The bent tube is excited and its vibration amplitude measured. The instrument measures mass flow rate directly, and thus is ideal for two-phase or liquid–solid flow measurements. Pressure drop of the Coriolis meter may be high, but its useful flow rate range is 100:1.

Traversing Methods

In situations such as in air handling or refrigeration equipment, it may be impractical or impossible to install a fixed flow meter, but it may be possible to measure flow rate using a traversing technique.[1] To

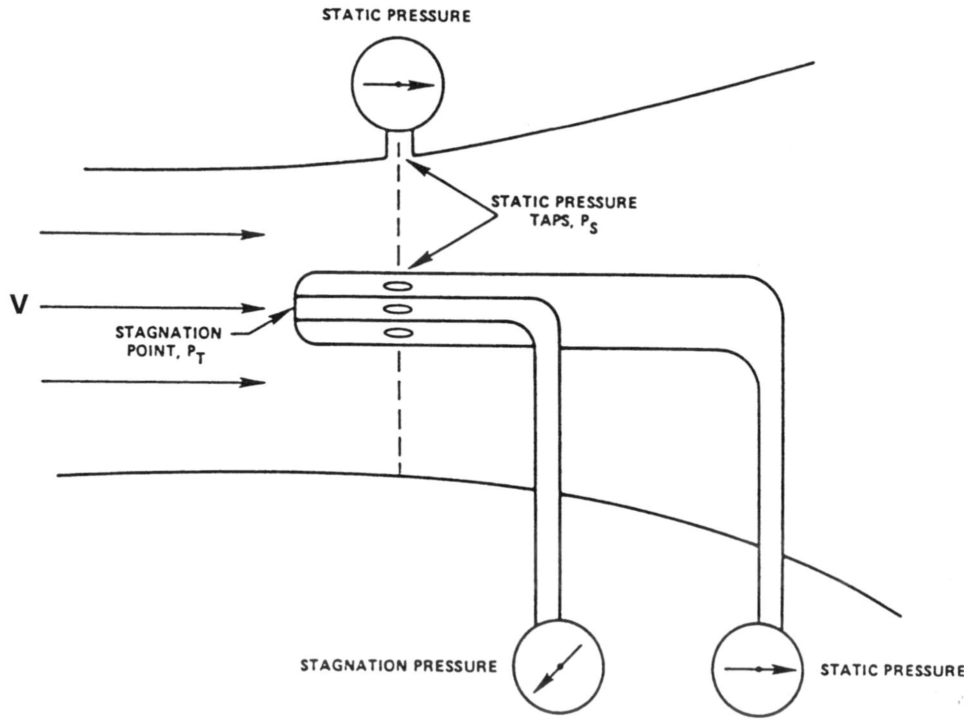

$$V = \sqrt{2(P_T - P_S)/\rho}$$

FIGURE 13.9 Pitot tube.

measure flow rate by **traverse**, the duct cross section is subdivided into segments of equal area. The fluid velocity is measured at the center of each area segment using a pitot tube (see Figure 13.9), a total head tube, or a suitable anemometer. The volume flow rate for each segment is approximated by the product of the measured velocity and segment area. Flow rate through the entire duct is the sum of these segmental flow rates. For details of recommended procedures see Reference 1 or 6.

Use of probes for traverse measurements requires direct access to the flow field. Pitot tubes give uncertain results when pressure gradients or streamline curvature are present, and they respond slowly. Two types of anemometers — **thermal anemometers** and **laser Doppler anemometers** — partially overcome these difficulties, although they introduce new complications.

Thermal anemometers use electrically heated tiny elements (either hot-wire or hot-film elements). Sophisticated feedback circuits are used to maintain the temperature of the element constant and to sense the input heating rate. The heating rate is related to the local flow velocity by calibration. The primary advantage of thermal anemometers is the small size of the sensing element. Sensors as small as 0.002 mm in diameter and 0.1 mm long are available commercially. Because the thermal mass of such tiny elements is extremely small, their response to fluctuations in flow velocity is rapid. Frequency responses to the 50-kHz range have been quoted.[3] Thus, thermal anemometers are ideal for measuring turbulence quantities. Insulating coatings may be applied to permit their use in conductive or corrosive gases or liquids.

Because of their fast response and small size, thermal anemometers are used extensively for research. Numerous schemes for treating the resulting data have been published. Digital processing techniques, including fast Fourier transforms, can be used to obtain mean values and moments, and to analyze signal frequency content and correlations.

Laser Doppler anemometers (LDAs) can be used for specialized applications where direct physical access to the flow field is difficult or impossible.[5] Laser beam(s) are focused to a small volume in the flow at the location of interest; laser light is scattered from particles present in the flow or introduced for this purpose. A frequency shift is caused by the local flow speed (Doppler effect). Scattered light and a reference beam are collected by receiving optics. The frequency shift is proportional to the flow speed; this relationship may be calculated, so there is no need for calibration. Since velocity is measured directly, the signal is unaffected by changes in temperature, density, or composition in the flow field.

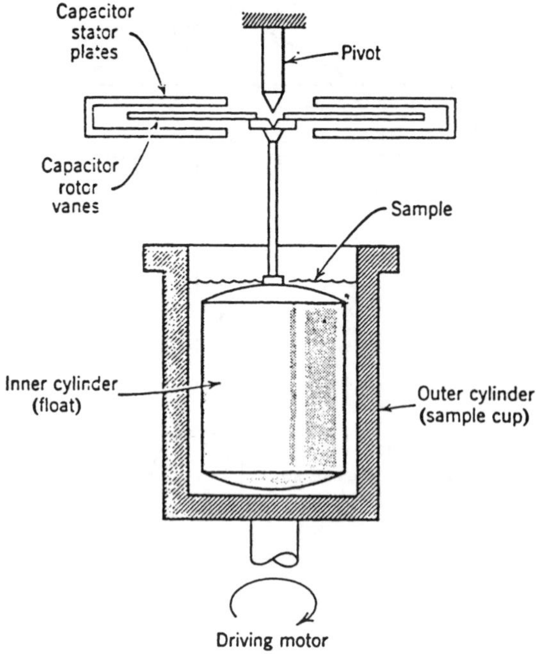

FIGURE 13.10 Rotational viscometer.

The primary disadvantages of LDAs are the expensive and fragile optical equipment and the need for careful alignment.

Viscosity Measurements

Viscometry is the technique of measuring the viscosity of a fluid. Viscometers are classified as rotational, capillary, or miscellaneous, depending on the technique employed. Rotational viscometers use the principle that a rotating body immersed in a liquid experiences a viscous drag which is a function of the viscosity of the liquid, the shape and size of the body, and the speed of its rotation. Rotational viscometers are widely used because measurements can be carried out for extended periods of time. Several types of viscometers are classified as rotational and Figure 13.10 is a schematic diagram illustrating a typical instrument of this type.

Capillary viscometry uses the principle that when a liquid passes in laminar flow through a tube, the viscosity of the liquid can be determined from measurements of the volume flow rate, the applied pressure, and the tube dimensions. Viscometers that cannot be classified either as rotational or capillary include the falling ball viscometer. Its method of operation is based on Stokes' law which relates the viscosity of a Newtonian fluid to the velocity of a sphere falling in it. Falling ball viscometers are often employed for reasonably viscous fluids. Rising bubble viscometers utilize the principle that the rise of an air bubble through a liquid medium gives a visual measurement of liquid viscosity. Because of their simplicity, rising bubble viscometers are commonly used to estimate the viscosity of varnish, lacquer, and other similar media.

Defining Terms

Flow meter: Device used to measure mass flow rate or volume flow rate of fluid flowing in a duct.

Restriction flow meter: Flow meter that causes flowing fluid to accelerate in a nozzle, creating a pressure change that can be measured and related to flow rate.

Thermal anemometer: Heated sensor used to infer local fluid velocity by sensing changes in heat transfer from a small electrically heated surface exposed to the fluid flow.

Traverse: Systematic procedure used to traverse a probe across a duct cross-section to measure flow rate through the duct.

References

1. *ASHRAE Handbook Fundamentals.* 1981. American Society of Heating, Refrigerating, and Air Conditioning Engineers, Atlanta, GA.
2. Baker, R.C. *An Introductory Guide to Flow Measurement.* 1989. Institution of Mechanical Engineers, London.
3. Bruun, H.H. 1995. *Hot-Wire Anemometry: Principles and Signal Analysis.* Oxford University Press, New York.
4. Fox, R.W. and McDonald, A.T. 1992. *Introduction to Fluid Mechanics*, 4th ed., John Wiley & Sons, New York.
5. Goldstein, R.J., Ed. 1996. *Fluid Mechanics Measurements*, 2nd ed., Taylor and Francis, Bristol, PA.
6. ISO 7145, *Determination of Flowrate of Fluids in Closed Conduits or Circular Cross Sections Method of Velocity Determination at One Point in the Cross Section*, ISO UDC 532.57.082.25:532.542. International Standards Organization, Geneva, 1982.
7. Miller, R.W. 1996. *Flow Measurement Engineering Handbook*, 3rd ed., McGraw-Hill, New York.
8. Spitzer, R.W., Ed. 1991. *Flow Measurement: A Practical Guide for Measurement and Control.* Instrument Society of America, Research Triangle Park, NC.
9. White, F.M. 1994. *Fluid Mechanics*, 3rd ed., McGraw-Hill, New York.

Further Information

This section presents only a survey of flow measurement methods. The references contain a wealth of further information. Baker[2] surveys the field and discusses precision, calibration, probe and tracer methods, and likely developments. Miller[7] is the most comprehensive and current reference available for information on restriction flow meters. Goldstein[5] treats a variety of measurement methods in his recently revised and updated book. Spitzer[8] presents an excellent practical discussion of flow measurement. Measurement of viscosity is extensively treated in *Viscosity and Flow Measurement: A Laboratory Handbook of Rheology*, by Van Wazer, J.R., Lyons, J.W., Kim, K.Y., and Colwell, R.E., Interscience Publishers, John Wiley & Sons, New York, 1963.

14

Micro/Nanotribology

Bharat Bhushan
Ohio State University

14 Micro/Nanotribology 211
Introduction • Experimental Techniques • Surface Roughness, Adhesion, and Friction • Scratching, Wear, and Indentation • Boundary Lubrication

Micro/Nanotribology

Bharat Bhushan

Introduction

The emerging field of micro/nanotribology is concerned with processes ranging from atomic and molecular scales to microscale, occurring during adhesion, friction, wear, and thin-film lubrication at sliding surfaces (Bhushan, 1995, 1997; Bhushan et al., 1995). The differences between conventional tribology or macrotribology and micro/nanotribology are contrasted in Figure 14.1. In macrotribology, tests are conducted on components with relatively large mass under heavily loaded conditions. In these tests, wear is inevitable and the bulk properties of mating components dominate the tribological performance. In micro/nanotribology, measurements are made on at least one of the mating components, with relatively small mass under lightly loaded conditions. In this situation, negligible wear occurs and the surface properties dominate the tribological performance.

Micro/nanotribological investigations are needed to develop fundamental understanding of interfacial phenomena on a small scale and to study interfacial phenomena in the micro- and nanostructures used in magnetic storage systems, microelectromechanical systems (MEMS), and other industrial applications (Bhushan, 1995, 1996, 1997). Friction and wear of lightly loaded micro/nanocomponents are highly dependent on the surface interactions (few atomic layers). These structures are generally lubricated with molecularly thin films. Micro- and nanotribological studies are also valuable in fundamental understand-

0-8493-0055-X/00/$0.00+$.50

Macrotribology	Micro/nanotribology
Large Mass Heavy Load ↓ Wear (Inevitable)	Small mass (μg) Light load (μg to mg) ↓ No Wear (Few atomic layers)
Bulk Material	Surface (few atomic layers)

FIGURE 14.1 Comparison between macrotribology and microtribology.

ing of interfacial phenomena in macrostructures to provide a bridge between science and engineering (Bowden and Tabor, 1950, 1964; Bhushan and Gupta, 1997; Bhushan, 1996).

In 1985, Binnig et al. (1986) developed an "atomic force microscope" (AFM) to measure ultrasmall forces (less than 1 μN) present between the AFM tip surface and the sample surface. AFMs can be used for measurement of *all engineering surfaces* of any surface roughness, which may be either electrically conducting or insulating. AFM has become a popular surface profiler for topographic measurements on micro- to nanoscale. These are also used for scratching, wear, and nanofabrication purposes. AFMs have been modified in order to measure both normal and friction forces and this instrument is generally called a friction force microscope (FFM) or a lateral force microscope (LFM). New transducers in conjunction with an AFM can be used for measurements of elastic/plastic mechanical properties (such as load-displacement curves, indentation hardness, and modulus of elasticity) (Bhushan et al., 1996). A surface force apparatus (SFA) was first developed in 1969 (Tabor and Winterton, 1969) to study both static and dynamic properties of molecularly thin liquid films sandwiched between two molecularly smooth surfaces. SFAs are being used to study rheology of molecularly thin liquid films; however, the liquid under study has to be confined between molecularly smooth surfaces with radii of curvature on the order of 1 mm (leading to poorer lateral resolution as compared with AFMs) (Bhushan, 1995). Only AFMs/FFMs can be used to study *engineering surfaces* in the *dry and wet conditions* with *atomic resolution.* The scope of this section is limited to the applications of AFMs/FFMs.

At most solid–solid interfaces of technological relevance, contact occurs at numerous asperities with a range of radii; a sharp AFM/FFM tip sliding on a surface simulates just one such contact. Surface roughness, adhesion, friction, wear, and lubrication at the interface between two solids with and without liquid films have been studied using the AFM and FFM. The status of current understanding of micro/nanotribology of engineering interfaces follows.

Experimental Techniques

An AFM relies on a scanning technique to produce very high resolution, three-dimensional images of sample surfaces. The AFM measures ultrasmall forces (less than 1 nN) present between the AFM tip surface and a sample surface. These small forces are measured by measuring the motion of a very flexible cantilever beam having an ultrasmall mass. The deflection can be measured to with ±0.02 nm, so for a typical cantilever force constant of 10 N/m, a force as low as 0.2 nN can be detected. An AFM is capable of investigating surfaces of both conductors and insulators on an atomic scale. In the operation of a high-resolution AFM, the sample is generally scanned; however, AFMs are also available where the tip is scanned and the sample is stationary. To obtain atomic resolution with an AFM, the spring constant of the cantilever should be weaker than the equivalent spring between atoms. A cantilever beam with a spring constant of about 1 N/m or lower is desirable. Tips have to be sharp as possible. Tips with a radius ranging from 10 to 100 nm are commonly available.

In the AFM/FFM shown in Figure 14.2, the sample is mounted on a PZT tube scanner which consists of separate electrodes to precisely scan the sample in the X–Y plane in a raster pattern and to move the sample in the vertical (Z) direction. A sharp tip at the end of a flexible cantilever is brought in contact

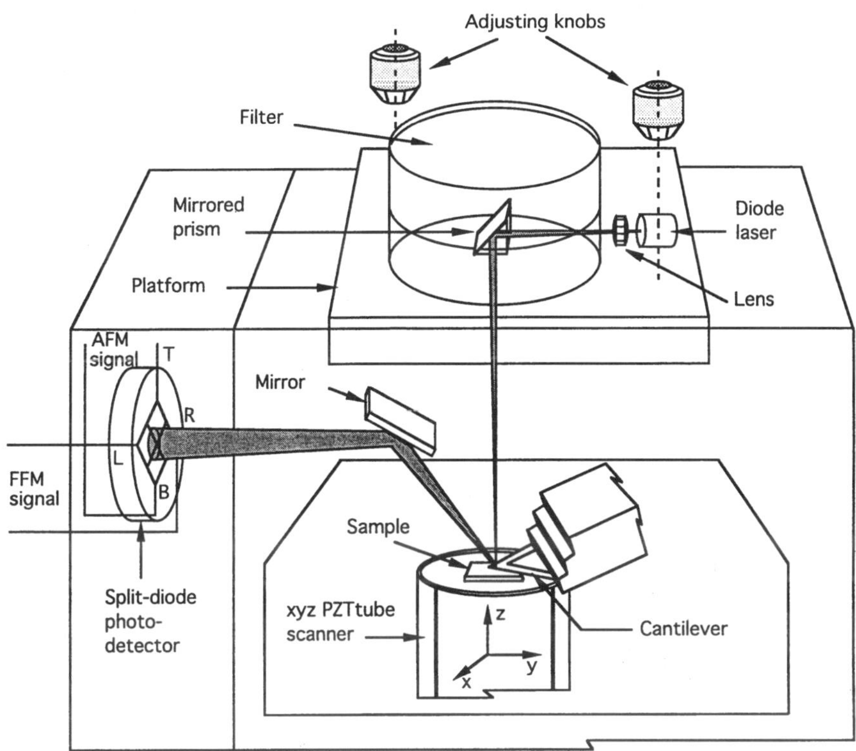

FIGURE 14.2 Schematic of a commercial AFM/FFM using laser beam deflection method.

with the sample. Normal and frictional forces being applied at the tip–sample interface are measured simultaneously, using a laser beam deflection technique.

Topographic measurements are typically made using a sharp tip on a cantilever beam with normal stiffness on the order of 0.5 N/m at a normal load of about 10 nN, and friction measurements are carried out in the load range of 10 to 150 nN. The tip is scanned in such a way that its trajectory on the sample forms a triangular pattern. Scanning speeds in the fast and slow scan directions depend on the scan area and scan frequency. A maximum scan size of 125 × 125 μm and scan rate of 122 Hz typically can be used. Higher scan rates are used for small scan lengths.

For nanoscale boundary lubrication studies, the samples are typically scanned over an area of 1 × 1 μm at a normal force of about 300 nN, in a direction orthogonal to the long axis of the cantilever beam (Bhushan, 1997). The samples are generally scanned with a scan rate of 1 Hz and the scanning speed of 2 μm/sec. The coefficient of friction is monitored during scanning for a desired number of cycles. After the scanning test, a larger area of 2 × 2 μm is scanned at a normal force of 40 nN to observe for any wear scar.

For microscale scratching, microscale wear, and nano-scale indentation hardness measurements, a sharp single-crystal natural diamond tip mounted on a stainless steel cantilever beam with a normal stiffness on the order of 25 N/m is used at relatively higher loads (1 to 150 μN). For wear studies, typically an area of 2 × 2 μm is scanned at various normal loads (ranging from 1 to 100 μN) for a selected number of cycles. For nanoindentation hardness measurements the scan size is set to zero and then normal load is applied to make the indents. During this procedure, the diamond tip is continuously pressed against the sample surface for about 2 sec at various indentation loads. The sample surface is scanned before and after the scratching, wear, or indentation to obtain the initial and the final surface topography, at a low normal load of about 0.3 μN using the same diamond tip.

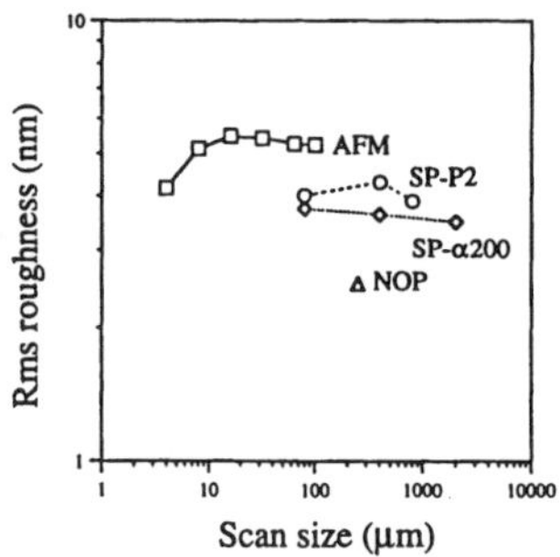

FIGURE 14.3 Scale dependence of standard deviation of surface heights (rms) for a glass–ceramic substrate, measured using an AFM, a stylus profiler (SP-P2 and SP-α 200), and a noncontact optical profiler (NOP).

An area larger than the indentation region is scanned to observe the indentation marks. Nanohardness is calculated by dividing the indentation load by the projected residual area of the indents.

In measurements using conventional AFMs, the hardness value is based on the projected residual area after imaging the incident. Identification of the boundary of the indentation mark is difficult to accomplish with great accuracy, which makes the direct measurement of contact area somewhat inaccurate. A capacitive transducer with the dual capability of depth sensing as well as *in situ* imaging is used in conjunction with an AFM (Bhushan et al., 1996). This indentation system, called nano/picoindention, is used to make load-displacement measurements and subsequently carry out *in situ* imaging of the indent, if necessary. Indenter displacement at a given load is used to calculate the projected indent area for calculation of the hardness value. Young's modulus of elasticity is obtained from the slope of the unloading portion of the load-displacement curve.

Surface Roughness, Adhesion, and Friction

Solid surfaces, irrespective of the method of formation, contain surface irregularities or deviations from the prescribed geometrical form. When two nominally flat surfaces are placed in contact, surface roughness causes contact to occur at discrete contact points. Deformation occurs in these points and may be either elastic or plastic, depending on the nominal stress, surface roughness, and material properties. The sum of the areas of all the contact points constitutes the real area that would be in contact, and for most materials at normal loads this will be only a small fraction of the area of contact if the surfaces were perfectly smooth. In general, real area of contact must be minimized to minimize adhesion, friction, and wear (Bhushan and Gupta, 1997; Bhushan, 1996). Characterizing surface roughness is therefore important for predicting and understanding the tribological properties of solids in contact.

Surface roughness most commonly refers to the variations in the height of the surface relative to a reference plane (Bowden and Tabor, 1950; Bhushan, 1996). Commonly measured roughness parameters, such as standard deviation of surface heights (rms), are found to be scale dependent and a function of the measuring instrument, for any given surface, Figure 14.3 (Poon and Bhushan, 1995). The topography of most engineering surfaces is fractal, possessing a self-similar structure over a range of scales. By using fractal analysis one can characterize the roughness of surfaces with two scale-independent fractal parameters D and C which provide information about roughness at all length scales (Ganti and Bhushan, 1995; Bhushan, 1995). These two parameters are instrument independent and are unique for each surface. D (generally ranging from 1 to 2) primarily relates to distribution of different frequencies in the surface profile, and C to the amplitude of the surface height variations at all frequencies. A fractal model of elastic plastic contact has been used to predict whether contacts experience elastic or plastic deformation and to predict the statistical distribution of contact points.

Based on atomic-scale friction measurements of a well-characterized freshly cleaved surface of highly oriented pyrolytic graphite (HOPG), the atomic-scale friction force of HOPG exhibits the same periodicity as that of corresponding topography (Figure 14.4(a)), but the peaks in friction and those in

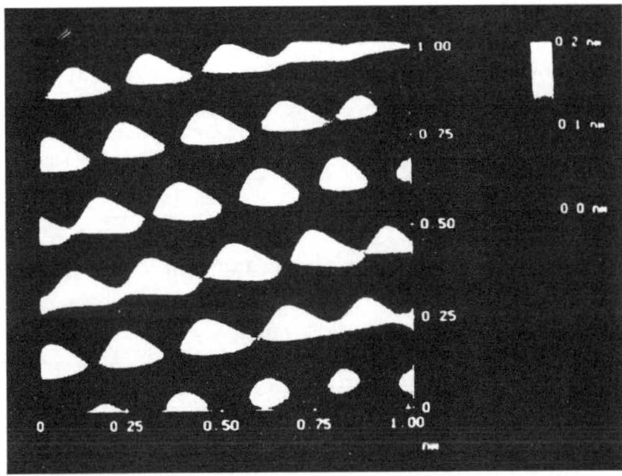

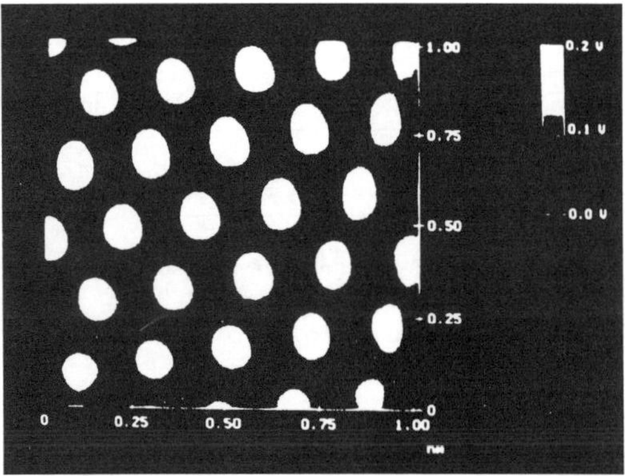

Topography Friction

a

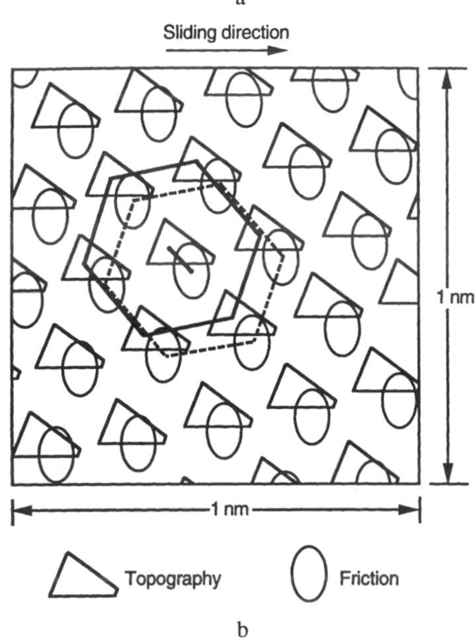

b

FIGURE 14.4 (a) Gray-scale plots of surface topography (left) and friction profiles (right) of a 1 × 1 nm area of freshly cleaved HOPG, showing the atomic-scale variation of topography and friction, (b) diagram of superimposed topography and friction profiles from (a); the symbols correspond to maxima. Note the spatial shift between the two profiles. (From Bhushan, B., *Handbook of Micro/Nanotribology*, CRC Press, Boca Raton, FL, 1995. With permission.)

topography were displaced relative to each other (Figure 14.4(b)). A Fourier expansion of the interatomic potential has been used to calculate the conservative interatomic forces between atoms of the FFM tip and those of the graphite surface. Maxima in the interatomic forces in the normal and lateral directions do not occur at the same location, which explains the observed shift between the peaks in the lateral force and those in the corresponding topography. Furthermore, the observed local variations in friction force were explained by variation in the intrinsic lateral force between the sample and the FFM tip, and these variations may not necessarily occur as a result of an atomic-scale stick–slip process.

Friction forces of HOPG have also been studied. Local variations in the microscale friction of cleaved graphite are observed, which arise from structural changes that occur during the cleaving process. The cleaved HOPG surface is largely atomically smooth, but exhibits line-shaped regions in which the coefficient of friction is more than an order of magnitude larger. Transmission electron microscopy

TABLE 14.1 Surface Roughness and Micro- and Macroscale Coefficients of Friction of Various Samples

Material	rms Roughness, nm	Microscale Coefficient of Friction vs. Si_3N_4Tip[a]	Macroscale Coefficient of Friction vs. Alumina Ball[b]
Si(111)	0.11	0.03	0.18
C^+-implanted Si	0.33	0.02	0.18

[a] Tip radius of about 50 nm in the load range of 10 to 150 nN (2.5 to 6.1 GPa), a scanning speed of 5 m/sec and scan area of 1 × 1 μm.

[b] Ball radius of 3 mm at a normal load of 0.1 N (0.3 GPa) and average sliding speed of 0.8 mm/sec.

indicates that the line-shaped regions consist of graphite planes of different orientation, as well as of amorphous carbon. Differences in friction can also be seen for organic mono- and multilayer films, which again seem to be the result of structural variations in the films. These measurements suggest that the FFM can be used for structural mapping of the surfaces. FFM measurements can be used to map chemical variations, as indicated by the use of the FFM with a modified probe tip to map the spatial arrangement of chemical functional groups in mixed organic monolayer films. Here, sample regions that had stronger interactions with the functionalized probe tip exhibited larger friction. For further details, see Bhushan (1995).

Local variations in the microscale friction of scratched surfaces can be significant and are seen to depend on the local surface slope rather than on the surface height distribution (Bhushan, 1995). Directionality in friction is sometimes observed on the macroscale; on the microscale this is the norm (Bhushan, 1995). This is because most "engineering" surfaces have asymmetric surface asperities so that the interaction of the FFM tip with the surface is dependent on the direction of the tip motion. Moreover, during surface-finishing processes material can be transferred preferentially onto one side of the asperities, which also causes asymmetry and directional dependence. Reduction in local variations and in the directionality of frictional properties therefore requires careful optimization of surface roughness distributions and of surface-finishing processes.

Table 14.1 shows the coefficient of friction measured for two surfaces on micro- and macroscales. The coefficient of friction is defined as the ratio of friction force to the normal load. The values on the microscale are much lower than those on the macroscale. When measured for the small contact areas and very low loads used in microscale studies, indentation hardness and modulus of elasticity are higher than at the macroscale. This reduces the degree of wear. In addition, the small apparent areas of contact reduce the number of particles trapped at the interface, and thus minimize the "ploughing" contribution to the friction force.

At higher loads (with contact stresses exceeding the hardness of the softer material), however, the coefficient of friction for microscale measurements increases toward values comparable with those obtained from macroscale measurements, and surface damage also increases (Bhushan et al., 1995; Bhushan and Kulkarni, 1996). Thus, Amontons' law of friction, which states that the coefficient of friction is independent of apparent contact area and normal load, does not hold for microscale measurements. These findings suggest that microcomponents sliding under lightly loaded conditions should experience very low friction and near zero wear.

Scratching, Wear, and Indentation

The AFM can be used to investigate how surface materials can be moved or removed on micro- to nanoscales, for example, in scratching and wear (Bhushan, 1995) (where these things are undesirable) and, in nanomachining/nanofabrication (where they are desirable). The AFM can also be used for measurements of mechanical properties on micro- to nanoscales. Figure 14.5 shows microscratches made on Si(111) at various loads after 10 cycles. As expected, the depth of scratch increases with load. Such microscratching measurements can be used to study failure mechanisms on the microscale and to evaluate the mechanical integrity (scratch resistance) of ultrathin films at low loads.

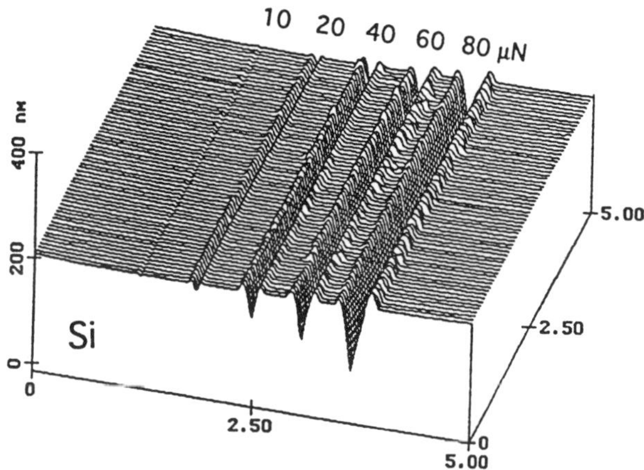

FIGURE 14.5 Surface profiles of Si(111) scratched at various loads. Note that the *x* and *y* axes are in micrometers and the *z* axis is in nanometers. (From Bhushan, B., *Handbook of Micro/Nanotribology*, CRC Press, Boca Raton, FL, 1995. With permission.)

By scanning the sample in two dimensions with the AFM, wear scars are generated on the surface. The evolution of wear of a diamond-like carbon coating on a polished aluminum substrate is showing in Figure 14.6 which illustrates how the microwear profile for a load of 20 μN develops as a function of the number of scanning cycles. Wear is not uniform, but is initiated at the nanoscratches indicating that surface defects (with high surface energy) act as initiation sites. Thus, scratch-free surfaces will be relatively resistant to wear.

Mechanical properties, such as load-displacement curves, hardness, and modulus of elasticity can be determined on micro- to picoscales using an AFM and its modifications (Bhushan, 1995; Bhushan et al., 1995, 1996). Indentability on the scale of picometers can be studied by monitoring the slope of cantilever deflection as a function of sample traveling distance after the tip is engaged and the sample is pushed against the tip. For a rigid sample, cantilever deflection equals the sample traveling distance; but the former quantity is smaller if the tip indents the sample. The indentation hardness on nanoscale of bulk materials and surface films with an indentation depth as small as 1 nm can be measured. An example of hardness data as a function of indentation depth is shown in Figure 14.7. A decrease in hardness with an increase in indentation depth can be rationalized on the basis that, as the volume of deformed materials increases, there is a higher probability of encountering material defects. AFM measurements on ion-implanted silicon surfaces show that ion implantation increases their hardness and, thus, their wear resistance (Bhushan, 1995). Formation of surface alloy films with improved mechanical properties by ion implantation is growing in technological importance as a means of improving the mechanical properties of materials.

Young's modulus of elasticity is calculated from the slope of the indentation curve during unloading (Bhushan, 1995; Bhushan et al., 1996). AFM can be used in a *force modulation mode* to measure surface elasticities: an AFM tip is scanned over the modulated sample surface with the feedback loop keeping the average force constant. For the same applied force, a soft area deforms more, and thus causes less cantilever deflection, than a hard area. The ratio of modulation amplitude to the local tip deflection is then used to create a *force modulation image*. The force modulation mode makes it easier to identify soft areas on hard substrates.

Detection of the transfer of material on a nanoscale is possible with the AFM. Indentation of C_{60}-rich fullerene films with an AFM tip has been shown to result in the transfer of fullerene molecules to the AFM tip, as indicated by discontinuities in the cantilever deflection as a function of sample traveling distance in subsequent indentation studies (Bhushan, 1995).

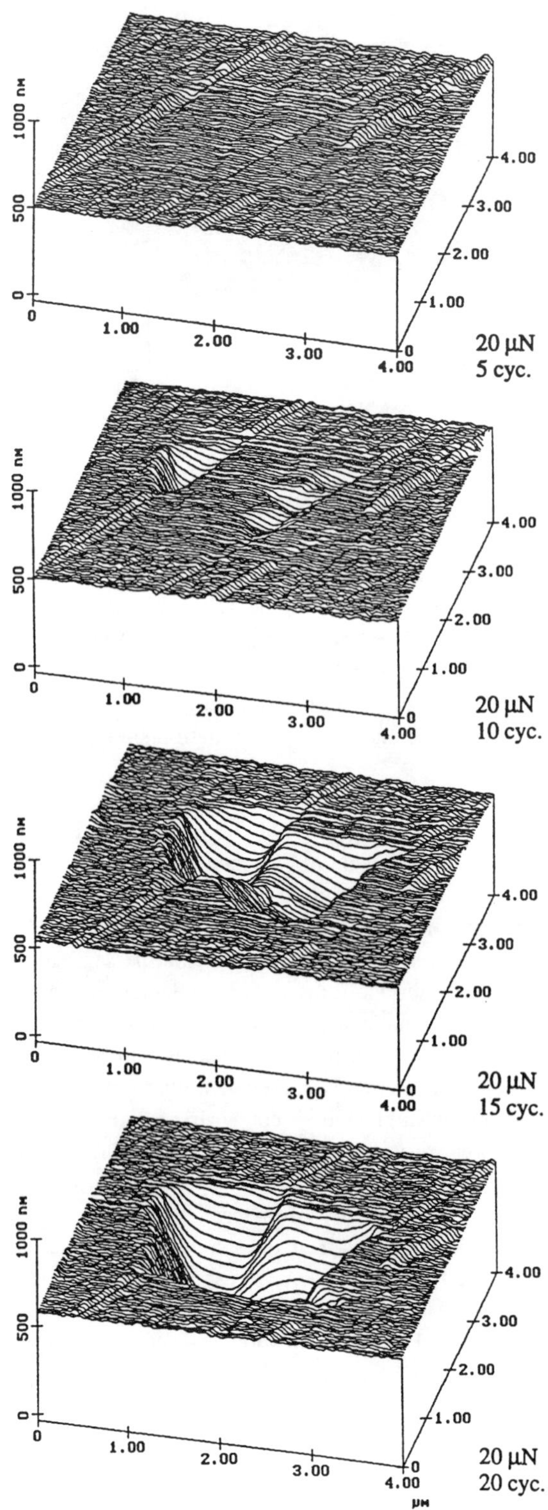

FIGURE 14.6 Surface profiles of diamond-like carbon-coated thin-film disk showing the worn region; the normal load and number of test cycles are indicated. (From Bhushan, B., *Handbook of Micro/Nanotribology*, CRC Press, Boca Raton, FL, 1995. With permission.)

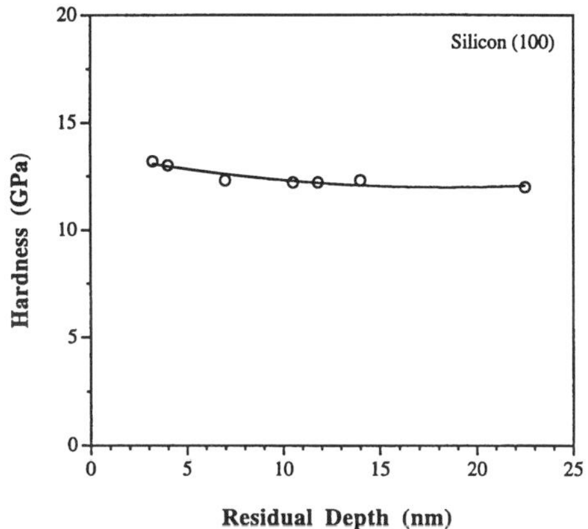

FIGURE 14.7 Indentation hardness as a function of residual indentation depth for Si(100). (From Bhushan, B. et al., *Philos. Mag.*, A74, 117–1128, 1996. With permission.)

Boundary Lubrication

The "classical" approach to lubrication uses freely supported multimolecular layers of liquid lubricants (Bowden and Tabor, 1950, 1964; Bhushan, 1996). The liquid lubricants are chemically bonded to improve their wear resistance (Bhushan, 1995, 1996). To study depletion of boundary layers, the microscale friction measurements are made as a function of the number of cycles. For an example of the data of virgin Si(100) surface and silicon surface lubricated with about 2-nm-thick Z-15 and Z-Dol perfluoropolyether (PEPE) lubricants, see Figure 14.8. Z-Dol is PFPE lubricant with hydroxyl end groups. Its lubricant film was thermally bonded.t In Figure 14.8, he unlubricated silicon sample shows a slight increase in friction force followed by a drop to a lower steady state value after some cycles. Depletion of native oxide and possible roughening of the silicon sample are responsible for the decrease in this friction force. The initial friction force for the Z-15-lubricated sample is lower than that of the unlubricated silicon and increases gradually to a friction force value comparable with that of the silicon after some cycles. This suggests the depletion of the Z-15 lubricant in the wear track. In the case of the Z-Dol-coated silicon sample, the friction force starts out to be low and remains low during the entire test. It suggests that Z-Dol does not get displaced/depleted as readily as Z-15. Additional studies of freely supported liquid lubricants showed that either increasing the film thickness or chemically bonding the molecules to the substrate with a mobile fraction improves the lubricationserformance (Bhushan, 1997).

For lubrication of microdevices, a more effect approach involves the deposition of organized, dense molecular layers of long-chain molecules on the surface contact. Such monolayers and thin films are

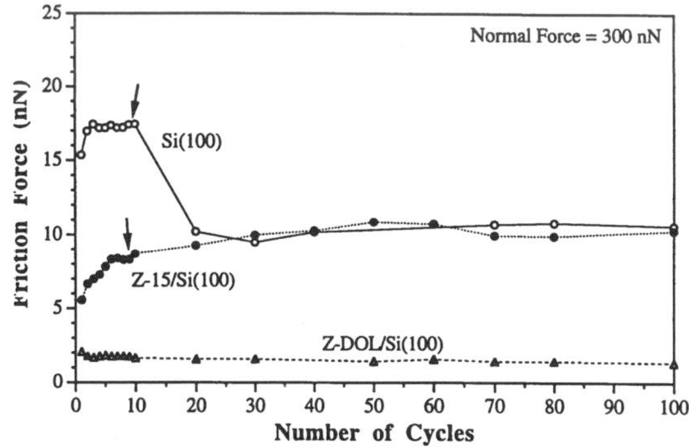

FIGURE 14.8 Friction force as a function of number of cycles using a silicon nitride tip at a normal force of 300 nN for the unlubricated and lubricated silicon samples. (From Bhushan, B., *Micro/Nanotribology and Its Applications*, Kluwer, Dordrecht, The Netherlands, 1997. With permission.)

commonly produced by Langmuir–Blodgett (LB) deposition and by chemical grafting of molecules into self-assembled monolayers (SAMs). Based on the measurements, SAMs of octodecyl (C_{18}) compounds based on aminosilanes on a oxidized silicon exhibited a lower coefficient of friction of (0.018) and greater durability than LB films of zinc arachidate adsorbed on a gold surface coated with octadecylthiol (ODT) (coefficient of friction of 0.03) (Bhushan et al., 1995). LB films are bonded to the substrate by weak van der Waals attraction, whereas SAMs are chemically bound via covalent bonds. Because of the choice of chain length and terminal linking group that SAMs offer, they hold great promise for boundary lubrication of microdevices.

Measurement of ultrathin lubricant films with nanometer lateral resolution can be made with the AFM (Bhushan, 1995). The lubricant thickness is obtained by measuring the force on the tip as it approaches, contacts, and pushes through the liquid film and ultimately contacts the substrate. The distance between the sharp "snap-in" (owing to the formation of a liquid of meniscus between the film and the tip) at the liquid surface and the hard repulsion at the substrate surface i a measure of the liquid film thickness. This technique is now used routinely in the information-storage industry for thickness measurements (with nanoscale spatial resolution) of lubricant films, a few nanometers thick, in rigid magnetic disks.

References

Bhushan, B. 1995. *Handbook of Micro/Nanotribology*, CRC Press, Boca Raton, FL.

Bhushan, B. 1996. *Tribology and Mechanics of Magnetic Storage Devices*, 2nd ed., Springer, New York.

Bhushan, B. 1997. *Micro/Nanotribiology and Its Applications*, NATO ASI Series E: Applied Sciences, Kluwer, Dordrecht, Netherlands.

Bhushan, B. and Gupta, B.K. 1997. *Handbook of Tribology: Materials, Coatings and Surface Treatments*, McGraw-Hill, New York (1991); Reprint with corrections, Kreiger, Malabar, FL.

Bhushan, B. and Kulkarni, A.V. 1996. Effect of normal load on microscale friction measurements, *Thin Solid Films*, 278, 49–56.

Bhushan, B., Israelachvili, J.N., and Landman, U. 1995. Nanotribology: friction, wear and lubrication at the atomic scale, *Nature*, 374, 607–616.

Bhushan, B., Kulkarni, A.V., Bonin, W., and Wyrobek, J.T. 1996. Nano-indentation and pico-indentation measurements using capacitive transducer system in atomic force microscopy, *Philos. Mag.*, A74, 1117–1128.

Binning, G., Quate, C.F., and Gerber, Ch. 1986. Atomic force microscopy, *Phys. Rev. Lett.*, 56, 930–933.

Bowden, F.P. and Tabor, D. 1950; 1964. *The Friction and Lubrication of Solids*, Parts I and II, Clarendon, Oxford.

Ganti, S. and Bhushan, B. 1995. Generalized fractal analysis and its applications to engineering surfaces, *Wear*, 180, 17–34.

Poon, C.Y. and Bhushan, B. 1995. Comparison of surface roughness measurements by stylus profiler, AFM and non-contact optical profiler, *Wear*, 190, 76–88.

Tabor, D. and Winterton, R.H.S. 1969. The direct measurement of normal and retarded van der Walls forces, *Proc. R. Soc. London*, A312, 435–450.

Nomenclature for Fluid Mechanics

Symbol	Quantity	Unit		Dimensions (*MLtT*)
		SI	English	
a	Velocity of sound	m/sec	ft/sec	Lt^{-1}
a	Acceleration	m/sec^2	ft/sec^2	Lt^{-2}
A	Area	m^2	ft^2	L^2
b	Distance, width	m	ft	L

Symbol	Quantity	Unit SI	Unit English	Dimensions ($MLtT$)
c_p	Specific heat, constant pressure	J/kg·K	ft·lb/lb$_m$·°R	$L^2t^{-2}T^{-1}$
c_v	Specific heat, constant volume	J/kg·K	ft·lb/lb$_m$·°R	$L^2t^{-2}T^{-1}$
C	Concentration	No./m^3	No./ft^3	L^{-3}
C	Coefficient	—	—	—
C	Empirical constant	—	—	—
D	Diameter	m	ft	L
D_H	Hydraulic diameter	m	ft	L
e	Total energy per unit mass	J/kg	ft·lb/lb$_m$	L^2t^{-2}
E	Total energy	J	ft·lb or Btu	ML^2t^{-2}
E	Modulus of leasticity	Pa	lb/ft^2	ML^1t^{-2}
Eu	Euler number	—	—	—
f	Friction factor	—	—	—
F	Force	N	lb	MLt^{-2}
Fr	Froude number	—	—	—
F_B	Buoyant force	N	lb	MLt^{-2}
g	Acceleration of gravity	m/sec^2	ft/sec^2	Lt^{-2}
g_0	Gravitation constant	kg·m/N·sec^2	lb$_m$·ft/lb·sec^2	—
G	Mass flow rate per unit area	kg/sec·m^2	lb$_m$/sec·ft^2	$ML^{-2}t^{-1}$
h	Head, vertical distance	m	ft	L
h	Enthalpy per unit mass	J/kg	ft·lb/lb$_m$	L^2t^{-2}
H	Head, elevation of hydraulic grade line	m	ft	L
I	Moment of inertia	m^4	ft^4	L^4
k	Specific heat ratio	—	—	—
K	Bulk modulus of elasticity	Pa	lb/ft^2	ML^1t^{-2}
K	Minor loss coefficient	—	—	—
L	Length	m	ft	L
L	Lift	N	lb	MLt^{-2}
l	Length, mixing length	m	ft	L
ln	Natural logarithm	—	—	—
m	Mass	kg	lb$_m$	M
$\dot{m}$	Strength of source	m^3/sec	ft^3/sec	L^3t^{-1}
m	Mass flow rate	kg/sec	lb$_m$/sec	Mt^{-1}
M	Molecular weight	—	—	—
$\dot{M}$	Momentum per unit time	N	lb	MLt^{-2}
M	Mach number	—	—	—
n	Exponent, constant	—	—	—
n	Normal direction	m	ft	L
n	Manning roughness factor	—	—	—
n	Number of moles	—	—	—
N	Rotation speed	1/sec	1/sec	t^{-1}
NPSH	Net positive suction head	m	ft	L
p	Pressure	Pa	lb/ft^2	$ML^{-1}t^{-2}$
P	Height of weir	m	ft	L
P	Wetted perimeter	m	ft	L
q	Discharge per unit width	m^2/sec	ft^2/sec	L^2t^{-1}
q	Heat transfer per unit time	J/sec	Btu	ML^2t^{-3}
r	Radial distance	m	ft	L
R	Gas constant	J/kg·K	ft·lb/lb$_m$·°R	$L^2t^{-2}T^{-1}$
Re	Reynolds number	—	—	—
s	Distance	m	ft	L

Symbol	Quantity	Unit SI	Unit English	Dimensions (*MLtT*)
s	Entropy per unit mass	J/kg·K	$ft \cdot lb/lb_m \cdot {}^\circ R$	$L^2t^{-2}T^{-1}$
S	Entropy	J/K	ft·lb/°R	$ML^2t^{-2}T^{-1}$
S	Specific gravity, slope	—	—	—
t	Time	sec	sec	t
t	Distance, thickness	m	ft	L
T	Temperature	K	°R	T
T	Torque	N·m	lb·ft	ML^2t^{-2}
u	Velocity, Velocity component	m/sec	ft/sec	Lt^{-1}
u	Peripheral speed	m/sec	ft/sec	Lt^{-1}
u	Internal energy per unit mass	J/kg	$ft \cdot lb/lb_m$	L^2t^{-2}
u_*	Shear stress velocity	m/sec	ft/sec	Lt^{-1}
U	Internal energy	J	Btu	ML^2t^{-2}
v	Velocity, velocity component	m/sec	ft/sec	Lt^{-1}
v_s	Specific volume	m^3/kg	ft^3/lb_m	$M^{-1}L^3$
V	Volume	m^3	ft^3	L^3
V	Volumetric flow rate	m^3/sec	ft^3/sec	L^3t^{-1}
V	Velocity	m/sec	ft/sec	Lt^{-1}
w	Velocity component	m/sec	ft/sec	Lt^{-1}
w	Work per unit mass	J/kg	$ft \cdot lb/lb_m$	L^2t^{-2}
W	Work per unit time	J/sec	ft·lb/sec	ML^2t^{-3}
W_s	Shaft work	m·N	ft·lb	ML^2t^{-2}
W	Weight	N	lb	MLt^{-2}
We	Weber number	—	—	—
x	Distance	m	ft	L
y	Distance, depth	m	ft	L
Y	Weir height	m	ft	L
z	Vertical distance	m	ft	L

Greek Symbols

Symbol	Quantity	SI	English	Dimensions
α	Angle, coefficient	—	—	—
β	Blade angle	—	—	—
Γ	Circulation	m^2	ft^2	L^2t^{-1}
υ	Vector operator	1/m	1/ft	L^{-1}
γ	Specific weight	N/m^3	lb/ft^3	$ML^{-2}t^{-2}$
δ	Boundary layer thickness	m	ft	L
ε	Kinematic eddy viscosity	m^2/sec	ft^2/sec	L^2t^{-1}
ε	Roughness height	m	ft	L
η	Eddy viscosity	$N \cdot sec/m^2$	$lb \cdot sec/ft^2$	$ML^{-1}t^{-1}$
η	Head ratio	—	—	—
η	Efficiency	—	—	—
Θ	Angle	—	—	—
κ	Universal constant	—	—	—
λ	Scale ratio, undetermined multiplier	—	—	—
μ	Viscosity	$N \cdot sec/m^2$	$lb \cdot sec/ft^2$	$ML^{-1}t^{-1}$
ν	Kinematic viscosity (= μ/ρ)	m^2/sec	ft^2/sec	L^2t^{-1}
Φ	Velocity potential	m^2/sec	ft^2/sec	L^2t^{-1}
Φ	Function	—	—	—
π	Constant	—	—	—
Π	Dimensionless constant	—	—	—
ρ	Density	kg/m^3	lb_m/ft^3	ML^{-3}
σ	Surface tension	N/m	lb/ft	Mt^{-2}
σ	Cavitation index	—	—	—
τ	Shear stress	Pa	lb/ft^2	ML^1t^2
ψ	Stream function, two dimensions	m/sec	ft/sec	L^2t^{-1}
ψ	Stokes' stream function	m^3/sec	ft^3/sec	L^3t^{-1}
ω	Angular velocity	rad/sec	rad/sec	t^{-1}

Subscripts

c	Critical condition
u	Unit quantities
c.s.	Control surface
c.v.	Control volume
o	Stagnation or standard state condition
1, 2	Inlet and outlet, respectively, of control volume or machine rotor
∞	Upstream condition far away from body, free stream
T	Total pressure
J	Static pressure

Appendices

Paul Norton

National Renewable Energy Laboratory

A. Properties of Gases and Vapors A-2
B. Properties of Liquids B-35
C. Properties of Solids C-38
D. SI Units D-74
E. Miscellaneous E-75

0-8493-0055-X/00/$0.00+$.50

Appendix A. Properties of Gases and Vapors

TABLE A.1 Properties of Dry Air at Atmospheric Pressure

Symbols and Units:

K = absolute temperature, degrees Kelvin
deg C = temperature, degrees Celsius
deg F = temperature, degrees Fahrenheit
ρ = density, kg/m^3
c_p = specific heat capacity, kJ/kg·K
c_p/c_v = specific heat capacity ratio, dimensionless
μ = viscosity, $N{\cdot}s/m^2 \times 10^6$ (For $N{\cdot}s/m^2$ (= kg/m·s) multiply tabulated values by 10^{-6})
k = thermal conductivity, $W/m{\cdot}k \times 10^3$ (For W/m·K multiply tabulated values by 10^{-3})
Pr = Prandtl number, dimensionless
h = enthalpy, kJ/kg
V_s = sound velocity, m/s

Temperature			Properties							
K	deg C	deg F	ρ	c_p	c_p/c_v	μ	k	Pr	h	V_s
100	−173.15	−280	3.598	1.028		6.929	9.248	.770	98.42	198.4
110	−163.15	−262	3.256	1.022	1.420 2	7.633	10.15	.768	108.7	208.7
120	−153.15	−244	2.975	1.017	1.416 6	8.319	11.05	.766	118.8	218.4
130	−143.15	−226	2.740	1.014	1.413 9	8.990	11.94	.763	129.0	227.6
140	−133.15	−208	2.540	1.012	1.411 9	9.646	12.84	.761	139.1	236.4
150	−123.15	−190	2.367	1.010	1.410 2	10.28	13.73	.758	149.2	245.0
160	−113.15	−172	2.217	1.009	1.408 9	10.91	14.61	.754	159.4	253.2
170	−103.15	−154	2.085	1.008	1.407 9	11.52	15.49	.750	169.4	261.0
180	−93.15	−136	1.968	1.007	1.407 1	12.12	16.37	.746	179.5	268.7
190	−83.15	−118	1.863	1.007	1.406 4	12.71	17.23	.743	189.6	276.2
200	−73.15	−100	1.769	1.006	1.405 7	13.28	18.09	.739	199.7	283.4
205	−68.15	−91	1.726	1.006	1.405 5	13.56	18.52	.738	204.7	286.9
210	−63.15	−82	1.684	1.006	1.405 3	13.85	18.94	.736	209.7	290.5
215	−58.15	−73	1.646	1.006	1.405 0	14.12	19.36	.734	214.8	293.9
220	−53.15	−64	1.607	1.006	1.404 8	14.40	19.78	.732	219.8	297.4
225	−48.15	−55	1.572	1.006	1.404 6	14.67	20.20	.731	224.8	300.8
230	−43.15	−46	1.537	1.006	1.404 4	14.94	20.62	.729	229.8	304.1
235	−38.15	−37	1.505	1.006	1.404 2	15.20	21.04	.727	234.9	307.4
240	−33.15	−28	1.473	1.005	1.404 0	15.47	21.45	.725	239.9	310.6
245	−28.15	−19	1.443	1.005	1.403 8	15.73	21.86	.724	244.9	313.8
250	−23.15	−10	1.413	1.005	1.403 6	15.99	22.27	.722	250.0	317.1
255	−18.15	−1	1.386	1.005	1.403 4	16.25	22.68	.721	255.0	320.2
260	−13.15	8	1.359	1.005	1.403 2	16.50	23.08	.719	260.0	323.4
265	−8.15	17	1.333	1.005	1.403 0	16.75	23.48	.717	265.0	326.5
270	−3.15	26	1.308	1.006	1.402 9	17.00	23.88	.716	270.1	329.6
275	+1.85	35	1.285	1.006	1.402 6	17.26	24.28	.715	275.1	332.6
280	6.85	44	1.261	1.006	1.402 4	17.50	24.67	.713	280.1	335.6
285	11.85	53	1.240	1.006	1.402 2	17.74	25.06	.711	285.1	338.5
290	16.85	62	1.218	1.006	1.402 0	17.98	25.47	.710	290.2	341.5
295	21.85	71	1.197	1.006	1.401 8	18.22	25.85	.709	295.2	344.4
300	26.85	80	1.177	1.006	1.401 7	18.46	26.24	.708	300.2	347.3
305	31.85	89	1.158	1.006	1.401 5	18.70	26.63	.707	305.3	350.2
310	36.85	98	1.139	1.007	1.401 3	18.93	27.01	.705	310.3	353.1
315	41.85	107	1.121	1.007	1.401 0	19.15	27.40	.704	315.3	355.8
320	46.85	116	1.103	1.007	1.400 8	19.39	27.78	.703	320.4	358.7

**Condensed and computed from: "Tables of Thermal Properties of Gases," National Bureau of Standards Circular 564, U.S. Government Printing Office, November 1955.

TABLE A.1 (continued) Properties of Dry Air at Atmospheric Pressure

Temperature			*Properties*							
K	*deg C*	*deg F*	ρ	c_p	c_p/c_v	μ	*k*	*Pr*	*h*	V_s
325	51.85	125	1.086	1.008	1.400 6	19.63	28.15	.702	325.4	361.4
330	56.85	134	1.070	1.008	1.400 4	19.85	28.53	.701	330.4	364.2
335	61.85	143	1.054	1.008	1.400 1	20.08	28.90	.700	335.5	366.9
340	66.85	152	1.038	1.008	1.399 9	20.30	29.28	.699	340.5	369.6
345	71.85	161	1.023	1.009	1.399 6	20.52	29.64	.698	345.6	372.3
350	76.85	170	1.008	1.009	1.399 3	20.75	30.03	.697	350.6	375.0
355	81.85	179	0.994 5	1.010	1.399 0	20.97	30.39	.696	355.7	377.6
360	86.85	188	0.980 5	1.010	1.398 7	21.18	30.78	.695	360.7	380.2
365	91.85	197	0.967 2	1.010	1.398 4	21.38	31.14	.694	365.8	382.8
370	96.85	206	0.953 9	1.011	1.398 1	21.60	31.50	.693	370.8	385.4
375	101.85	215	0.941 3	1.011	1.397 8	21.81	31.86	.692	375.9	388.0
380	106.85	224	0.928 8	1.012	1.397 5	22.02	32.23	.691	380.9	390.5
385	111.85	233	0.916 9	1.012	1.397 1	22.24	32.59	.690	386.0	393.0
390	116.85	242	0.905 0	1.013	1.396 8	22.44	32.95	.690	391.0	395.5
395	121.85	251	0.893 6	1.014	1.396 4	22.65	33.31	.689	396.1	398.0
400	126.85	260	0.882 2	1.014	1.396 1	22.86	33.65	.689	401.2	400.4
410	136.85	278	0.860 8	1.015	1.395 3	23.27	34.35	.688	411.3	405.3
420	146.85	296	0.840 2	1.017	1.394 6	23.66	35.05	.687	421.5	410.2
430	156.85	314	0.820 7	1.018	1.393 8	24.06	35.75	.686	431.7	414.9
440	166.85	332	0.802 1	1.020	1.392 9	24.45	36.43	.684	441.9	419.6
450	176.85	350	0.784 2	1.021	1.392 0	24.85	37.10	.684	452.1	424.2
460	186.85	368	0.767 7	1.023	1.391 1	25.22	37.78	.683	462.3	428.7
470	196.85	386	0.750 9	1.024	1.390 1	25.58	38.46	.682	472.5	433.2
480	206.85	404	0.735 1	1.026	1.389 2	25.96	39.11	.681	482.8	437.6
490	216.85	422	0.720 1	1.028	1.388 1	26.32	39.76	.680	493.0	442.0
500	226.85	440	0.705 7	1.030	1.387 1	26.70	40.41	.680	503.3	446.4
510	236.85	458	0.691 9	1.032	1.386 1	27.06	41.06	.680	513.6	450.6
520	246.85	476	0.678 6	1.034	1.385 1	27.42	41.69	.680	524.0	454.9
530	256.85	494	0.665 8	1.036	1.384 0	27.78	42.32	.680	534.3	459.0
540	266.85	512	0.653 5	1.038	1.382 9	28.14	42.94	.680	544.7	463.2
550	276.85	530	0.641 6	1.040	1.381 8	28.48	43.57	.680	555.1	467.3
560	286.85	548	0.630 1	1.042	1.380 6	28.83	44.20	.680	565.5	471.3
570	296.85	566	0.619 0	1.044	1.379 5	29.17	44.80	.680	575.9	475.3
580	306.85	584	0.608 4	1.047	1.378 3	29.52	45.41	.680	586.4	479.2
590	316.85	602	0.598 0	1.049	1.377 2	29.84	46.01	.680	596.9	483.2
600	326.85	620	0.588 1	1.051	1.376 0	30.17	46.61	.680	607.4	486.9
620	346.85	656	0.569 1	1.056	1.373 7	30.82	47.80	.681	628.4	494.5
640	366.85	692	0.551 4	1.061	1.371 4	31.47	48.96	.682	649.6	502.1
660	386.85	728	0.534 7	1.065	1.369 1	32.09	50.12	.682	670.9	509.4
680	406.85	764	0.518 9	1.070	1.366 8	32.71	51.25	.683	692.2	516.7
700	426.85	800	0.504 0	1.075	1.364 6	33.32	52.36	.684	713.7	523.7
720	446.85	836	0.490 1	1.080	1.362 3	33.92	53.45	.685	735.2	531.0
740	466.85	872	0.476 9	1.085	1.360 1	34.52	54.53	.686	756.9	537.6
760	486.85	908	0.464 3	1.089	1.358 0	35.11	55.62	.687	778.6	544.6
780	506.85	944	0.452 4	1.094	1.355 9	35.69	56.68	.688	800.5	551.2
800	526.85	980	0.441 0	1.099	1.354	36.24	57.74	.689	822.4	557.8
850	576.85	1 070	0.415 2	1.110	1.349	37.63	60.30	.693	877.5	574.1
900	626.85	1 160	0.392 0	1.121	1.345	38.97	62.76	.696	933.4	589.6
950	676.85	1 250	0.371 4	1.132	1.340	40.26	65.20	.699	989.7	604.9
1 000	726.85	1 340	0.352 9	1.142	1.336	41.53	67.54	.702	1 046	619.5
1 100	826.85	1 520	0.320 8	1.161	1.329	43.96			1 162	648.0
1 200	926.85	1 700	0.294 1	1.179	1.322	46.26			1 279	675.2
1 300	1 026.85	1 880	0.271 4	1.197	1.316	48.46			1 398	701.0
1 400	1 126.85	2 060	0.252 1	1.214	1.310	50.57			1 518	725.9
1 500	1 220.85	2 240	0.235 3	1.231	1.304	52.61			1 640	749.4
1 600	1 326.85	2 420	0.220 6	1.249	1.299	54.57			1 764	772.6
1 800	1 526.85	2 780	0.196 0	1.288	1.288	58.29			2 018	815.7
2 000	1 726.85	3 140	0.176 4	1.338	1.274				2 280	855.5
2 400	2 126.85	3 860	0.146 7	1.574	1.238				2 853	924.4
2 800	2 526.85	4 580	0.124 5	2.259	1.196				3 599	983.1

TABLE A.2 Ideal Gas Properties of Nitrogen, Oxygen, and Carbon Dioxide

Symbols and Units:

T = absolute temperature, degrees Kelvin
$\bar{h}$ = enthalpy, kJ/kmol
$\bar{u}$ = internal energy, kJ/kmol
$\bar{s}°$ = absolute entropy at standard reference pressure, kJ/kmol K
[$\bar{h}$ = enthalpy of formation per mole at standard state = 0 kJ/kmol]

Part a. Ideal Gas Properties of Nitrogen, N_2

T	$\bar{h}$	$\bar{u}$	$\bar{s}°$	T	$\bar{h}$	$\bar{u}$	$\bar{s}°$
0	0	0	0	600	17,563	12,574	212.066
220	6,391	4,562	182.639	610	17,864	12,792	212.564
230	6,683	4,770	183.938	620	18,166	13,011	213.055
240	6,975	4,979	185.180	630	18,468	13,230	213.541
250	7,266	5,188	186.370	640	18,772	13,450	214.018
260	7,558	5,396	187.514	650	19,075	13,671	214.489
270	7,849	5,604	188.614	660	19,380	13,892	214.954
280	8,141	5,813	189.673	670	19,685	14,114	215.413
290	8,432	6,021	190.695	680	19,991	14,337	215.866
298	8,669	6,190	191.502	690	20,297	14,560	216.314
300	8,723	6,229	191.682	700	20,604	14,784	216.756
310	9,014	6,437	192.638	710	20,912	15,008	217.192
320	9,306	6,645	193.562	720	21,220	15,234	217.624
330	9,597	6,853	194.459	730	21,529	15,460	218.059
340	9,888	7,061	195.328	740	21,839	15,686	218.472
350	10,180	7,270	196.173	750	22,149	15,913	218.889
360	10,471	7,478	196.995	760	22,460	16,141	219.301
370	10,763	7,687	197.794	770	22,772	16,370	219.709
380	11,055	7,895	198.572	780	23,085	16,599	220.113
390	11,347	8,104	199.331	790	23,398	16,830	220.512
400	11,640	8,314	200.071	800	23,714	17,061	220.907
410	11,932	8,523	200.794	810	24,027	17,292	221.298
420	12,225	8,733	201.499	820	24,342	17,524	221.684
430	12,518	8,943	202.189	830	24,658	17,757	222.067
440	12,811	9,153	202.863	840	24,974	17,990	222.447
450	13,105	9,363	203.523	850	25,292	18,224	222.822
460	13,399	9,574	204.170	860	25,610	18,459	223.194
470	13,693	9,786	204.803	870	25,928	18,695	223.562
480	13,988	9,997	205.424	880	26,248	18,931	223.927
490	14,285	10,210	206.033	890	26,568	19,168	224.288
500	14,581	10,423	206.630	900	26,890	19,407	224.647
510	14,876	10,635	207.216	910	27,210	19,644	225.002
520	15,172	10,848	207.792	920	27,532	19,883	225.353
530	15,469	11,062	208.358	930	27,854	20,122	225.701
540	15,766	11,277	208.914	940	28,178	20,362	226.047
550	16,064	11,492	209.461	950	28,501	20,603	226.389
560	16,363	11,707	209.999	960	28,826	20,844	226.728
570	16,662	11,923	210.528	970	29,151	21,086	227.064
580	16,962	12,139	211.049	980	29,476	21,328	227.398
590	17,26_	12,356	211.562	990	29,803	21,571	227.728

Source: Adapted from M.J. Moran and H.N. Shapiro, *Fundamentals of Engineering Thermodynamics*, 3rd. ed., Wiley, New York, 1995, as presented in K. Wark. *Thermodynamics*, 4th ed., McGraw-Hill, New York, 1983, based on the *JANAF Thermochemical Tables*, NSRDS-NBS-37, 1971.

TABLE A.2 (continued) Ideal Gas Properties of Nitrogen, Oxygen, and Carbon Dioxide

T	$\bar{h}$	$\bar{u}$	$\bar{s}°$	T	$\bar{h}$	$\bar{u}$	$\bar{s}°$
1000	30,129	21,815	228.057	1760	56,227	41.594	247.396
1020	30,784	22,304	228.706	1780	56,938	42,139	247.798
1040	31,442	22,795	229.344	1800	57,651	42,685	248.195
1060	32,101	23,288	229.973	1820	58,363	43,231	248.589
1080	32,762	23,782	230.591	1840	59,075	43,777	248.979
1100	33,426	24,280	231.199	1860	59,790	44,324	249.365
1120	34,092	24,780	231.799	1880	60,504	44,873	249.748
1140	34,760	25,282	232.391	1900	61,220	45,423	250.128
1160	35,430	25,786	232.973	1920	61,936	45,973	250.502
1180	36,104	26,291	233.549	1940	62,654	46,524	250.874
1200	36,777	26,799	234.115	1960	63,381	47,075	251.242
1220	37,452	27,308	234.673	1980	64,090	47,627	251.607
1240	38,129	27,819	235.223	2000	64,810	48,181	251.969
1260	38,807	28,331	235.766	2050	66,612	49,567	252.858
1280	39,488	28,845	236.302	2100	68,417	50,957	253.726
1300	40,170	29,361	236.831	2150	70,226	52,351	254.578
1320	40,853	29,878	237.353	2200	72,040	53,749	255.412
1340	41,539	30,398	237.867	2250	73,856	55,149	256.227
1360	42,227	30,919	238.376	2300	75,676	56,553	257.027
1380	42,915	31,441	238.878	2350	77,496	57,958	257.810
1400	43,605	31,964	239.375	2400	79,320	59,366	258.580
1420	44,295	32,489	239.865	2450	81,149	60,779	259.332
1440	44,988	33,014	240.350	2500	82,981	62,195	260.073
1460	45,682	33,543	240.827	2550	84,814	63,613	260.799
1480	46,377	34,071	241.301	2600	86,650	65,033	261.512
1500	47,073	34,601	241.768	2650	88,488	66,455	262.213
1520	47,771	35,133	242.228	2700	90,328	67,880	262.902
1540	48,470	35,665	242.685	2750	92,171	69,306	263.577
1560	49,168	36,197	243.137	2800	94,014	70,734	264.241
1580	49,869	36,732	243.585	2850	95,859	72,163	264.895
1600	50,571	37,268	244.028	2900	97,705	73,593	265.538
1620	51,275	37,806	244.464	2950	99,556	75,028	266.170
1640	51,980	38,344	244.896	3000	101,407	76,464	266.793
1660	52,686	38,884	245.324	3050	103,260	77,902	267.404
1680	53,393	39,424	245.747	3100	105,115	79,341	268.007
1700	54,099	39,965	246.166	3150	106,972	80,782	268.601
1720	54,807	40,507	246.580	3200	108,830	82,224	269.186
1740	55,516	41,049	246.990	3250	110,690	83,668	269.763

TABLE A.2 (continued) Ideal Gas Properties of Nitrogen, Oxygen, and Carbon Dioxide

Part b. Ideal Gas Properties of Oxygen, O_2

T	$\bar{h}$	$\bar{u}$	$\bar{s}°$	T	$\bar{h}$	$\bar{u}$	$\bar{s}°$
0	0	0	0	600	17,929	12,940	226.346
220	6,404	4,575	196.171	610	18,250	13,178	226.877
230	6,694	4,782	197.461	620	18,572	13,417	227.400
240	6,984	4,989	198.696	630	18,895	13,657	227.918
250	7,275	5,197	199.885	640	19,219	13,898	228.429
260	7,566	5,405	201.027	650	19,544	14,140	228.932
270	7,858	5,613	202.128	660	19,870	14,383	229.430
280	8,150	5,822	203.191	670	20,197	14,626	229.920
290	8,443	6,032	204.218	680	20,524	14,871	230.405
298	8,682	6,203	205.033	690	20,854	15,116	230.885
300	8,736	6,242	205.213	700	21,184	15,364	231.358
310	9,030	6,453	206.177	710	21,514	15,611	231.827
320	9,325	6,664	207.112	720	21,845	15,859	232.291
330	9,620	6,877	208.020	730	22,177	16,107	232.748
340	9,916	7,090	208.904	740	22,510	16,357	233.201
350	10,213	7,303	209.765	750	22,844	16,607	233.649
360	10,511	7,518	210.604	760	23,178	16,859	234.091
370	10,809	7,733	211.423	770	23,513	17,111	234.528
380	11,109	7,949	212.222	780	23,850	17,364	234.960
390	11,409	8,166	213.002	790	24,186	17,618	235.387
400	11,711	8,384	213.765	800	24,523	17,872	235.810
410	12,012	8,603	214.510	810	24,861	18,126	236.230
420	12,314	8,822	215.241	820	25,199	18,382	236.644
430	12,618	9,043	215.955	830	25,537	18,637	237.055
440	12,923	9,264	216.656	840	25,877	18,893	237.462
450	13,228	9,487	217.342	850	26,218	19,150	237.864
460	13,535	9,710	218.016	860	26,559	19,408	238.264
470	13,842	9,935	218.676	870	26,899	19,666	238.660
480	14,151	10,160	219.326	880	27,242	19,925	239.051
490	14,460	10,386	219.963	890	27,584	20,185	239.439
500	14,770	10,614	220.589	900	27,928	20,445	239.823
510	15,082	10,842	221.206	910	28,272	20,706	240.203
520	15,395	11,071	221.812	920	28,616	20,967	240.580
530	15,708	11,301	222.409	930	28,960	21,228	240.953
540	16,022	11,533	222.997	940	29,306	21,491	241.323
550	16,338	11,765	223.576	950	29,652	21,754	241.689
560	16,654	11,998	224.146	960	29,999	22,017	242.052
570	16,971	12,232	224.708	970	30,345	22,280	242.411
580	17,290	12,467	225.262	980	30,692	22,544	242.768
590	17,609	12,703	225.808	990	31,041	22,809	243.120

TABLE A.2 (continued) Ideal Gas Properties of Nitrogen, Oxygen, and Carbon Dioxide

T	$\bar{h}$	$\bar{u}$	$\bar{s}^\circ$	T	$\bar{h}$	$\bar{u}$	$\bar{s}^\circ$
1000	31,389	23,075	243.471	1760	58,880	44,247	263.861
1020	32,088	23,607	244.164	1780	59,624	44,825	264.283
1040	32,789	24,142	244.844	1800	60,371	45,405	264.701
1060	33,490	24,677	245.513	1820	61,118	45,986	265.113
1080	34,194	25,214	246.171	1840	61,866	46,568	265.521
1100	34,899	25,753	246.818	1860	62,616	47,151	265.925
1120	35,606	26,294	247.454	1880	63,365	47,734	266.326
1140	36,314	26,836	248.081	1900	64,116	48,319	266.722
1160	37,023	27,379	248.698	1920	64,868	48,904	267.115
1180	37,734	27,923	249.307	1940	65,620	49,490	267.505
1200	38,447	28,469	249.906	1960	66,374	50,078	267.891
1220	39,162	29,018	250.497	1980	67,127	50,665	268.275
1240	39,877	29,568	251.079	2000	67,881	51,253	268.655
1260	40,594	30,118	251.653	2050	69,772	52,727	269.588
1280	41,312	30,670	252.219	2100	71,668	54,208	270.504
1300	42,033	31,224	252.776	2150	73,573	55,697	271.399
1320	42,753	31,778	253.325	2200	75,484	57,192	272.278
1340	43,475	32,334	253.868	2250	77,397	58,690	273.136
1360	44,198	32,891	254.404	2300	79,316	60,193	273.981
1380	44,923	33,449	254.932	2350	81,243	61,704	274.809
1400	45,648	34,008	255.454	2400	83,174	63,219	275.625
1420	46,374	34,567	255.968	2450	85,112	64,742	276.424
1440	47,102	35,129	256.475	2500	87,057	66,271	277.207
1460	47,831	35,692	256.978	2550	89,004	67,802	277.979
1480	48,561	36,256	257.474	2600	90,956	69,339	278.738
1500	49,292	36,821	257.965	2650	92,916	70,883	279.485
1520	50,024	37,387	258.450	2700	94,881	72,433	280.219
1540	50,756	37,952	258.928	2750	96,852	73,987	280.942
1560	51,490	38,520	259.402	2800	98,826	75,546	281.654
1580	52,224	39,088	259.870	2850	100,808	77,112	282.357
1600	52,961	39,658	260.333	2900	102,793	78,682	283.048
1620	53,696	40,227	260.791	2950	104,785	80,258	283.728
1640	54,434	40,799	261.242	3000	106,780	81,837	284.399
1660	55,172	41,370	261.690	3050	108,778	83,419	285.060
1680	55,912	41,944	262.132	3100	110,784	85,009	285.713
1700	56,652	42,517	262.571	3150	112,795	86,601	286.355
1720	57,394	43,093	263.005	3200	114,809	88,203	286.989
1740	58,136	43,669	263.435	3250	116,827	89,804	287.614

TABLE A.2 (continued) Ideal Gas Properties of Nitrogen, Oxygen, and Carbon Dioxide

Part c. Ideal Gas Properties of Carbon Dioxide, CO_2

T	$\bar{h}$	$\bar{u}$	$\bar{s}^\circ$	T	$\bar{h}$	$\bar{u}$	$\bar{s}^\circ$
0	0	0	0	600	22,280	17,291	243.199
220	6,601	4,772	202.966	610	22,754	17,683	243.983
230	6,938	5,026	204.464	620	23,231	18,076	244.758
240	7,280	5,285	205.920	630	23,709	18,471	245.524
250	7,627	5,548	207.337	640	24,190	18,869	246.282
260	7,979	5,817	208.717	650	24,674	19,270	247.032
270	8,335	6,091	210.062	660	25,160	19,672	247.773
280	8,697	6,369	211.376	670	25,648	20,078	248.507
290	9,063	6,651	212.660	680	26,138	20,484	249.233
298	9,364	6,885	213.685	690	26,631	20,894	249.952
300	9,431	6,939	213.915	700	27,125	21,305	250.663
310	9,807	7,230	215.146	710	27,622	21,719	251.368
320	10,186	7,526	216.351	720	28,121	22,134	252.065
330	10,570	7,826	217.534	730	28,622	22,552	252.755
340	10,959	8,131	218.694	740	29,124	22,972	253.439
350	11,351	8,439	219.831	750	29,629	23,393	254.117
360	11,748	8,752	220.948	760	30,135	23,817	254.787
370	12,148	9,068	222.044	770	30,644	24,242	255.452
380	12,552	9,392	223.122	780	31,154	24,669	256.110
390	12,960	9,718	224.182	790	31,665	25,097	256.762
400	13,372	10,046	225.225	800	32,179	25,527	257.408
410	13,787	10,378	226.250	810	32,694	25,959	258.048
420	14,206	10,714	227.258	820	33,212	26,394	258.682
430	14,628	11,053	228.252	830	33,730	26,829	259.311
440	15,054	11,393	229.230	840	34,251	27,267	259.934
450	15,483	11,742	230.194	850	34,773	27,706	260.551
460	15,916	12,091	231.144	860	35,296	28,125	261.164
470	16,351	12,444	232.080	870	35,821	28,588	261.770
480	16,791	12,800	233.004	880	36,347	29,031	262.371
490	17,232	13,158	233.916	890	36,876	29,476	262.968
500	17,678	13,521	234.814	900	37,405	29,922	263.559
510	18,126	13,885	235.700	910	37,935	30,369	264.146
520	18,576	14,253	236.575	920	38,467	30,818	264.728
530	19,029	14,622	237.439	930	39,000	31,268	265.304
540	19,485	14,996	238.292	940	39,535	31,719	265.877
550	19,945	15,372	239.135	950	40,070	32,171	266.444
560	20,407	15,751	239.962	960	40,607	32,625	267.007
570	20,870	16,131	240.789	970	41,145	33,081	267.566
580	21,337	16,515	241.602	980	41,685	33,537	268.119
590	21,807	16,902	242.405	990	42,226	33,995	268.670

TABLE A.2 (continued) Ideal Gas Properties of Nitrogen, Oxygen, and Carbon Dioxide

T	$\bar{h}$	$\bar{u}$	$\bar{s}^\circ$	T	$\bar{h}$	$\bar{u}$	$\bar{s}^\circ$
1000	42,769	34,455	269.215	1760	86,420	71,787	301.543
1020	43,859	35,378	270.293	1780	87,612	72,812	302.271
1040	44,953	36,306	271.354	1800	88,806	73,840	302.884
1060	46,051	37,238	272.400	1820	90,000	74,868	303.544
1080	47,153	38,174	273.430	1840	91,196	75,897	304.198
1100	48,258	39,112	274.445	1860	92,394	76,929	304.845
1120	49,369	40,057	275.444	1880	93,593	77,962	305.487
1140	50,484	41,006	276.430	1900	94,793	78,996	306.122
1160	51,602	41,957	277.403	1920	95,995	80,031	306.751
1180	52,724	42,913	278.362	1940	97,197	81,067	307.374
1200	53,848	43,871	279.307	1960	98,401	82,105	307.992
1220	54,977	44,834	280.238	1980	99,606	83,144	308.604
1240	56,108	45,799	281.158	2000	100,804	84,185	309.210
1260	57,244	46,768	282.066	2050	103,835	86,791	310.701
1280	58,381	47,739	282.962	2100	106,864	89,404	312.160
1300	59,522	48,713	283.847	2150	109,898	92,023	313.589
1320	60,666	49,691	284.722	2200	112,939	94,648	314.988
1340	61,813	50,672	285.586	2250	115,984	97,277	316.356
1360	62,963	51,656	286.439	2300	119,035	99,912	317.695
1380	64,116	52,643	287.283	2350	122,091	102,552	319.011
1400	65,271	53,631	288.106	2400	125,152	105,197	320.302
1420	66,427	54,621	288.934	2450	128,219	107,849	321.566
1440	67,586	55,614	289.743	2500	131,290	110,504	322.808
1460	68,748	56,609	290.542	2550	134,368	113,166	324.026
1480	69,911	57,606	291.333	2600	137,449	115,832	325.222
1500	71,078	58,606	292.114	2650	140,533	118,500	326.396
1520	72,246	59,609	292.888	2700	143,620	121,172	327.549
1540	73,417	60,613	292.654	2750	146,713	123,849	328.684
1560	74,590	61,620	294.411	2800	149,808	126,528	329.800
1580	76,767	62,630	295.161	2850	152,908	129,212	330.896
1600	76,944	63,741	295.901	2900	156,009	131,898	331.975
1620	78,123	64,653	296.632	2950	159,117	134,589	333.037
1640	79,303	65,668	297.356	3000	162,226	137,283	334.084
1660	80,486	66,592	298.072	3050	165,341	139,982	335.114
1680	81,670	67,702	298.781	3100	168,456	142,681	336.126
1700	82,856	68,721	299.482	3150	171,576	145,385	337.124
1720	84,043	69,742	300.177	3200	174,695	148,089	338.109
1740	85,231	70,764	300.863	3250	177,822	150,801	339.069

TABLE A.3 Psychrometric Table: Properties of Moist Air at 101 325 N/m²

Symbols and Units:

P_s = pressure of water vapor at saturation, N/m²
W_s = humidity ratio at saturation, mass of water vapor associated with unit mass of dry air
V_a = specific volume of dry air, m³/kg
V_s = specific volume of saturated mixture, m³/kg dry air
h_a^a = specific enthalpy of dry air, kJ/kg
h_s = specific enthalpy of saturated mixture, kJ/kg dry air
s_s = specific entropy of saturated mixture, J/K·kg dry air

Temperature			*Properties*						
C	*K*	*F*	P_s	W_s	V_a	V_s	h_a	h_s	s_s
−40	233.15	−40	12.838	0.000 079 25	0.659 61	0.659 68	−22.35	−22.16	−90.659
−30	243.15	−22	37.992	0.000 234 4	0.688 08	0.688 33	−12.29	−11.72	−46.732
−25	248.15	−13	63.248	0.000 390 3	0.702 32	0.702 75	−7.265	−6.306	−24.706
−20	253.15	−4	103.19	0.000 637 1	0.716 49	0.717 24	−2.236	−0.6653	−2.2194
−15	258.15	+5	165.18	0.001 020	0.730 72	0.731 91	+2.794	5.318	21.189
−10	263.15	14	259.72	0.001 606	0.744 95	0.746 83	7.823	11.81	46.104
−5	268.15	23	401.49	0.002 485	0.759 12	0.762 18	12.85	19.04	73.365
0	273.15	32	610.80	0.003 788	0.773 36	0.778 04	17.88	27.35	104.14
5	278.15	41	871.93	0.005 421	0.787 59	0.794 40	22.91	36.52	137.39
10	283.15	50	1 227.2	0.007 658	0.801 76	0.811 63	27.94	47.23	175.54
15	288.15	59	1 704.4	0.010 69	0.816 00	0.829 98	32.97	59.97	220.22
20	293.15	68	2 337.2	0.014 75	0.830 17	0.849 83	38.00	75.42	273.32
25	298.15	77	3 167.0	0.020 16	0.844 34	0.871 62	43.03	94.38	337.39
30	303.15	86	4 242.8	0.027 31	0.858 51	0.896 09	48.07	117.8	415.65
35	308.15	95	5 623.4	0.036 73	0.872 74	0.924 06	53.10	147.3	512.17
40	313.15	104	7 377.6	0.049 11	0.886 92	0.956 65	58.14	184.5	532.31
45	318.15	113	9 584.8	0.065 36	0.901 15	0.995 35	63.17	232.0	783.06
50	323.15	122	12 339	0.086 78	0.915 32	1.042 3	68.21	293.1	975.27
55	328.15	131	15 745	0.115 2	0.929 49	1.100 7	73.25	372.9	1 221.5
60	333.15	140	19 925	0.153 4	0.943 72	1.174 8	78.29	478.5	1 543.5
65	338.15	149	25 014	0.205 5	0.957 90	1.272 1	83.33	621.4	1 973.6
70	343.15	158	31 167	0.278 8	0.972 07	1.404 2	88.38	820.5	2 564.8
75	348.15	167	38 554	0.385 8	0.986 30	1.592 4	93.42	1 110	3 412.8
80	353.15	176	47 365	0.551 9	1.000 5	1.879 1	98.47	1 557	4 710.9
85	358.15	185	57 809	0.836 3	1.014 6	2.363 2	103.5	2 321	6 892.6
90	363.15	194	70 112	1.416	1.028 8	3.340 9	108.6	3 876	11 281

Note: The P_s column in this table gives the vapor pressure of pure water at temperature intervals of five degrees Celsius. For the latest data on vapor pressures at intervals of 0.1 deg C, see "Vapor Pressure Equation for Water," A. Wexler and L. Greenspan, *J. Res. Nat. Bur. Stand.*, 75A(3):213-229, May-June 1971.

* Fpr very low barometric pressures and high wet-bulb temperatures, the values of h_s in this table are somewhat low; for corrections, see "ASHRAE Handbook of Fundamentals."

*Computed from: Psychrometric Tables, in "ASHRAE Handbook of Fundamentals," American Society of Heating, refrigerating and Air-Conditioning Engineers, 1972.

TABLE A.4 Water Vapor at Low Pressures: Perfect Gas Behavior pv/T = R = 0.461 51 kJ/kg·K

Symbols and Units:

t = thermodynamic temperature, deg C
T = thermodynamic temperature, K
pv = RT, kJ/kg
u_o = specific internal energy at zero pressure, kJ/kg
h_o = specific enthalpy at zero pressure, kJ/kg
s_l = specific entropy of semiperfect vapor at 0.1 MN/m², kJ/kg·K
ψ_l = specific Helmholtz free energy of semiperfect vapor at 0.1 MN/m², kJ/kg
ψ_l = specific Helmholtz free energy of semiperfect vapor at 0.1 MN/m², kJ/kg
ζ_l = specific Gibbs free energy of semiperfect vapor at 0.1 MN/m², kJ/kg
p_r = relative pressure, pressure of semiperfect vapor at zero entropy, TN/m²
v_r = relative specific volume, specific volume of semiperfect vapor at zero entropy, mm³/kg
c_{po} = specific heat capacity at constant pressure for zero pressure, kJ/kg·K
c_{vo} = specific heat capacity at constant volume for zero pressure, kJ/kg·K
k = c_{po}/c_{vo} = isentropic exponent, $-(\partial \log p/\partial \log v)_s$

t	T	pv	u_o	h_o	s_l	ψ_l	ζ_l	p_r	v_r	c_{po}	c_{vo}	k
0	273.15	126.06	2 375.5	2 501.5	6.804 2	516.9	643.0	.252 9	498.4	1.858 4	1.396 9	1.330 4
10	283.15	130.68	2 389.4	2 520.1	6.871 1	443.9	574.6	.292 3	447.0	1.860 1	1.398 6	1.330 0
20	293.15	135.29	2 403.4	2 538.7	6.935 7	370.2	505.5	.336 3	402.4	1.862 2	1.400 7	1.329 5
30	303.15	139.91	2 417.5	2 557.4	6.998 2	296.0	435.9	.385 0	363.4	1.864 7	1.403 1	1.328 9
40	313.15	144.52	2 431.5	2 576.0	7.058 7	221.1	365.6	.439 0	329.2	1.867 4	1.405 9	1.328 3
50	323.15	149.14	2 445.6	2 594.7	7.117 5	145.6	294.7	.498 6	299.1	1.870 5	1.409 0	1.327 5
60	333.15	153.75	2 459.7	2 613.4	7.174 5	69.5	223.2	.564 2	272.5	1.873 8	1.412 3	1.326 8
70	343.15	158.37	2 473.8	2 632.2	7.230 0	−7.2	151.2	.636 3	248.9	1.877 4	1.415 9	1.325 9
80	353.15	162.98	2 488.0	2 651.0	7.284 0	−84.3	78.6	.715 2	227.9	1.881 2	1.419 7	1.325 1
90	363.15	167.60	2 502.2	2 669.8	7.336 6	−162.1	5.5	.801 5	209.1	1.885 2	1.423 7	1.324 2
100	373.15	172.21	2 516.5	2 688.7	7.387 8	−240.3	−68.1	.895 7	192.26	1.889 4	1.427 9	1.323 2
120	393.15	181.44	2 545.1	2 726.6	7.486 7	−398.3	−216.8	1.109 7	163.50	1.898 3	1.436 7	1.321 2
140	413.15	190.67	2 573.9	2 764.6	7.581 1	−558.2	−367.5	1.361 7	140.03	1.907 7	1.446 2	1.319 1
160	433.15	199.90	2 603.0	2 802.9	7.671 5	−720.0	−520.1	1.656 4	120.69	1.917 7	1.456 2	1.316 9
180	453.15	209.13	2 632.2	2 841.3	7.758 3	−883.5	−674.4	1.999 1	104.61	1.928 1	1.466 6	1.314 7
200	473.15	218.4	2 661.6	2 880.0	7.841 8	−1 048.7	−830.4	2.396	91.15	1.938 9	1.477 4	1.312 4
300	573.15	264.5	2 812.3	3 076.8	8.218 9	−1 898.4	−1 633.9	5.423	48.77	1.997 5	1.536 0	1.300 5
400	673.15	310.7	2 969.0	3 279.7	8.545 1	−2 783.1	−2 472.5	10.996	28.25	2.061 4	1.599 9	1.288 5
500	773.15	356.8	3 132.4	3 489.2	8.835 2	−3 699	−3 342	20.61	17.310	2.128 7	1.667 2	1.276 8
600	873.15	403.0	3 302.5	3 705.5	9.098 2	−4 642	−4 239	36.45	11.056	2.198 0	1.736 5	1.265 8
700	973.15	449.1	3 479.7	3 928.8	9.340 3	−5 610	−5 161	61.58	7.293	2.268 3	1.806 8	1.255 4
800	1 073.15	495.3	3 663.9	4 159.2	9.565 5	−6 601	−6 106	100.34	4.936	2.338 7	1.877 1	1.245 9
900	1 173.15	541.4	3 855.1	4 396.5	9.776 9	−7 615	−7 073	158.63	3.413	2.407 8	1.946 2	1.237 1
1 000	1 273.15	587.6	4 053.1	4 640.6	9.976 6	−8 649	−8 061	244.5	2.403	2.474 4	2.012 8	1.299 3
1 100	1 373.15	633.7	4 257.5	4 891.2	10.166 1	−9 702	−9 068	368.6	1.719	2.536 9	2.075 4	1.222 4
1 200	1 473.15	679.9	4 467.9	5 147.8	10.346 4	−10 774	−10 094	544.9	1.248	2.593 8	2.132 3	1.216 4
1 300	1 573.15	726.0	4 683.7	5 409.7	10.518 4	−11 863	−11 137	791.0	.918	2.643 1	2.181 6	1.211 5

*Adapted from: "Steam Tables," J. H. Keenan, F. G. Keyes, P. G. Hill, and J. G. Moore, John Wiley & Sons, Inc. 1969 (International Edition - Metric Units).

REFERENCE

For other steam tables in metric units, see "Steam Tables in SI Units," Ministry of Technology, London, 1970.

TABLE A.5 Properties of Saturated Water and Steam

Part a. Temperature Table

Temp. °C	Press. bars	Specific Volume m^3/kg Sat. Liquid $v_f \times 10^3$	Sat. Vapor v_g	Internal Energy kJ/kg Sat. Liquid u_f	Sat. Vapor u_g	Enthalpy kJ/kg Sat. Liquid h_f	Evap. h_{fg}	Sat. Vapor h_g	Entropy kJ/kg · K Sat. Liquid s_f	Sat. Vapor s_g	Temp. °C
.01	0.00611	1.0002	206.136	0.00	2375.3	0.01	2501.3	2501.4	0.0000	9.1562	.01
4	0.00813	1.0001	157.232	16.77	2380.9	16.78	2491.9	2508.7	0.0610	9.0514	4
5	0.00872	1.0001	147.120	20.97	2382.3	20.98	2489.6	2510.6	0.0761	9.0257	5
6	0.00935	1.0001	137.734	25.19	2383.6	25.20	2487.2	2512.4	0.0912	9.0003	6
8	0.01072	1.0002	120.917	33.59	2386.4	33.60	2482.5	2516.1	0.1212	8.9501	8
10	0.01228	1.0004	106.379	42.00	2389.2	42.01	2477.7	2519.8	0.1510	8.9008	10
11	0.01312	1.0004	99.857	46.20	2390.5	46.20	2475.4	2521.6	0.1658	8.8765	11
12	0.01402	1.0005	93.784	50.41	2391.9	50.41	2473.0	2523.4	0.1806	8.8524	12
13	0.01497	1.0007	88.124	54.60	2393.3	54.60	2470.7	2525.3	0.1953	8.8285	13
14	0.01598	1.0008	82.848	58.79	2394.7	58.80	2468.3	2527.1	0.2099	8.8048	14
15	0.01705	1.0009	77.926	62.99	2396.1	62.99	2465.9	2528.9	0.2245	8.7814	15
16	0.01818	1.0011	73.333	67.18	2397.4	67.19	2463.6	2530.8	0.2390	8.7582	16
17	0.01938	1.0012	69.044	71.38	2398.8	71.38	2461.2	2532.6	0.2535	8.7351	17
18	0.02064	1.0014	65.038	75.57	2400.2	75.58	2458.8	2534.4	0.2679	8.7123	18
19	0.02198	1.0016	61.293	79.76	2401.6	79.77	2456.5	2536.2	0.2823	8.6897	19
20	0.02339	1.0018	57.791	83.95	2402.9	83.96	2454.1	2538.1	0.2966	8.6672	20
21	0.02487	1.0020	54.514	88.14	2404.3	88.14	2451.8	2539.9	0.3109	8.6450	21
22	0.02645	1.0022	51.447	92.32	2405.7	92.33	2449.4	2541.7	0.3251	8.6229	22
23	0.02810	1.0024	48.574	96.51	2407.0	96.52	2447.0	2543.5	0.3393	8.6011	23
24	0.02985	1.0027	45.883	100.70	2408.4	100.70	2444.7	2545.4	0.3534	8.5794	24
25	0.03169	1.0029	43.360	104.88	2409.8	104.89	2442.3	2547.2	0.3674	8.5580	25
26	0.03363	1.0032	40.994	109.06	2411.1	109.07	2439.9	2549.0	0.3814	8.5367	26
27	0.03567	1.0035	38.774	113.25	2412.5	113.25	2437.6	2550.8	0.3954	8.5156	27
28	0.03782	1.0037	36.690	117.42	2413.9	117.43	2435.2	2552.6	0.4093	8.4946	28
29	0.04008	1.0040	34.733	121.60	2415.2	121.61	2432.8	2554.5	0.4231	8.4739	29
30	0.04246	1.0043	32.894	125.78	2416.6	125.79	2430.5	2556.3	0.4369	8.4533	30
31	0.04496	1.0046	31.165	129.96	2418.0	129.97	2428.1	2558.1	0.4507	8.4329	31
32	0.04759	1.0050	29.540	134.14	2419.3	134.15	2425.7	2559.9	0.4644	8.4127	32
33	0.05034	1.0053	28.011	138.32	2420.7	138.33	2423.4	2561.7	0.4781	8.3927	33
34	0.05324	1.0056	26.571	142.50	2422.0	142.50	2421.0	2563.5	0.4917	8.3728	34
35	0.05628	1.0060	25.216	146.67	2423.4	146.68	2418.6	2565.3	0.5053	8.3531	35
36	0.05947	1.0063	23.940	150.85	2424.7	150.86	2416.2	2567.1	0.5188	8.3336	36
38	0.06632	1.0071	21.602	159.20	2427.4	159.21	2411.5	2570.7	0.5458	8.2950	38
40	0.07384	1.0078	19.523	167.56	2430.1	167.57	2406.7	2574.3	0.5725	8.2570	40
45	0.09593	1.0099	15.258	188.44	2436.8	188.45	2394.8	2583.2	0.6387	8.1648	45

TABLE A.5 (continued) Properties of Saturated Water and Steam

		Specific Volume m³/kg		Internal Energy kJ/kg		Enthalpy kJ/kg			Entropy kJ/kg · K		
Temp. °C	Press. bars	Sat. Liquid $v_f \times 10^3$	Sat. Vapor v_g	Sat. Liquid u_f	Sat. Vapor u_g	Sat. Liquid h_f	Evap. h_{fg}	Sat. Vapor h_g	Sat. Liquid s_f	Sat. Vapor s_g	Temp. °C
50	.1235	1.0121	12.032	209.32	2443.5	209.33	2382.7	2592.1	.7038	8.0763	50
55	.1576	1.0146	9.568	230.21	2450.1	230.23	2370.7	2600.9	.7679	7.9913	55
60	.1994	1.0172	7.671	251.11	2456.6	251.13	2358.5	2609.6	.8312	7.9096	60
65	.2503	1.0199	6.197	272.02	2463.1	272.06	2346.2	2618.3	.8935	7.8310	65
70	.3119	1.0228	5.042	292.95	2469.6	292.98	2333.8	2626.8	.9549	7.7553	70
75	.3858	1.0259	4.131	313.90	2475.9	313.93	2321.4	2635.3	1.0155	7.6824	75
80	.4739	1.0291	3.407	334.86	2482.2	334.91	2308.8	2643.7	1.0753	7.6122	80
85	.5783	1.0325	2.828	355.84	2488.4	355.90	2296.0	2651.9	1.1343	7.5445	85
90	.7014	1.0360	2.361	376.85	2494.5	376.92	2283.2	2660.1	1.1925	7.4791	90
95	.8455	1.0397	1.982	397.88	2500.6	397.96	2270.2	2668.1	1.2500	7.4159	95
100	1.014	1.0435	1.673	418.94	2506.5	419.04	2257.0	2676.1	1.3069	7.3549	100
110	1.433	1.0516	1.210	461.14	2518.1	461.30	2230.2	2691.5	1.4185	7.2387	110
120	1.985	1.0603	0.8919	503.50	2529.3	503.71	2202.6	2706.3	1.5276	7.1296	120
130	2.701	1.0697	0.6685	546.02	2539.9	546.31	2174.2	2720.5	1.6344	7.0269	130
140	3.613	1.0797	0.5089	588.74	2550.0	589.13	2144.7	2733.9	1.7391	6.9299	140
150	4.758	1.0905	0.3928	631.68	2559.5	632.20	2114.3	2746.5	1.8418	6.8379	150
160	6.178	1.1020	0.3071	674.86	2568.4	675.55	2082.6	2758.1	1.9427	6.7502	160
170	7.917	1.1143	0.2428	718.33	2576.5	719.21	2049.5	2768.7	2.0419	6.6663	170
180	10.02	1.1274	0.1941	762.09	2583.7	763.22	2015.0	2778.2	2.1396	6.5857	180
190	12.54	1.1414	0.1565	806.19	2590.0	807.62	1978.8	2786.4	2.2359	6.5079	190
200	15.54	1.1565	0.1274	850.65	2595.3	852.45	1940.7	2793.2	2.3309	6.4323	200
210	19.06	1.1726	0.1044	895.53	2599.5	897.76	1900.7	2798.5	2.4248	6.3585	210
220	23.18	1.1900	0.08619	940.87	2602.4	943.62	1858.5	2802.1	2.5178	6.2861	220
230	27.95	1.2088	0.07158	986.74	2603.9	990.12	1813.8	2804.0	2.6099	6.2146	230
240	33.44	1.2291	0.05976	1033.2	2604.0	1037.3	1766.5	2803.8	2.7015	6.1437	240
250	39.73	1.2512	0.05013	1080.4	2602.4	1085.4	1716.2	2801.5	2.7927	6.0730	250
260	46.88	1.2755	0.04221	1128.4	2599.0	1134.4	1662.5	2796.6	2.8838	6.0019	260
270	54.99	1.3023	0.03564	1177.4	2593.7	1184.5	1605.2	2789.7	2.9751	5.9301	270
280	64.12	1.3321	0.03017	1227.5	2586.1	1236.0	1543.6	2779.6	3.0668	5.8571	280
290	74.36	1.3656	0.02557	1278.9	2576.0	1289.1	1477.1	2766.2	3.1594	5.7821	290
300	85.81	1.4036	0.02167	1332.0	2563.0	1344.0	1404.9	2749.0	3.2534	5.7045	300
320	112.7	1.4988	0.01549	1444.6	2525.5	1461.5	1238.6	2700.1	3.4480	5.5362	320
340	145.9	1.6379	0.01080	1570.3	2464.6	1594.2	1027.9	2622.0	3.6594	5.3357	340
360	186.5	1.8925	0.006945	1725.2	2351.5	1760.5	720.5	2481.0	3.9147	5.0526	360
374.14	220.9	3.155	0.003155	2029.6	2029.6	2099.3	0	2099.3	4.4298	4.4298	374.14

TABLE A.5 (continued) Properties of Saturated Water and Steam

Part b. Pressure Table

		Specific Volume m³/kg		Internal Energy kJ/kg		Enthalpy kJ/kg			Entropy kJ/kg · K		
Press. bars	Temp. °C	Sat. Liquid $v_f \times 10^3$	Sat. Vapor v_g	Sat. Liquid u_f	Sat. Vapor u_g	Sat. Liquid h_f	Evap. h_{fg}	Sat. Vapor h_g	Sat. Liquid s_f	Sat. Vapor s_g	Press. bars
0.04	28.96	1.0040	34.800	121.45	2415.2	121.46	2432.9	2554.4	0.4226	8.4746	0.04
0.06	36.16	1.0064	23.739	151.53	2425.0	151.53	2415.9	2567.4	0.5210	8.3304	0.06
0.08	41.51	1.0084	18.103	173.87	2432.2	173.88	2403.1	2577.0	0.5926	8.2287	0.08
0.10	45.81	1.0102	14.674	191.82	2437.9	191.83	2392.8	2584.7	0.6493	8.1502	0.10
0.20	60.06	1.0172	7.649	251.38	2456.7	251.40	2358.3	2609.7	0.8320	7.9085	0.20
0.30	69.10	1.0223	5.229	289.20	2468.4	289.23	2336.1	2625.3	0.9439	7.7686	0.30
0.40	75.87	1.0265	3.993	317.53	2477.0	317.58	2319.2	2636.8	1.0259	7.6700	0.40
0.50	81.33	1.0300	3.240	340.44	2483.9	340.49	2305.4	2645.9	1.0910	7.5939	0.50
0.60	85.94	1.0331	2.732	359.79	2489.6	359.86	2293.6	2653.5	1.1453	7.5320	0.60
0.70	89.95	1.0360	2.365	376.63	2494.5	376.70	2283.3	2660.0	1.1919	7.4797	0.70
0.80	93.50	1.0380	2.087	391.58	2498.8	391.66	2274.1	2665.8	1.2329	7.4346	0.80
0.90	96.71	1.0410	1.869	405.06	2502.6	405.15	2265.7	2670.9	1.2695	7.3949	0.90
1.00	99.63	1.0432	1.694	417.36	2506.1	417.46	2258.0	2675.5	1.3026	7.3594	1.00
1.50	111.4	1.0528	1.159	466.94	2519.7	467.11	2226.5	2693.6	1.4336	7.2233	1.50
2.00	120.2	1.0605	0.8857	504.49	2529.5	504.70	2201.9	2706.7	1.5301	7.1271	2.00
2.50	127.4	1.0672	0.7187	535.10	2537.2	535.37	2181.5	2716.9	1.6072	7.0527	2.50
3.00	133.6	1.0732	0.6058	561.15	2543.6	561.47	2163.8	2725.3	1.6718	6.9919	3.00
3.50	138.9	1.0786	0.5243	583.95	2546.9	584.33	2148.1	2732.4	1.7275	6.9405	3.50
4.00	143.6	1.0836	0.4625	604.31	2553.6	604.74	2133.8	2738.6	1.7766	6.8959	4.00
4.50	147.9	1.0882	0.4140	622.25	2557.6	623.25	2120.7	2743.9	1.8207	6.8565	4.50
5.00	151.9	1.0926	0.3749	639.68	2561.2	640.23	2108.5	2748.7	1.8607	6.8212	5.00
6.00	158.9	1.1006	0.3157	669.90	2567.4	670.56	2086.3	2756.8	1.9312	6.7600	6.00
7.00	165.0	1.1080	0.2729	696.44	2572.5	697.22	2066.3	2763.5	1.9922	6.7080	7.00
8.00	170.4	1.1148	0.2404	720.22	2576.8	721.11	2048.0	2769.1	2.0462	6.6628	8.00
9.00	175.4	1.1212	0.2150	741.83	2580.5	742.83	2031.1	2773.9	2.0946	6.6226	9.00
10.0	179.9	1.1273	0.1944	761.68	2583.6	762.81	2015.3	2778.1	2.1387	6.5863	10.0
15.0	198.3	1.1539	0.1318	843.16	2594.5	844.84	1947.3	2792.2	2.3150	6.4448	15.0
20.0	212.4	1.1767	0.09963	906.44	2600.3	908.79	1890.7	2799.5	2.4474	6.3409	20.0
25.0	224.0	1.1973	0.07998	959.11	2603.1	962.11	1841.0	2803.1	2.5547	6.2575	25.0
30.0	233.9	1.2165	0.06668	1004.8	2604.1	1008.4	1795.7	2804.2	2.6457	6.1869	30.0
35.0	242.6	1.2347	0.05707	1045.4	2603.7	1049.8	1753.7	2803.4	2.7253	6.1253	35.0
40.0	250.4	1.2522	0.04978	1082.3	2602.3	1087.3	1714.1	2801.4	2.7964	6.0701	40.0
45.0	257.5	1.2692	0.04406	1116.2	2600.1	1121.9	1676.4	2798.3	2.8610	6.0199	45.0
50.0	264.0	1.2859	0.03944	1147.8	2597.1	1154.2	1640.1	2794.3	2.9202	5.9734	50.0
60.0	275.6	1.3187	0.03244	1205.4	2589.7	1213.4	1571.0	2784.3	3.0267	5.8892	60.0
70.0	285.9	1.3513	0.02737	1257.6	2580.5	1267.0	1505.1	2772.1	3.1211	5.8133	70.0
80.0	295.1	1.3842	0.02352	1305.6	2569.8	1316.6	1441.3	2758.0	3.2068	5.7432	80.0
90.0	303.4	1.4178	0.02048	1350.5	2557.8	1363.3	1378.9	2742.1	3.2858	5.6772	90.0
100.	311.1	1.4524	0.01803	1393.0	2544.4	1407.6	1317.1	2724.7	3.3596	5.6141	100.
110.	318.2	1.4886	0.01599	1433.7	2529.8	1450.1	1255.5	2705.6	3.4295	5.5527	110.
120.	324.8	1.5267	0.01426	1473.0	2513.7	1491.3	1193.6	2684.9	3.4962	5.4924	120.
130.	330.9	1.5671	0.01278	1511.1	2496.1	1531.5	1130.7	2662.2	3.5606	5.4323	130.
140.	336.8	1.6107	0.01149	1548.6	2476.8	1571.1	1066.5	2637.6	3.6232	5.3717	140.
150.	342.2	1.6581	0.01034	1585.6	2455.5	1610.5	1000.0	2610.5	3.6848	5.3098	150.
160.	347.4	1.7107	0.009306	1622.7	2431.7	1650.1	930.6	2580.6	3.7461	5.2455	160.
170.	352.4	1.7702	0.008364	1660.2	2405.0	1690.3	856.9	2547.2	3.8079	5.1777	170.
180.	357.1	1.8397	0.007489	1698.9	2374.3	1732.0	777.1	2509.1	3.8715	5.1044	180.
190.	361.5	1.9243	0.006657	1739.9	2338.1	1776.5	688.0	2464.5	3.9388	5.0228	190.
200.	365.8	2.036	0.005834	1785.6	2293.0	1826.3	583.4	2409.7	4.0139	4.9269	200.
220.9	374.1	3.155	0.003155	2029.6	2029.6	2099.3	0	2099.3	4.4298	4.4298	220.9

Source: Adapted from M.J. Moran and H.N. Shapiro, *Fundamentals of Engineering Thermodynamics*, 3rd. ed., Wiley, New York, 1995, as extracted from J.H. Keenan, F.G. Keyes, P.G. Hill, and J.G. Moore, *Steam Tables*, Wiley, New York, 1969.

TABLE A.6 Properties of Superheated Steam

Symbols and Units:

T = temperature, °C
T_{sat} = Saturation temperature, °C
v = Specific volume, m³/kg
u = internal energy, kJ/kg
h = enthalpy, kJ/kg
S = entropy, kJ/kg·K
p = pressure, bar and μPa

T °C	v m³/kg	u kJ/kg	h kJ/kg	s kJ/kg · K	v m³/kg	u kJ/kg	h kJ/kg	s kJ/kg · K
	p = 0.06 bar = 0.006 MPa (T_{sat} = 36.16°C)				p = 0.35 bar = 0.035 MPa (T_{sat} = 72.69°C)			
Sat.	23.739	2425.0	2567.4	8.3304	4.526	2473.0	2631.4	7.7158
80	27.132	2487.3	2650.1	8.5804	4.625	2483.7	2645.6	7.7564
120	30.219	2544.7	2726.0	8.7840	5.163	2542.4	2723.1	7.9644
160	33.302	2602.7	2802.5	8.9693	5.696	2601.2	2800.6	8.1519
200	36.383	2661.4	2879.7	9.1398	6.228	2660.4	2878.4	8.3237
240	39.462	2721.0	2957.8	9.2982	6.758	2720.3	2956.8	8.4828
280	42.540	2781.5	3036.8	9.4464	7.287	2780.9	3036.0	8.6314
320	45.618	2843.0	3116.7	9.5859	7.815	2842.5	3116.1	8.7712
360	48.696	2905.5	3197.7	9.7180	8.344	2905.1	3197.1	8.9034
400	51.774	2969.0	3279.6	9.8435	8.872	2968.6	3279.2	9.0291
440	54.851	3033.5	3362.6	9.9633	9.400	3033.2	3362.2	9.1490
500	59.467	3132.3	3489.1	10.1336	10.192	3132.1	3488.8	9.3194
	p = 0.70 bar = 0.07 MPa (T_{sat} = 89.95°C)				p = 1.0 bar = 0.10 MPa (T_{sat} = 99.63°C)			
Sat.	2.365	2494.5	2660.0	7.4797	1.694	2506.1	2675.5	7.3594
100	2.434	2509.7	2680.0	7.5341	1.696	2506.7	2676.2	7.3614
120	2.571	2539.7	2719.6	7.6375	1.793	2537.3	2716.6	7.4668
160	2.841	2599.4	2798.2	7.8279	1.984	2597.8	2796.2	7.6597
200	3.108	2659.1	2876.7	8.0012	2.172	2658.1	2875.3	7.8343
240	3.374	2719.3	2955.5	8.1611	2.359	2718.5	2954.5	7.9949
280	3.640	2780.2	3035.0	8.3162	2.546	2779.6	3034.2	8.1445
320	3.905	2842.0	3115.3	8.4504	2.732	2841.5	3114.6	8.2849
360	4.170	2904.6	3196.5	8.5828	2.917	2904.2	3195.9	8.4175
400	4.434	2968.2	3278.6	8.7086	3.103	2967.9	3278.2	8.5435
440	4.698	3032.9	3361.8	8.8286	3.288	3032.6	3361.4	8.6636
500	5.095	3131.8	3488.5	8.9991	3.565	3131.6	3488.1	8.8342
	p = 1.5 bars = 0.15 MPa (T_{sat} = 111.37°C)				p = 3.0 bars = 0.30 MPa (T_{sat} = 133.55°C)			
Sat.	1.159	2519.7	2693.6	7.2233	0.606	2543.6	2725.3	6.9919
120	1.188	2533.3	2711.4	7.2693				
160	1.317	2595.2	2792.8	7.4665	0.651	2587.1	2782.3	7.1276
200	1.444	2656.2	2872.9	7.6433	0.716	2650.7	2865.5	7.3115
240	1.570	2717.2	2952.7	7.8052	0.781	2713.1	2947.3	7.4774
280	1.695	2778.6	3032.8	7.9555	0.844	2775.4	3028.6	7.6299
320	1.819	2840.6	3113.5	8.0964	0.907	2838.1	3110.1	7.7722
360	1.943	2903.5	3195.0	8.2293	0.969	2901.4	3192.2	7.9061
400	2.067	2967.3	3277.4	8.3555	1.032	2965.6	3275.0	8.0330
440	2.191	3032.1	3360.7	8.4757	1.094	3030.6	3358.7	8.1538
500	2.376	3131.2	3487.6	8.6466	1.187	3130.0	3486.0	8.3251
600	2.685	3301.7	3704.3	8.9101	1.341	3300.8	3703.2	8.5892

TABLE A.6 (continued) Properties of Superheated Steam

Symbols and Units:

T = temperature, °C
T_{sat} = Saturation temperature, °C
v = Specific volume, m³/kg
u = internal energy, kJ/kg
h = enthalpy, kJ/kg
S = entropy, kJ/kg·K
p = pressure, bar and μPa

T °C	v m³/kg	u kJ/kg	h kJ/kg	s kJ/kg · K	v m³/kg	u kJ/kg	h kJ/kg	s kJ/kg . k
	p = 5.0 bars = 0.50 MPa (T_{sat} = 151.86°C)				p = 7.0 bars = 0.70 MPa (T_{sat} = 164.97°C)			
Sat.	0.3749	2561.2	2748.7	6.8213	0.2729	2572.5	2763.5	6.7080
180	0.4045	2609.7	2812.0	6.9656	0.2847	2599.8	2799.1	6.7880
200	0.4249	2642.9	2855.4	7.0592	0.2999	2634.8	2844.8	6.8865
240	0.4646	2707.6	2939.9	7.2307	0.3292	2701.8	2932.2	7.0641
280	0.5034	2771.2	3022.9	7.3865	0.3574	2766.9	3017.1	7.2233
320	0.5416	2834.7	3105.6	7.5308	0.3852	2831.3	3100.9	7.3697
360	0.5796	2898.7	3188.4	7.6660	0.4126	2895.8	3184.7	7.5063
400	0.6173	2963.2	3271.9	7.7938	0.4397	2960.9	3268.7	7.6350
440	0.6548	3028.6	3356.0	7.9152	0.4667	3026.6	3353.3	7.7571
500	0.7109	3128.4	3483.9	8.0873	0.5070	3126.8	3481.7	7.9299
600	0.8041	3299.6	3701.7	8.3522	0.5738	3298.5	3700.2	8.1956
700	0.8969	3477.5	3925.9	8.5952	0.6403	3476.6	3924.8	8.4391

T °C	v m³/kg	u kJ/kg	h kJ/kg	s kJ/kg · K	v m³/kg	u kJ/kg	h kJ/kg	s kJ/kg . k
	p = 10.0 bars = 1.0 MPa (T_{sat} = 179.91°C)				p = 15.0 bars = 1.5 MPa (T_{sat} = 198.32°C)			
Sat.	0.1944	2583.6	2778.1	6.5865	0.1318	2594.5	2792.2	6.4448
200	0.2060	2621.9	2827.9	6.6940	0.1325	2598.1	2796.8	6.4546
240	0.2275	2692.9	2920.4	6.8817	0.1483	2676.9	2899.3	6.6628
280	0.2480	2760.2	3008.2	7.0465	0.1627	2748.6	2992.7	6.8381
320	0.2678	2826.1	3093.9	7.1962	0.1765	2817.1	3081.9	6.9938
360	0.2873	2891.6	3178.9	7.3349	0.1899	2884.4	3169.2	7.1363
400	0.3066	2957.3	3263.9	7.4651	0.2030	2951.3	3255.8	7.2690
440	0.3257	3023.6	3349.3	7.5883	0.2160	3018.5	3342.5	7.3940
500	0.3541	3124.4	3478.5	7.7622	0.2352	3120.3	3473.1	7.5698
540	0.3729	3192.6	3565.6	7.8720	0.2478	3189.1	3560.9	7.6805
600	0.4011	3296.8	3697.9	8.0290	0.2668	3293.9	3694.0	7.8385
640	0.4198	3367.4	3787.2	8.1290	0.2793	3364.8	3783.8	7.9391

T °C	v m³/kg	u kJ/kg	h kJ/kg	s kJ/kg · K	v m³/kg	u kJ/kg	h kJ/kg	s kJ/kg . k
	p = 20.0 bars = 2.0 MPa (T_{sat} = 212.42°C)				p = 30.0 bars = 3.0 MPa (T_{sat} = 233.90°C)			
Sat.	0.0996	2600.3	2799.5	6.3409	0.0667	2604.1	2804.2	6.1869
240	0.1085	2659.6	2876.5	6.4952	0.0682	2619.7	2824.3	6.2265
280	0.1200	2736.4	2976.4	6.6828	0.0771	2709.9	2941.3	6.4462
320	0.1308	2807.9	3069.5	6.8452	0.0850	2788.4	3043.4	6.6245
360	0.1411	2877.0	3159.3	6.9917	0.0923	2861.7	3138.7	6.7801
400	0.1512	2945.2	3247.6	7.1271	0.0994	2932.8	3230.9	6.9212
440	0.1611	3013.4	3335.5	7.2540	0.1062	3002.9	3321.5	7.0520
500	0.1757	3116.2	3467.6	7.4317	0.1162	3108.0	3456.5	7.2338
540	0.1853	3185.6	3556.1	7.5434	0.1227	3178.4	3546.6	7.3474
600	0.1996	3290.9	3690.1	7.7024	0.1324	3285.0	3682.3	7.5085
640	0.2091	3362.2	3780.4	7.8035	0.1388	3357.0	3773.5	7.6106
700	0.2232	3470.9	3917.4	7.9487	0.1484	3466.5	3911.7	7.7571

TABLE A.6 (continued) Properties of Superheated Steam

Symbols and Units:

T = temperature, °C
T_{sat} = Saturation temperature, °C
v = Specific volume, m^3/kg
u = internal energy, kJ/kg
h = enthalpy, kJ/kg
S = entropy, kJ/kg·K
p = pressure, bar and μPa

T °C	v m^3/kg	u kJ/kg	h kJ/kg	s kJ/kg · K	v m^3/kg	u kJ/kg	h kJ/kg	s kJ/kg · K
	p = 40 bars = 4.0 MPa (T_{sat} = 250.4°C)				p = 60 bars = 6.0 MPa (T_{sat} = 275.64°C)			
Sat.	0.04978	2602.3	2801.4	6.0701	0.03244	2589.7	2784.3	5.8892
280	0.05546	2680.0	2901.8	6.2568	0.03317	2605.2	2804.2	5.9252
320	0.06199	2767.4	3015.4	6.4553	0.03876	2720.0	2952.6	6.1846
360	0.06788	2845.7	3117.2	6.6215	0.04331	2811.2	3071.1	6.3782
400	0.07341	2919.9	3213.6	6.7690	0.04739	2892.9	3177.2	6.5408
440	0.07872	2992.2	3307.1	6.9041	0.05122	2970.0	3277.3	6.6853
500	0.08643	3099.5	3445.3	7.0901	0.05665	3082.2	3422.2	6.8803
540	0.09145	3171.1	3536.9	7.2056	0.06015	3156.1	3517.0	6.9999
600	0.09885	3279.1	3674.4	7.3688	0.06525	3266.9	3658.4	7.1677
640	0.1037	3351.8	3766.6	7.4720	0.06859	3341.0	3752.6	7.2731
700	0.1110	3462.1	3905.9	7.6198	0.07352	3453.1	3894.1	7.4234
740	0.1157	3536.6	3999.6	7.7141	0.07677	3528.3	3989.2	7.5190

T °C	v m^3/kg	u kJ/kg	h kJ/kg	s kJ/kg · K	v m^3/kg	u kJ/kg	h kJ/kg	s kJ/kg · K
	p = 80 bars = 8.0 MPa (T_{sat} = 295.06°C)				p = 100 bars = 10.0 MPa (T_{sat} = 311.06°C)			
Sat.	0.02352	2569.8	2758.0	5.7432	0.01803	2544.4	2724.7	5.6141
320	0.02682	2662.7	2877.2	5.9489	0.01925	2588.8	2781.3	5.7103
360	0.03089	2772.7	3019.8	6.1819	0.02331	2729.1	2962.1	6.0060
400	0.03432	2863.8	3138.3	6.3634	0.02641	2832.4	3096.5	6.2120
440	0.03742	2946.7	3246.1	6.5190	0.02911	2922.1	3213.2	6.3805
480	0.04034	3025.7	3348.4	6.6586	0.03160	3005.4	3321.4	6.5282
520	0.04313	3102.7	3447.7	6.7871	0.03394	3085.6	3425.1	6.6622
560	0.04582	3178.7	3545.3	6.9072	0.03619	3164.1	3526.0	6.7864
600	0.04845	3254.4	3642.0	7.0206	0.03837	3241.7	3625.3	6.9029
640	0.05102	3330.1	3738.3	7.1283	0.04048	3318.9	3723.7	7.0131
700	0.05481	3443.9	3882.4	7.2812	0.04358	3434.7	3870.5	7.1687
740	0.05729	3520.4	3978.7	7.3782	0.04560	3512.1	3968.1	7.2670

T °C	v m^3/kg	u kJ/kg	h kJ/kg	s kJ/kg · K	v m^3/kg	u kJ/kg	h kJ/kg	s kJ/kg · K
	p = 120 bars = 12.0 MPa (T_{sat} = 324.75°C)				p = 140 bars = 14.0 MPa (T_{sat} = 336.75°C)			
Sat.	0.01426	2513.7	2684.9	5.4924	0.01149	2476.8	2637.6	5.3717
360	0.01811	2678.4	2895.7	5.8361	0.01422	2617.4	2816.5	5.6602
400	0.02108	2798.3	3051.3	6.0747	0.01722	2760.9	3001.9	5.9448
440	0.02355	2896.1	3178.7	6.2586	0.01954	2868.6	3142.2	6.1474
480	0.02576	2984.4	3293.5	6.4154	0.02157	2962.5	3264.5	6.3143
520	0.02781	3068.0	3401.8	6.5555	0.02343	3049.8	3377.8	6.4610
560	0.02977	3149.0	3506.2	6.6840	0.02517	3133.6	3486.0	6.5941
600	0.03164	3228.7	3608.3	6.8037	0.02683	3215.4	3591.1	6.7172
640	0.03345	3307.5	3709.0	6.9164	0.02843	3296.0	3694.1	6.8326
700	0.03610	3425.2	3858.4	7.0749	0.03075	3415.7	3846.2	6.9939
740	0.03781	3503.7	3957.4	7.1746	0.03225	3495.2	3946.7	7.0952

TABLE A.7 Chemical, Physical, and Thermal Properties of Gases: Gases and Vapors, Including Fuels and Refrigerants, English and Metric Units

Common name(s) / *Chemical formula*	*Acetylene (Ethyne)* C_2H_2	*Air [mixture]*	*Ammonia, anhyd.* NH_3	*Argon* *Ar*
Refrigerant number	—	729	717	740
CHEMICAL AND PHYSICAL PROPERTIES				
Molecular weight	26.04	28.966	17.02	39.948
Specific gravity, air = 1	0.90	1.00	0.59	1.38
Specific volume, ft^3/lb	14.9	13.5	23.0	9.80
Specific volume, m^3/kg	0.93	0.842	1.43	0.622
Density of liquid (at atm bp), lb/ft^3	43.0	54.6	42.6	87.0
Density of liquid (at atm bp), kg/m^3	693.	879.	686.	1 400.
Vapor pressure at 25 deg C, psia			145.4	
Vapor pressure at 25 deg C, MN/m^2			1.00	
Viscosity (abs), lbm/ft·sec	6.72×10^{-6}	12.1×10^{-6}	6.72×10^{-6}	13.4×10^{-6}
Viscosity (abs), centipoises[a]	0.01	0.018	0.010	0.02
Sound velocity in gas, m/sec	343	346	415	322
THERMAL AND THERMODYNAMIC PROPERTIES				
Specific heat, c_p, Btu/lb·deg F or cal/g·deg C	0.40	0.240 3	0.52	0.125
Specific heat, c_p, J/kg·K	1 674.	1 005.	2 175.	523.
Specific heat ratio, c_p/c_v	1.25	1.40	1.3	1.67
Gas constant R, ft-lb/lb·deg R	59.3	53.3	90.8	38.7
Gas constant R, J/kg·deg C	319	286.8	488.	208.
Thermal conductivity, Btu/hr·ft·deg F	0.014	0.015 1	0.015	0.010 2
Thermal conductivity, W/m·deg C	0.024	0.026	0.026	0.017 2
Boiling point (sat 14.7 psia), deg F	−103	−320	−28.	−303.
Boiling point (sat 760 mm), deg C	−75	−195	−33.3	−186
Latent heat of evap (at bp), Btu/lb	264	88.2	589.3	70.
Latent heat of evap (at bp), J/kg	614 000	205 000.	1 373 000	163 000
Freezing (melting) point, deg F (1 atm)	−116	−357.2	−107.9	−308.5
Freezing (melting) point, deg C (1 atm)	−82.2	−216.2	−77.7	−189.2
Latent heat of fusion, Btu/lb	23.	10.0	143.0	
Latent heat of fusion, J/kg	53 500	23 200	332 300	
Critical temperature, deg F	97.1	−220.5	271.4	−187.6
Critical temperature, deg C	36.2	−140.3	132.5	−122
Critical pressure, psia	907.	550.	1 650.	707.
Critical pressure, MN/m^2	6.25	3.8	11.4	4.87
Critical volume, ft^3/lb		0.050	0.068	0.029 9
Critical volume, m^3/kg		0.003	0.004 24	0.001 86
Flammable (yes or no)	Yes	No	No	No
Heat of combustion, Btu/ft^3	1 450	—	—	—
Heat of combustion, Btu/lb	21 600	—	—	—
Heat of combustion, kJ/kg	50 200	—	—	—

[a]For N·sec/m^2 divide by 1 000.

Note: The properties of pure gases are given at 25°C (77°F, 298 K) and atmospheric pressure (except as stated).

TABLE A.7 (continued) Chemical, Physical, and Thermal Properties of Gases: Gases and Vapors, Including Fuels and Refrigerants, English and Metric Units

Common name(s)	*Butadiene*	*n-Butane*	*Isobutane (2-Methyl propane)*	*1-Butene (Butylene)*
Chemical formula	C_4H_6	C_4H_{10}	C_4H_{10}	C_4H_8
Refrigerant number	—	600	600a	—
CHEMICAL AND PHYSICAL PROPERTIES				
Molecular weight	54.09	58.12	58.12	56.108
Specific gravity, air = 1	1.87	2.07	2.07	1.94
Specific volume, ft^3/lb	7.1	6.5	6.5	6.7
Specific volume, m^3/kg	0.44	0.405	0.418	0.42
Density of liquid (at atm bp), lb/ft^3		37.5	37.2	
Density of liquid (at atm bp), kg/m^3		604.	599.	
Vapor pressure at 25 deg C, psia		35.4	50.4	
Vapor pressure at 25 deg C, MN/m^2		0.024 4	0.347	
Viscosity (abs), lbm/ft·sec		4.8×10^{-6}		
Viscosity (abs), centipoises[a]		0.007		
Sound velocity in gas, m/sec	226	216	216	222
THERMAL AND THERMODYNAMIC PROPERTIES				
Specific heat, c_p, Btu/lb·deg F or cal/g·deg C	0.341	0.39	0.39	0.36
Specific heat, c_p, J/kg·K	1 427.	1 675.	1 630.	1 505.
Specific heat ratio, c_p/c_v	1.12	1.096	1.10	1.112
Gas constant R, ft-lb/lb·deg F	28.55	26.56	26.56	27.52
Gas constant R, J/kg·deg C	154.	143.	143.	148.
Thermal conductivity, Btu/hr·ft·deg F		0.01	0.01	
Thermal conductivity, W/m·deg C		0.017	0.017	
Boiling point (sat 14.7 psia), deg F	24.1	31.2	10.8	20.6
Boiling point (sat 760 mm), deg C	−4.5	−0.4	−11.8	−6.3
Latent heat of evap (at bp), Btu/lb		165.6	157.5	167.9
Latent heat of evap (at bp), J/kg		386 000	366 000	391 000
Freezing (melting) point, deg F (1 atm)	−164.	−217.	−229	−301.6
Freezing (melting) point, deg C (1 atm)	−109.	−138	−145	−185.3
Latent heat of fusion, Btu/lb		19.2		16.4
Latent heat of fusion, J/kg		44 700		38 100
Critical temperature, deg F		306	273.	291.
Critical temperature, deg C	171.	152.	134.	144.
Critical pressure, psia	652.	550.	537.	621.
Critical pressure, MN/m^2		3.8	3.7	4.28
Critical volume, ft^3/lb		0.070		0.068
Critical volume, m^3/kg		0.004 3		0.004 2
Flammable (yes or no)	Yes	Yes	Yes	Yes
Heat of combustion, Btu/ft^3	2 950	3 300	3 300	3 150
Heat of combustion, Btu/lb	20 900	21 400	21 400	21 000
Heat of combustion, kJ/kg	48 600	49 700	49 700	48 800

[a] For $N \cdot sec/m^2$ divide by 1 000.

TABLE A.7 (continued) Chemical, Physical, and Thermal Properties of Gases: Gases and Vapors, Including Fuels and Refrigerants, English and Metric Units

Common name(s)	*cis-2-Butene*	*trans-2-Butene*	*Isobutene*	*Carbon dioxide*
Chemical formula	C_4H_8	C_4H_8	C_4H_8	CO_2
Refrigerant number	-	—	—	744
CHEMICAL AND PHYSICAL PROPERTIES				
Molecular weight	56.108	56.108	56.108	44.01
Specific gravity, air = 1	1.94	1.94	1.94	1.52
Specific volume, ft³/lb	6.7	6.7	6.7	8.8
Specific volume, m³/kg	0.42	0.42	0.42	0.55
Density of liquid (at atm bp), lb/ft³				—
Density of liquid (at atm bp), kg/m³				—
Vapor pressure at 25 deg C, psia				931.
Vapor pressure at 25 deg C, MN/m²				6.42
Viscosity (abs), lbm/ft·sec				9.4×10^{-6}
Viscosity (abs), centipoises[a]				0.014
Sound velocity in gas, m/sec	223.	221.	221.	270.
THERMAL AND THERMODYNAMIC PROPERTIES				
Specific heat, c_p, Btu/lb·deg F or cal/g·deg C	0.327	0.365	0.37	0.205
Specific heat, c_p, J/kg·K	1 368.	1 527.	1 548.	876.
Specific heat ratio, c_p/c_v	1.121	1.107	1.10	1.30
Gas constant R, ft-lb/lb·deg F				35.1
Gas constant R, J/kg·deg C				189.
Thermal conductivity, Btu/hr·ft·deg F				0.01
Thermal conductivity, W/m·deg C				0.017
Boiling point (sat 14.7 psia), deg F	38.6	33.6	19.2	−109.4[b]
Boiling point (sat 760 mm), deg C	3.7	0.9	−7.1	−78.5
Latent heat of evap (at bp), Btu/lb	178.9	174.4	169.	246.
Latent heat of evap (at bp), J/kg	416 000.	406 000.	393 000.	572 000.
Freezing (melting) point, deg F (1 atm)	−218.	−158.		
Freezing (melting) point, deg C (1 atm)	−138.9	−105.5		
Latent heat of fusion, Btu/lb	31.2	41.6	25.3	—
Latent heat of fusion, J/kg	72 600.	96 800.	58 800.	—
Critical temperature, deg F				88.
Critical temperature, deg C	160.	155.		31.
Critical pressure, psia	595.	610.		1 072.
Critical pressure, MN/m²	4.10	4.20		7.4
Critical volume, ft³/lb				
Critical volume, m³/kg				
Flammable (yes or no)	Yes	Yes	Yes	No
Heat of combustion, Btu/ft³	3 150.	3 150.	3 150.	—
Heat of combustion, Btu/lb	21 000.	21 000.	21 000.	—
Heat of combustion, kJ/kg	48 800.	48 800.	48 800.	—

[a] For N·sec/m² divide by 1 000.
[b] Sublimes.

TABLE A.7 (continued) Chemical, Physical, and Thermal Properties of Gases: Gases and Vapors, Including Fuels and Refrigerants, English and Metric Units

Common name(s)	*Carbon monoxide*	*Chlorine*	*Deuterium*	*Ethane*
Chemical formula	CO	Cl_2	D_2	C_2H_6
Refrigerant number	—	—	—	*170*
CHEMICAL AND PHYSICAL PROPERTIES				
Molecular weight	28.011	70.906	2.014	30.070
Specific gravity, air = 1	0.967	2.45	0.070	1.04
Specific volume, ft^3/lb	14.0	5.52	194.5	13.025
Specific volume, m^3/kg	0.874	0.344	12.12	0.815
Density of liquid (at atm bp), lb/ft^3		97.3		28.
Density of liquid (at atm bp), kg/m^3		1 559.		449.
Vapor pressure at 25 deg C, psia			0.756	
Vapor pressure at 25 deg C, MN/m^2			0.005 2	
Viscosity (abs), lbm/ft·sec	12.1×10^{-6}	9.4×10^{-6}	8.75×10^{-6}	$64. \times 10^{-6}$
Viscosity (abs), centipoises[a]	0.018	0.014	0.013	0.095
Sound velocity in gas, m/sec	352.	215.	930.	316.
THERMAL AND THERMODYNAMIC PROPERTIES				
Specific heat, c_p, Btu/lb·deg F or cal/g·deg C	0.25	0.114	1.73	0.41
Specific heat, c_p, J/kg·K	1 046.	477.	7 238.	1 715.
Specific heat ratio, c_p/c_v	1.40	1.35	1.40	1.20
Gas constant R, ft-lb/lb·deg F	55.2	21.8	384.	51.4
Gas constant R, J/kg·deg C	297.	117.	2 066.	276.
Thermal conductivity, Btu/hr·ft·deg F	0.014	0.005	0.081	0.010
Thermal conductivity, W/m·deg C	0.024	0.008 7	0.140	0.017
Boiling point (sat 14.7 psia), deg F	−312.7	−29.2		−127.
Boiling point (sat 760 mm), deg C	−191.5	−34.		−88.3
Latent heat of evap (at bp), Btu/lb	92.8	123.7		210.
Latent heat of evap (at bp), J/kg	216 000.	288 000.		488 000.
Freezing (melting) point, deg F (1 atm)	−337.	−150.		−278.
Freezing (melting) point, deg C (1 atm)	−205.	−101.		−172.2
Latent heat of fusion, Btu/lb	12.8	41.0		41.
Latent heat of fusion, J/kg		95 400.		95 300.
Critical temperature, deg F	−220.	291.	−390.6	90.1
Critical temperature, deg C	−140.	144.	−234.8	32.2
Critical pressure, psia	507.	1 120.	241.	709.
Critical pressure, MN/m^2	3.49	7.72	1.66	4.89
Critical volume, ft^3/lb	0.053	0.028	0.239	0.076
Critical volume, m^3/kg	0.003 3	0.001 75	0.014 9	0.004 7
Flammable (yes or no)	Yes	No		Yes
Heat of combustion, Btu/ft^3	310.	—		
Heat of combustion, Btu/lb	4 340.	—		22 300.
Heat of combustion, kJ/kg	10 100.	—		51 800.

[a] For $N \cdot sec/m^2$ divide by 1 000.

TABLE A.7 (continued) Chemical, Physical, and Thermal Properties of Gases: Gases and Vapors, Including Fuels and Refrigerants, English and Metric Units

Common name(s)	*Ethyl chloride*	*Ethylene (Ethene)*	*Fluorine*
Chemical formula	C_2H_5Cl	C_2H_4	F_2
Refrigerant number	*160*	*1 150*	—
CHEMICAL AND PHYSICAL PROPERTIES			
Molecular weight	64.515	28.054	37.996
Specific gravity, air = 1	2.23	0.969	1.31
Specific volume, ft^3/lb	6.07	13.9	10.31
Specific volume, m^3/kg	0.378	0.87	0.706
Density of liquid (at atm bp), lb/ft^3	56.5	35.5	
Density of liquid (at atm bp), kg/m^3	905.	569.	
Vapor pressure at 25 deg C, psia			
Vapor pressure at 25 deg C, MN/m^2			
Viscosity (abs), lbm/ft·sec		6.72×10^{-6}	16.1×10^{-6}
Viscosity (abs), centipoises[a]		0.010	0.024
Sound velocity in gas, m/sec	204.	331.	290.
THERMAL AND THERMODYNAMIC PROPERTIES			
Specific heat, c_p, Btu/lb·deg F or cal/g·deg C	0.27	0.37	0.198
Specific heat, c_p, J/kg·K	1 130.	1 548.	828.
Specific heat ratio, c_p/c_v	1.13	1.24	1.35
Gas constant *R*, ft-lb/lb·deg F	24.0	55.1	40.7
Gas constant *R*, J/kg·deg C	129.	296.	219.
Thermal conductivity, Btu/hr·ft·deg F		0.010	0.016
Thermal conductivity, W/m·deg C		0.017	0.028
Boiling point (sat 14.7 psia), deg F	54.	−155.	−306.4
Boiling point (sat 760 mm), deg C	12.2	−103.8	−188.
Latent heat of evap (at bp), Btu/lb	166.	208.	74.
Latent heat of evap (at bp), J/kg	386 000.	484 000.	172 000.
Freezing (melting) point, deg F (1 atm)	−218.	−272.	−364.
Freezing (melting) point, deg C (1 atm)	−138.9	−169.	−220.
Latent heat of fusion, Btu/lb	29.3	51.5	11.
Latent heat of fusion, J/kg	68 100.	120 000.	25 600.
Critical temperature, deg F	368.6	49.	−200
Critical temperature, deg C	187.	9.5	−129.
Critical pressure, psia	764.	741.	810.
Critical pressure, MN/m^2	5.27	5.11	5.58
Critical volume, ft^3/lb	0.049	0.073	
Critical volume, m^3/kg	0.003 06	0.004 6	
Flammable (yes or no)	No	Yes	
Heat of combustion, Btu/ft^3	—	1 480.	
Heat of combustion, Btu/lb	—	20 600.	
Heat of combustion, kJ/kg	—	47 800.	

[a]For N·sec/m^2 divide by 1 000.

TABLE A.7 (continued) Chemical, Physical, and Thermal Properties of Gases: Gases and Vapors, Including Fuels and Refrigerants, English and Metric Units

Common name(s)	*Fluorocarbons*			
Chemical formula	CCl_3F	CCl_2F_2	$CClF_3$	$CBrF_3$
Refrigerant number	*11*	*12*	*13*	*13B1*
CHEMICAL AND PHYSICAL PROPERTIES				
Molecular weight	137.37	120.91	104.46	148.91
Specific gravity, air = 1	4.74	4.17	3.61	5.14
Specific volume, ft^3/lb	2.74	3.12	3.58	2.50
Specific volume, m^3/kg	0.171	0.195	0.224	0.975
Density of liquid (at atm bp), lb/ft^3	92.1	93.0	95.0	124.4
Density of liquid (at atm bp), kg/m^3	1 475.	1 490.	1 522.	1 993.
Vapor pressure at 25 deg C, psia		94.51	516.	234.8
Vapor pressure at 25 deg C, MN/m^2		0.652	3.56	1.619
Viscosity (abs), lbm/ft·sec	7.39×10^{-6}	8.74×10^{-6}		
Viscosity (abs), centipoises[a]	0.011	0.013		
Sound velocity in gas, m/sec				
THERMAL AND THERMODYNAMIC PROPERTIES				
Specific heat, c_p, Btu/lb·deg F or cal/g·deg C	0.14	0.146	0.154	
Specific heat, c_p, J/kg·K	586.	611.	644.	
Specific heat ratio, c_p/c_v	1.14	1.14	1.145	
Gas constant R, ft-lb/lb·deg F				
Gas constant R, J/kg·deg C				
Thermal conductivity, Btu/hr·ft·deg F	0.005	0.006		
Thermal conductivity, W/m·deg C	0.008 7	0.010 4		
Boiling point (sat 14.7 psia), deg F	74.9	−21.8	−114.6	−72.
Boiling point (sat 760 mm), deg C	23.8	−29.9	−81.4	−57.8
Latent heat of evap (at bp), Btu/lb	77.5	71.1	63.0	51.1
Latent heat of evap (at bp), J/kg	180 000.	165 000.	147 000.	119 000.
Freezing (melting) point, deg F (1 atm)	−168.	−252.	−294.	−270.
Freezing (melting) point, deg C (1 atm)	−111.	−157.8	−181.1	−167.8
Latent heat of fusion, Btu/lb				
Latent heat of fusion, J/kg				
Critical temperature, deg F	388.4	233.	83.9	152.
Critical temperature, deg C	198.	111.7	28.8	66.7
Critical pressure, psia	635.	582.	559.	573.
Critical pressure, MN/m^2	4.38	4.01	3.85	3.95
Critical volume, ft^3/lb	0.028 9	0.287	0.027 7	0.021 5
Critical volume, m^3/kg	0.001 80	0.018	0.001 73	0.001 34
Flammable (yes or no)	No	No	No	No
Heat of combustion, Btu/ft^3	—	—	—	—
Heat of combustion, Btu/lb	—	—	—	—
Heat of combustion, kJ/kg	—	—	—	—

[a]For $N \cdot sec/m^2$ divide by 1 000.

TABLE A.7 (continued) Chemical, Physical, and Thermal Properties of Gases: Gases and Vapors, Including Fuels and Refrigerants, English and Metric Units

Common name(s)	*Fluorocarbons*			
Chemical formula	CF_4	$CHCl_2F$	$CHClF_2$	$C_2Cl_2F_4$
Refrigerant number	*14*	*21*	*22*	*114*
CHEMICAL AND PHYSICAL PROPERTIES				
Molecular weight	88.00	102.92	86.468	170.92
Specific gravity, air = 1	3.04	3.55	2.99	5.90
Specific volume, ft^3/lb	4.34	3.7	4.35	2.6
Specific volume, m^3/kg	0.271	0.231	0.271	0.162
Density of liquid (at atm bp), lb/ft^3	102.0	87.7	88.2	94.8
Density of liquid (at atm bp), kg/m^3	1 634.	1 405.	1 413.	1 519.
Vapor pressure at 25 deg C, psia		26.4	151.4	30.9
Vapor pressure at 25 deg C, MN/m^2		0.182	1.044	0.213
Viscosity (abs), lbm/ft·sec		8.06×10^{-6}	8.74×10^{-6}	8.06×10^{-6}
Viscosity (abs), centipoises[a]		0.012	0.013	0.012
Sound velocity in gas, m/sec				
THERMAL AND THERMODYNAMIC PROPERTIES				
Specific heat, c_p, Btu/lb·deg F or cal/g·deg C		0.139	0.157	0.158
Specific heat, c_p, J/kg·K		582.	657.	661.
Specific heat ratio, c_p/c_v		1.18	1.185	1.09
Gas constant R, ft-lb/lb·deg F				
Gas constant R, J/kg·deg C				
Thermal conductivity, Btu/hr·ft·deg F			0.007	0.006
Thermal conductivity, W/m·deg C			0.012	0.010
Boiling point (sat 14.7 psia), deg F	−198.2	48.1	−41.3	38.4
Boiling point (sat 760 mm), deg C	−127.9	9.0	−40.7	3.55
Latent heat of evap (at bp), Btu/lb	58.5	104.1	100.4	58.4
Latent heat of evap (at bp), J/kg	136 000.	242 000.	234 000.	136 000.
Freezing (melting) point, deg F (1 atm)	−299.	−211.	−256.	−137.
Freezing (melting) point, deg C (1 atm)	−183.8	−135.	−160.	−93.8
Latent heat of fusion, Btu/lb	2.53			
Latent heat of fusion, J/kg	5 880.			
Critical temperature, deg F	−49.9	353.3	204.8	294.
Critical temperature, deg C	−45.5	178.5	96.5	
Critical pressure, psia	610.	750.	715.	475.
Critical pressure, MN/m^2	4.21	5.17	4.93	3.28
Critical volume, ft^3/lb	0.025	0.030 7	0.030 5	0.027 5
Critical volume, m^3/kg	0.001 6	0.001 91	0.001 90	0.001 71
Flammable (yes or no)	No	No	No	No
Heat of combustion, Btu/ft^3	—	—	—	—
Heat of combustion, Btu/lb	—	—	—	—
Heat of combustion, kJ/kg	—	—	—	—

[a]For $N \cdot sec/m^2$ divide by 1 000.

TABLE A.7 (continued) Chemical, Physical, and Thermal Properties of Gases: Gases and Vapors, Including Fuels and Refrigerants, English and Metric Units

Common name(s)	*Fluorocarbons*			*Helium*
Chemical formula	C_2ClF_5	$C_2H_3ClF_2$	$C_2H_4F_2$	*He*
Refrigerant number	*115*	*142b*	*152a*	*704*
CHEMICAL AND PHYSICAL PROPERTIES				
Molecular weight	154.47	100.50	66.05	4.002 6
Specific gravity, air = 1	5.33	3.47	2.28	0.138
Specific volume, ft^3/lb	2.44	3.7	5.9	97.86
Specific volume, m^3/kg	0.152	0.231	0.368	6.11
Density of liquid (at atm bp), lb/ft^3	96.5	74.6	62.8	7.80
Density of liquid (at atm bp), kg/m^3	1 546.	1 195.	1 006.	125.
Vapor pressure at 25 deg C, psia	132.1	49.1	86.8	
Vapor pressure at 25 deg C, MN/m^2	0.911	0.338 5	0.596	
Viscosity (abs), lbm/ft·sec				13.4×10^{-6}
Viscosity (abs), centipoises[a]				0.02
Sound velocity in gas, m/sec				1 015.
THERMAL AND THERMODYNAMIC PROPERTIES				
Specific heat, c_p, Btu/lb·deg F or cal/g·deg C	0.161			1.24
Specific heat, c_p, J/kg·K	674.			5 188.
Specific heat ratio, c_p/c_v	1.091			1.66
Gas constant R, ft-lb/lb·deg F				386.
Gas constant R, J/kg·deg C				2 077.
Thermal conductivity, Btu/hr·ft·deg F				0.086
Thermal conductivity, W/m·deg C				0.149
Boiling point (sat 14.7 psia), deg F	−38.0	14.	−13.	−452.
Boiling point (sat 760 mm), deg C	−38.9	−10.0	−25.0	4.22 K
Latent heat of evap (at bp), Btu/lb	53.4	92.5	137.1	10.0
Latent heat of evap (at bp), J/kg	124 000.	215 000.	319 000.	23 300.
Freezing (melting) point, deg F (1 atm)	−149.			[b]
Freezing (melting) point, deg C (1 atm)	−100.6			—
Latent heat of fusion, Btu/lb				—
Latent heat of fusion, J/kg				—
Critical temperature, deg F	176.		387.	−450.3
Critical temperature, deg C				5.2 K
Critical pressure, psia	457.6			33.22
Critical pressure, MN/m^2	3.155			
Critical volume, ft^3/lb	0.026 1			0.231
Critical volume, m^3/kg	0.001 63			0.014 4
Flammable (yes or no)	No	No	No	No
Heat of combustion, Btu/ft^3	—	—	—	—
Heat of combustion, Btu/lb	—	—	—	—
Heat of combustion, kJ/kg	—	—	—	—

[a] For $N \cdot sec/m^2$ divide by 1 000.
[b] Helium cannot be solidified at atmospheric pressure.

TABLE A.7 (continued) Chemical, Physical, and Thermal Properties of Gases: Gases and Vapors, Including Fuels and Refrigerants, English and Metric Units

Common name(s)	*Hydrogen*	*Hydrogen chloride*	*Hydrogen sulfide*	*Krypton*
Chemical formula	H_2	HCl	H_2S	Kr
Refrigerant number	702	—	—	—
CHEMICAL AND PHYSICAL PROPERTIES				
Molecular weight	2.016	36.461	34.076	83.80
Specific gravity, air = 1	0.070	1.26	1.18	2.89
Specific volume, ft^3/lb	194.	10.74	11.5	4.67
Specific volume, m^3/kg	12.1	0.670	0.093 0	0.291
Density of liquid (at atm bp), lb/ft^3	4.43	74.4	62.	150.6
Density of liquid (at atm bp), kg/m^3	71.0	1 192.	993.	2 413.
Vapor pressure at 25 deg C, psia				
Vapor pressure at 25 deg C, MN/m^2				
Viscosity (abs), lbm/ft·sec	6.05×10^{-6}	10.1×10^{-6}	8.74×10^{-6}	16.8×10^{-6}
Viscosity (abs), centipoises[a]	0.009	0.015	0.013	0.025
Sound velocity in gas, m/sec	1 315.	310.	302.	223.
THERMAL AND THERMODYNAMIC PROPERTIES				
Specific heat, c_p, Btu/lb·deg F or cal/g·deg C	3.42	0.194	0.23	0.059
Specific heat, c_p, J/kg·K	14 310.	812.	962.	247.
Specific heat ratio, c_p/c_v	1.405	1.39	1.33	1.68
Gas constant R, ft-lb/lb·deg F	767.	42.4	45.3	18.4
Gas constant R, J/kg·deg C	4 126.	228.	244.	99.0
Thermal conductivity, Btu/hr·ft·deg F	0.105	0.008	0.008	0.005 4
Thermal conductivity, W/m·deg C	0.018 2	0.014	0.014	0.009 3
Boiling point (sat 14.7 psia), deg F	−423.	−121.	−76.	−244.
Boiling point (sat 760 mm), deg C	20.4 K	−85.	−60.	−153.
Latent heat of evap (at bp), Btu/lb	192.	190.5	234.	46.4
Latent heat of evap (at bp), J/kg	447 000.	443 000.	544 000.	108 000.
Freezing (melting) point, deg F (1 atm)	−434.6	−169.6	−119.2	−272.
Freezing (melting) point, deg C (1 atm)	−259.1	−112.	−84.	−169.
Latent heat of fusion, Btu/lb	25.0	23.4	30.2	4.7
Latent heat of fusion, J/kg	58 000.	54 400.	70 200.	10 900.
Critical temperature, deg F	−399.8	124.	213.	
Critical temperature, deg C	−240.0	51.2	100.4	−63.8
Critical pressure, psia	189.	1 201.	1 309.	800.
Critical pressure, MN/m^2	1.30	8.28	9.02	5.52
Critical volume, ft^3/lb	0.53	0.038	0.046	0.017 7
Critical volume, m^3/kg	0.033	0.002 4	0.002 9	0.001 1
Flammable (yes or no)	Yes	No	Yes	No
Heat of combustion, Btu/ft^3	320.	—	700.	—
Heat of combustion, Btu/lb	62 050.	—	8 000.	—
Heat of combustion, kJ/kg	144 000.	—	18 600.	—

[a]For $N \cdot sec/m^2$ divide by 1 000.

TABLE A.7 (continued) Chemical, Physical, and Thermal Properties of Gases: Gases and Vapors, Including Fuels and Refrigerants, English and Metric Units

Common name(s)	*Methane*	*Methyl chloride*	*Neon*	*Nitric oxide*
Chemical formula	CH_4	CH_3Cl	*Ne*	*NO*
Refrigerant number	*50*	*40*	*720*	—
CHEMICAL AND PHYSICAL PROPERTIES				
Molecular weight	16.044	50.488	20.179	30.006
Specific gravity, air = 1	0.554	1.74	0.697	1.04
Specific volume, ft³/lb	24.2	7.4	19.41	13.05
Specific volume, m³/kg	1.51	0.462	1.211	0.814
Density of liquid (at atm bp), lb/ft³	26.3	62.7	75.35	
Density of liquid (at atm bp), kg/m³	421.	1 004.	1 207.	
Vapor pressure at 25 deg C, psia		82.2		
Vapor pressure at 25 deg C, MN/m²		0.567		
Viscosity (abs), lbm/ft·sec	7.39×10^{-6}	7.39×10^{-6}	21.5×10^{-6}	12.8×10^{-6}
Viscosity (abs), centipoises[a]	0.011	0.011	0.032	0.019
Sound velocity in gas, m/sec	446.	251.	454.	341.
THERMAL AND THERMODYNAMIC PROPERTIES				
Specific heat, c_p, Btu/lb·deg F or cal/g·deg C	0.54	0.20	0.246	0.235
Specific heat, c_p, J/kg·K	2 260.	837.	1 030.	983.
Specific heat ratio, c_p/c_v	1.31	1.28	1.64	1.40
Gas constant R, ft-lb/lb·deg F	96.	30.6	76.6	51.5
Gas constant R, J/kg·deg C	518.	165.	412.	277.
Thermal conductivity, Btu/hr·ft·deg F	0.02	0.006	0.028	0.015
Thermal conductivity, W/m·deg C	0.035	0.010	0.048	0.026
Boiling point (sat 14.7 psia), deg F	−259.	−10.7	−410.9	−240.
Boiling point (sat 760 mm), deg C	−434.2	−23.7	−246.	−151.5
Latent heat of evap (at bp), Btu/lb	219.2	184.1	37.	
Latent heat of evap (at bp), J/kg	510 000.	428 000.	86 100.	
Freezing (melting) point, deg F (1 atm)	−296.6	−144.	−415.6	−258.
Freezing (melting) point, deg C (1 atm)	−182.6	−97.8	−248.7	−161.
Latent heat of fusion, Btu/lb	14.	56.	6.8	32.9
Latent heat of fusion, J/kg	32 600.	130 000.	15 800.	76 500.
Critical temperature, deg F	−116.	289.4	−379.8	−136.
Critical temperature, deg C	−82.3	143.	−228.8	−93.3
Critical pressure, psia	673.	968.	396.	945.
Critical pressure, MN/m²	4.64	6.67	2.73	6.52
Critical volume, ft³/lb	0.099	0.043	0.033	0.033 2
Critical volume, m³/kg	0.006 2	0.002 7	0.002 0	0.002 07
Flammable (yes or no)	Yes	Yes	No	No
Heat of combustion, Btu/ft³	985.		—	—
Heat of combustion, Btu/lb	2 290.		—	—
Heat of combustion, kJ/kg			—	—

[a]For N·sec/m² divide by 1 000.

TABLE A.7 (continued) Chemical, Physical, and Thermal Properties of Gases: Gases and Vapors, Including Fuels and Refrigerants, English and Metric Units

Common name(s)	*Nitrogen*	*Nitrous oxide*	*Oxygen*	*Ozone*
Chemical formula	N_2	N_2O	O_2	O_3
Refrigerant number	*728*	*744A*	*732*	—
CHEMICAL AND PHYSICAL PROPERTIES				
Molecular weight	28.013 4	44.012	31.998 8	47.998
Specific gravity, air = 1	0.967	1.52	1.105	1.66
Specific volume, ft^3/lb	13.98	8.90	12.24	8.16
Specific volume, m^3/kg	0.872	0.555	0.764	0.509
Density of liquid (at atm bp), lb/ft^3	50.46	76.6	71.27	
Density of liquid (at atm bp), kg/m^3	808.4	1 227.	1 142.	
Vapor pressure at 25 deg C, psia				
Vapor pressure at 25 deg C, MN/m^2				
Viscosity (abs), lbm/ft·sec	12.1×10^{-6}	10.1×10^{-6}	13.4×10^{-6}	8.74×10^{-6}
Viscosity (abs), centipoises[a]	0.018	0.015	0.020	0.013
Sound velocity in gas, m/sec	353.	268.	329.	
THERMAL AND THERMODYNAMIC PROPERTIES				
Specific heat, c_p, Btu/lb·deg F or cal/g·deg C	0.249	0.21	0.220	0.196
Specific heat, c_p, J/kg·K	1 040.	879.	920.	820.
Specific heat ratio, c_p/c_v	1.40	1.31	1.40	
Gas constant R, ft-lb/lb·deg F	55.2	35.1	48.3	32.2
Gas constant R, J/kg·deg C	297.	189.	260.	173.
Thermal conductivity, Btu/hr·ft·deg F	0.015	0.010	0.015	0.019
Thermal conductivity, W/m·deg C	0.026	0.017	0.026	0.033
Boiling point (sat 14.7 psia), deg F	−320.4	−127.3	−297.3	−170.
Boiling point (sat 760 mm), deg C	−195.8	−88.5	−182.97	−112.
Latent heat of evap (at bp), Btu/lb	85.5	161.8	91.7	
Latent heat of evap (at bp), J/kg	199 000.	376 000.	213 000.	
Freezing (melting) point, deg F (1 atm)	−346.	−131.5	−361.1	−315.5
Freezing (melting) point, deg C (1 atm)	−210.	−90.8	−218.4	−193.
Latent heat of fusion, Btu/lb	11.1	63.9	5.9	97.2
Latent heat of fusion, J/kg	25 800.	149 000.	13 700.	226 000.
Critical temperature, deg F	−232.6	97.7	−181.5	16.
Critical temperature, deg C	−147.	36.5	−118.6	−9.
Critical pressure, psia	493.	1 052.	726.	800.
Critical pressure, MN/m^2	3.40	7.25	5.01	5.52
Critical volume, ft^3/lb	0.051	0.036	0.040	0.029 8
Critical volume, m^3/kg	0.003 18	0.002 2	0.002 5	0.001 86
Flammable (yes or no)	No	No	No	No
Heat of combustion, Btu/ft^3	—	—	—	—
Heat of combustion, Btu/lb	—	—	—	—
Heat of combustion, kJ/kg	—	—	—	—

[a]For N·sec/m^2 divide by 1 000.

TABLE A.7 (continued) Chemical, Physical, and Thermal Properties of Gases: Gases and Vapors, Including Fuels and Refrigerants, English and Metric Units

Common name(s)	*Propane*	*Propylene (Propene)*	*Sulfur dioxide*	*Xenon*
Chemical formula	C_3H_8	C_3H_6	SO_2	*Xe*
Refrigerant number	*290*	*1 270*	*764*	—
CHEMICAL AND PHYSICAL PROPERTIES				
Molecular weight	44.097	42.08	64.06	131.30
Specific gravity, air = 1	1.52	1.45	2.21	4.53
Specific volume, ft³/lb	8.84	9.3	6.11	2.98
Specific volume, m³/kg	0.552	0.58		
Density of liquid (at atm bp), lb/ft³	36.2	37.5	42.8	190.8
Density of liquid (at atm bp), kg/m³	580.	601.	585.	3 060.
Vapor pressure at 25 deg C, psia	135.7	166.4	56.6	
Vapor pressure at 25 deg C, MN/m²	0.936	1.147	0.390	
Viscosity (abs), lbm/ft·sec	53.8×10^{-6}	57.1×10^{-6}	8.74×10^{-6}	15.5×10^{-6}
Viscosity (abs), centipoises[a]	0.080	0.085	0.013	0.023
Sound velocity in gas, m/sec	253.	261.	220.	177.
THERMAL AND THERMODYNAMIC PROPERTIES				
Specific heat, c_p, Btu/lb·deg F or cal/g·deg C	0.39	0.36	0.11	0.115
Specific heat, c_p, J/kg·K	1 630.	1 506.	460.	481.
Specific heat ratio, c_p/c_v	1.2	1.16	1.29	1.67
Gas constant *R*, ft-lb/lb·deg F	35.0	36.7	24.1	11.8
Gas constant *R*, J/kg·deg C	188.	197.	130.	63.5
Thermal conductivity, Btu/hr·ft·deg F	0.010	0.010	0.006	0.003
Thermal conductivity, W/m·deg C	0.017	0.017	0.010	0.005 2
Boiling point (sat 14.7 psia), deg F	−44.	−54.	14.0	−162.5
Boiling point (sat 760 mm), deg C	−42.2	−48.3	−10.	−108.
Latent heat of evap (at bp), Btu/lb	184.	188.2	155.5	41.4
Latent heat of evap (at bp), J/kg	428 000.	438 000.	362 000.	96 000.
Freezing (melting) point, deg F (1 atm)	−309.8	−301.	−104.	−220.
Freezing (melting) point, deg C (1 atm)	−189.9	−185.	−75.5	−140.
Latent heat of fusion, Btu/lb	19.1		58.0	10.
Latent heat of fusion, J/kg	44 400.		135 000.	23 300.
Critical temperature, deg F	205.	197.	315.5	61.9
Critical temperature, deg C	96.	91.7	157.6	16.6
Critical pressure, psia	618.	668.	1 141.	852.
Critical pressure, MN/m²	4.26	4.61	7.87	5.87
Critical volume, ft³/lb	0.073	0.069	0.03	0.014 5
Critical volume, m³/kg	0.004 5	0.004 3	0.001 9	0.000 90
Flammable (yes or no)	Yes	Yes	No	No
Heat of combustion, Btu/ft³	2 450.	2 310.	—	—
Heat of combustion, Btu/lb	21 660.	21 500.	—	—
Heat of combustion, kJ/kg	50 340.	50 000.	—	—

[a]For N·sec/m² divide by 1 000.

TABLE A.8 Ideal Gas Properties of Air

Part a. SI Units

T(K), h and u(kJ/kg), s°(kJ/kg·K)

T	h	p_r	u	v_r	s°	T	h	p_r	u	v_r	s°
200	199.97	0.3363	142.56	1707.	1.29559	450	451.80	5.775	322.62	223.6	2.11161
210	209.97	0.3987	149.69	1512.	1.34444	460	462.02	6.245	329.97	211.4	2.13407
220	219.97	0.4690	156.82	1346.	1.39105	470	472.24	6.742	337.32	200.1	2.15604
230	230.02	0.5477	164.00	1205.	1.43557	480	482.49	7.268	344.70	189.5	2.17760
240	240.02	0.6355	171.13	1084.	1.47824	490	492.74	7.824	352.08	179.7	2.19876
250	250.05	0.7329	178.28	979.	1.51917	500	503.02	8.411	359.49	170.6	2.21952
260	260.09	0.8405	185.45	887.8	1.55848	510	513.32	9.031	366.92	162.1	2.23993
270	270.11	0.9590	192.60	808.0	1.59634	520	523.63	9.684	374.36	154.1	2.25997
280	280.13	1.0889	199.75	738.0	1.63279	530	533.98	10.37	381.84	146.7	2.27967
285	285.14	1.1584	203.33	706.1	1.65055	540	544.35	11.10	389.34	139.7	2.29906
290	290.16	1.2311	206.91	676.1	1.66802	550	554.74	11.86	396.86	133.1	2.31809
295	295.17	1.3068	210.49	647.9	1.68515	560	565.17	12.66	404.42	127.0	2.33685
300	300.19	1.3860	214.07	621.2	1.70203	570	575.59	13.50	411.97	121.2	2.35531
305	305.22	1.4686	217.67	596.0	1.71865	580	586.04	14.38	419.55	115.7	2.37348
310	310.24	1.5546	221.25	572.3	1.73498	590	596.52	15.31	427.15	110.6	2.39140
315	315.27	1.6442	224.85	549.8	1.75106	600	607.02	16.28	434.78	105.8	2.40902
320	320.29	1.7375	228.42	528.6	1.76690	610	617.53	17.30	442.42	101.2	2.42644
325	325.31	1.8345	232.02	508.4	1.78249	620	628.07	18.36	450.09	96.92	2.44356
330	330.34	1.9352	235.61	489.4	1.79783	630	638.63	19.84	457.78	92.84	2.46048
340	340.42	2.149	242.82	454.1	1.82790	640	649.22	20.64	465.50	88.99	2.47716
350	350.49	2.379	250.02	422.2	1.85708	650	659.84	21.86	473.25	85.34	2.49364
360	360.58	2.626	257.24	393.4	1.88543	660	670.47	23.13	481.01	81.89	2.50985
370	370.67	2.892	264.46	367.2	1.91313	670	681.14	24.46	488.81	78.61	2.52589
380	380.77	3.176	271.69	343.4	1.94001	680	691.82	25.85	496.62	75.50	2.54175
390	390.88	3.481	278.93	321.5	1.96633	690	702.52	27.29	504.45	72.56	2.55731
400	400.98	3.806	286.16	301.6	1.99194	700	713.27	28.80	512.33	69.76	2.57277
410	411.12	4.153	293.43	283.3	2.01699	710	724.04	30.38	520.23	67.07	2.58810
420	421.26	4.522	300.69	266.6	2.04142	720	734.82	32.02	528.14	64.53	2.60319
430	431.43	4.915	307.99	251.1	2.06533	730	745.62	33.72	536.07	62.13	2.61803
440	441.61	5.332	315.30	236.8	2.08870	740	756.44	35.50	544.02	59.82	2.63280

TABLE A.8 (continued) Ideal Gas Properties of Air

T(K), h and u(kJ/kg), $s°$(kJ/kg·K)

T	h	p_r	u	v_r	$s°$	T	h	p_r	u	v_r	$s°$
750	767.29	37.35	551.99	57.63	2.64737	1300	1395.97	330.9	1022.82	11.275	3.27345
760	778.18	39.27	560.01	55.54	2.66176	1320	1419.76	352.5	1040.88	10.747	3.29160
770	789.11	41.31	568.07	53.39	2.67595	1340	1443.60	375.3	1058.94	10.247	3.30959
780	800.03	43.35	576.12	51.64	2.69013	1360	1467.49	399.1	1077.10	9.780	3.32724
790	810.99	45.55	584.21	49.86	2.70400	1380	1491.44	424.2	1095.26	9.337	3.34474
800	821.95	47.75	592.30	48.08	2.71787	1400	1515.42	450.5	1113.52	8.919	3.36200
820	843.98	52.59	608.59	44.84	2.74504	1420	1539.44	478.0	1131.77	8.526	3.37901
840	866.08	57.60	624.95	41.85	2.77170	1440	1563.51	506.9	1150.13	8.153	3.39586
860	888.27	63.09	641.40	39.12	2.79783	1460	1587.63	537.1	1168.49	7.801	3.41247
880	910.56	68.98	657.95	36.61	2.82344	1480	1611.79	568.8	1186.95	7.468	3.42892
900	932.93	75.29	674.58	34.31	2.84856	1500	1635.97	601.9	1205.41	7.152	3.44516
920	955.38	82.05	691.28	32.18	2.87324	1520	1660.23	636.5	1223.87	6.854	3.46120
940	977.92	89.28	708.08	30.22	2.89748	1540	1684.51	672.8	1242.43	6.569	3.47712
960	1000.55	97.00	725.02	28.40	2.92128	1560	1708.82	710.5	1260.99	6.301	3.49276
980	1023.25	105.2	741.98	26.73	2.94468	1580	1733.17	750.0	1279.65	6.046	3.50829
1000	1046.04	114.0	758.94	25.17	2.96770	1600	1757.57	791.2	1298.30	5.804	3.52364
1020	1068.89	123.4	776.10	23.72	2.99034	1620	1782.00	834.1	1316.96	5.574	3.53879
1040	1091.85	133.3	793.36	22.39	3.01260	1640	1806.46	878.9	1335.72	5.355	3.55381
1060	1114.86	143.0	810.62	21.14	3.03449	1660	1830.96	925.6	1354.48	5.147	3.56867
1080	1137.89	155.2	827.88	19.98	3.05608	1680	1855.50	974.2	1373.24	4.949	3.58335
1100	1161.07	167.1	845.33	18.896	3.07732	1700	1880.1	1025	1392.7	4.761	3.5979
1120	1184.28	179.7	862.79	17.886	3.09825	1750	1941.6	1161	1439.8	4.328	3.6336
1140	1207.57	193.1	880.35	16.946	3.11883	1800	2003.3	1310	1487.2	3.944	3.6684
1160	1230.92	207.2	897.91	16.064	3.13916	1850	2065.3	1475	1534.9	3.601	3.7023
1180	1254.34	222.2	915.57	15.241	3.15916	1900	2127.4	1655	1582.6	3.295	3.7354
1200	1277.79	238.0	933.33	14.470	3.17888	1950	2189.7	1852	1630.6	3:022	3.7677
1220	1301.31	254.7	951.09	13.747	3.19834	2000	2252.1	2068	1678.7	2.776	3.7994
1240	1324.93	272.3	968.95	13.069	3.21751	2050	2314.6	2303	1726.8	2.555	3.8303
1260	1348.55	290.8	986.90	12.435	3.23638	2100	2377.4	2559	1775.3	2.356	3.8605
1280	1372.24	310.4	1004.76	11.835	3.25510	2150	2440.3	2837	1823.8	2.175	3.8901
						2200	2503.2	3138	1872.4	2.012	3.9191
						2250	2566.4	3464	1921.3	1.864	3.9474

TABLE A.8 (continued) Ideal Gas Properties of Air

Part b. English Units

T(°R), h and u(Btu/lb), s°(Btu/lb · °R)

T	h	p_r	u	v_r	s°	T	h	p_r	u	v_r	s°
360	85.97	0.3363	61.29	396.6	0.50369	940	226.11	9.834	161.68	35.41	0.73509
380	90.75	0.4061	64.70	346.6	0.51663	960	231.06	10.61	165.26	33.52	0.74030
400	95.53	0.4858	68.11	305.0	0.52890	980	236.02	11.43	168.83	31.76	0.74540
420	100.32	0.5760	71.52	270.1	0.54058	1000	240.98	12.30	172.43	30.12	0.75042
440	105.11	0.6776	74.93	240.6	0.55172	1040	250.95	14.18	179.66	27.17	0.76019
460	109.90	0.7913	78.36	215.33	0.56235	1080	260.97	16.28	186.93	24.58	0.76964
480	114.69	0.9182	81.77	193.65	0.57255	1120	271.03	18.60	194.25	22.30	0.77880
500	119.48	1.0590	85.20	174.90	0.58233	1160	281.14	21.18	201.63	20.29	0.78767
520	124.27	1.2147	88.62	158.58	0.59172	1200	291.30	24.01	209.05	18.51	0.79628
537	128.34	1.3593	91.53	146.34	0.59945	1240	301.52	27.13	216.53	16.93	0.80466
540	129.06	1.3860	92.04	144.32	0.60078	1280	311.79	30.55	224.05	15.52	0.81280
560	133.86	1.5742	95.47	131.78	0.60950	1320	322.11	34.31	231.63	14.25	0.82075
580	138.66	1.7800	98.90	120.70	0.61793	1360	332.48	38.41	239.25	13.12	0.82848
600	143.47	2.005	102.34	110.88	0.62607	1400	342.90	42.88	246.93	12.10	0.83604
620	148.28	2.249	105.78	102.12	0.63395	1440	353.37	47.75	254.66	11.17	0.84341
640	153.09	2.514	109.21	94.30	0.64159	1480	363.89	53.04	262.44	10.34	0.85062
660	157.92	2.801	112.67	87.27	0.64902	1520	374.47	58.78	270.26	9.578	0.85767
680	162.73	3.111	116.12	80.96	0.65621	1560	385.08	65.00	278.13	8.890	0.86456
700	167.56	3.446	119.58	75.25	0.66321	1600	395.74	71.73	286.06	8.263	0.87130
720	172.39	3.806	123.04	70.07	0.67002	1650	409.13	80.89	296.03	7.556	0.87954
740	177.23	4.193	126.51	65.38	0.67665	1700	422.59	90.95	306.06	6.924	0.88758
760	182.08	4.607	129.99	61.10	0.68312	1750	436.12	101.98	316.16	6.357	0.89542
780	186.94	5.051	133.47	57.20	0.68942	1800	449.71	114.0	326.32	5.847	0.90308
800	191.81	5.526	136.97	53.63	0.69558	1850	463.37	127.2	336.55	5.388	0.91056
820	196.69	6.033	140.47	50.35	0.70160	1900	477.09	141.5	346.85	4.974	0.91788
840	201.56	6.573	143.98	47.34	0.70747	1950	490.88	157.1	357.20	4.598	0.92504
860	206.46	7.149	147.50	44.57	0.71323	2000	504.71	174.0	367.61	4.258	0.93205
880	211.35	7.761	151.02	42.01	0.71886	2050	518.61	192.3	378.08	3.949	0.93891
900	216.26	8.411	154.57	39.64	0.72438	2100	532.55	212.1	388.60	3.667	0.94564
920	221.18	9.102	158.12	37.44	0.72979	2150	546.54	233.5	399.17	3.410	0.95222

TABLE A.8 (continued) Ideal Gas Properties of Air

T(°R), h and u(Btu/lb), $s°$(Btu/lb · °R)

T	h	p_r	u	v_r	$s°$	T	h	p_r	u	v_r	$s°$
2200	560.59	256.6	409.78	3.176	0.95868	3700	998.11	2330	744.48	.5882	1.10991
2250	574.69	281.4	420.46	2.961	0.96501	3750	1013.1	2471	756.04	.5621	1.11393
2300	588.82	308.1	431.16	2.765	0.97123	3800	1028.1	2618	767.60	.5376	1.11791
2350	603.00	336.8	441.91	2.585	0.97732	3850	1043.1	2773	779.19	.5143	1.12183
2400	617.22	367.6	452.70	2.419	0.98331	3900	1058.1	2934	790.80	.4923	1.12571
2450	631.48	400.5	463.54	2.266	0.98919	3950	1073.2	3103	802.43	.4715	1.12955
2500	645.78	435.7	474.40	2.125	0.99497	4000	1088.3	3280	814.06	.4518	1.13334
2550	660.12	473.3	485.31	1.996	1.00064	4050	1103.4	3464	825.72	.4331	1.13709
2600	674.49	513.5	496.26	1.876	1.00623	4100	1118.5	3656	837.40	.4154	1.14079
2650	688.90	556.3	507.25	1.765	1.01172	4150	1133.6	3858	849.09	.3985	1.14446
2700	703.35	601.9	518.26	1.662	1.01712	4200	1148.7	4067	860.81	.3826	1.14809
2750	717.83	650.4	529.31	1.566	1.02244	4300	1179.0	4513	884.28	.3529	1.15522
2800	732.33	702.0	540.40	1.478	1.02767	4400	1209.4	4997	907.81	.3262	1.16221
2850	746.88	756.7	551.52	1.395	1.03282	4500	1239.9	5521	931.39	.3019	1.16905
2900	761.45	814.8	562.66	1.318	1.03788	4600	1270.4	6089	955.04	.2799	1.17575
2950	776.05	876.4	573.84	1.247	1.04288	4700	1300.9	6701	978.73	.2598	1.18232
3000	790.68	941.4	585.04	1.180	1.04779	4800	1331.5	7362	1002.5	.2415	1.18876
3050	805.34	1011	596.28	1.118	1.05264	4900	1362.2	8073	1026.3	.2248	1.19508
3100	820.03	1083	607.53	1.060	1.05741	5000	1392.9	8837	1050.1	.2096	1.20129
3150	834.75	1161	618.82	1.006	1.06212	5100	1423.6	9658	1074.0	.1956	1.20738
3200	849.48	1242	630.12	.9546	1.06676	5200	1454.4	10539	1098.0	.1828	1.21336
3250	864.24	1328	641.46	.9069	1.07134	5300	1485.3	11481	1122.0	.1710	1.21923
3300	879.02	1418	652.81	.8621	1.07585						
3350	893.83	1513	664.20	.8202	1.08031						
3400	908.66	1613	675.60	.7807	1.08470						
3450	923.52	1719	687.04	.7436	1.08904						
3500	938.40	1829	698.48	.7087	1.09332						
3550	953.30	1946	709.95	.6759	1.09755						
3600	968.21	2068	721.44	.6449	1.10172						
3650	983.15	2196	732.95	.6157	1.10584						

Source: Adapted from M.J. Moran and H.N. Shapiro, *Fundamentals of Engineering Thermodynamics*, 3rd. ed., Wiley, New York, 1995, as based on J.H. Keenan and J. Kaye, *Gas Tables*, Wiley, New York, 1945.

TABLE A.9 Equations for Gas Properties

Gas	Molar Mass M kg/kmol	Gas Constant R kJ/kg·K	Specific Heats at 25°C c_p kJ/kg·K	c_v kJ/kg·K	k	Temperature Range	Equation Coefficients for $c_p/R = a + bT + cT^2 + dT^3 + eT^4$: a	$b \times 10^3$ K^{-1}	$c \times 10^6$ K^{-2}	$d \times 10^{10}$ K^{-3}	$e \times 10^{13}$ K^{-4}	Critical State Properties p_c MPa	T_c K	Redlich–Kwong Constants a kPa·m^6·K$^{0.5}$/kmol^2	b m^3/kmol	Gas
Acetylene, C_2H_2	26.04	0.319	1.69	1.37	1.232	300–1000K	0.8021	23.51	−35.95	286.1	−87.64	6.14	308	8030	0.0362	Acetylene, C_2H_2
						1000–3000K	3.825	6.767	−3.014	6.931	−0.6469					
Air	28.97	0.287	1.01	0.718	1.400	300–1000K	3.721	−1.874	4.719	−34.45	8.531	3.77	132	1580	0.0253	Air
						1000–3000K	2.786	1.925	−0.9465	2.321	−0.2229					
Argon, Ar	39.95	0.208	0.520	0.312	1.667		2.50	0	0	0	0	4.90	151	1680	0.0222	Argon, Ar
Butane, C_4H_{10}	58.12	0.143	1.67	1.53	1.094	300–1500K	0.4756	44.65	−22.04	42.07	0	3.80	425	29000	0.0806	Butane, C_4H_{10}
Carbon Dioxide CO_2	44.01	0.189	0.844	0.655	1.289	300–1000K	2.227	9.992	−9.802	53.97	−12.81	7.38	304	6450	0.0297	Carbon Dioxide CO_2
						1000–3000K	3.247	5.847	−3.412	9.469	−1.009					
Carbon Monoxide CO	28.01	0.297	1.04	0.744	1.399	300–1000K	3.776	−2.093	4.880	−32.71	6.984	3.50	133	1720	0.0274	Carbon Monoxide, CO
						1000–3000K	2.654	2.226	−1.146	2.851	−0.2762					
Ethane, C_2H_6	30.07	0.276	1.75	1.48	1.187	300–1500K	0.8293	20.75	-7.704	8.756	0	4.88	306	9860	0.0450	Ethane, C_2H_6
Ethylene, C_2H_4	28.05	0.296	1.53	1.23	1.240	300–1000K	1.575	10.19	11.25	−199.1	81.98	5.03	282	7860	0.0404	Ethylene, C_2H_4
						1000–3000K	0.2530	18.67	−9.978	26.03	−2.668					
Helium, He	4.003	2.08	5.19	3.12	1.667		2.50	0	0	0	0	0.228	5.20	8.00	0.0165	Helium, He
Hydrogen, H_2	2.016	4.12	14.3	10.2	1.405	300–1000K	2.892	3.884	−8.850	86.94	−29.88	1.31	33.2	143	0.0182	Hydrogen, H_2
						1000–3000K	3.717	−0.9220	1.221	−4.328	0.5202					
Hydrogen, H	1.008	8.25	20.6	12.4	1.667	300–1000K	2.496	0.02977	−0.07655	0.8238	−0.3158					Hydrogen, H
						1000–3000K	2.567	−0.1509	0.1219	−0.4184	0.05182					
Hydroxyl, OH	17.01	0.489	1.76	1.27	1.384	300–1000K	3.874	−1.349	1.670	−5.670	0.6189					Hydroxyl, OH
						1000–3000K	3.229	0.2014	0.4357	−2.043	0.2696					
Methane, CH_4	16.04	0.518	2.22	1.70	1.304	300–1000K	4.503	−8.965	37.38	−364.9	122.2	4.60	191	3210	0.0298	Methane, CH_4
						1000–3000K	−0.6992	15.31	−7.695	18.96	−1.849					
Neon, Ne	20.18	0.412	1.03	0.618	1.667		2.50	0	0	0	0	2.65	44.4	146	0.0120	Neon, Ne
Nitric Oxide, N(	30.01	0.277	0.995	0.718	1.386	300–1000K	4.120	−4.225	10.77	−97.64	31.85	6.48	180	1980	0.0200	Nitric Oxide, NO
						1000–3000K	2.730	2.372	−1.338	3.604	−0.3743					
Nitrogen, N_2	28.01	0.297	1.04	0.743	1.400	300–1000K	3.725	−1.562	3.208	−15.54	1.154	3.39	126	1550	0.0267	Nitrogen, N_2
						1000–3000K	2.469	2.467	−1.312	3.401	−0.3454					
Nitrogen, N	14.01	0.594	1.48	0.890	1.667	300–1000K	2.496	0.02977	−0.07655	0.8238	−0.3158					Nitrogen, N
						1000–3000K	2.483	0.03033	−0.01517	0.001879	0.009657					
Oxygen, O_2	32.00	0.260	0.919	0.659	1.395	300–1000K	3.837	−3.420	10.99	−109.6	37.47	5.04	155	1740	0.0221	Oxygen, O_2
						1000–3000K	3.156	1.809	−1.052	3.190	−0.3629					
Oxygen, O	16.00	0.520	1.37	0.850	1.612	300–1000K	3.020	−2.176	3.793	−30.62	9.402					Oxygen, O
						1000–3000K	2.662	−0.3051	0.2250	−0.7447	0.09383					
Propane, C_3H_8	44.10	0.189	1.67	1.48	1.127	300-1500K	-0.4861	36.63	-18.91	38.14	0	4.26	370	18300	0.0626	Propane, C_3H_8
Water, H_2O	18.02	0.462	1.86	1.40	1.329	300–1000K	4.132	−1.559	5.315	−42.09	12.84	22.1	647	14300	0.0211	Water, H_2O
						1000–3000K	2.798	2.693	−0.5392	−0.01783	0.09027					

Source: Adapted from J.B. Jones and R.E. Dugan, *Engineering Thermodynamics*, Prentice-Hall, Englewood Cliffs, NJ 1996 from various sources: *JANAF Thermochemical Tables*, 3rd ed., published by the American Chemical Society and the American Institute of Physics for the National Bureau of Standards, 1986. Data for butane, ethane, and propane from K.A. Kobe and E.G. Long, "Thermochemistry for the Petrochemical Industry, Part II — Paraffinic Hydrocarbons, C_1-C_{16}" *Petroleum Refiner,* Vol. 28, No. 2, 1949, pp. 113-116.

Appendix B. Properties of Liquids

TABLE B.1 Properties of Liquid Water*

Symbols and Units:

ρ = density, lbm/ft³. For g/cm³ multiply by 0.016018. For kg/m³ multiply by 16.018.

c_p = specific heat, Btu/lbm·deg R = cal/g·K. For J/kg·K multiply by 4186.8

μ = viscosity. For lbf·sec/ft² = slugs/sec·ft, multiply by 10^{-7}. For lbm·sec·ft multiply by 10^{-7} and by 32.174. For g/sec·cm (poises) multiply by 10^{-7} and by 478.80. For N·sec/m² multiply by 10^{-7} and by 478.880.

k = thermal conductivity, Btu/hr·ft·deg R. For W/m·K multiply by 1.7307.

Temp, °F	At 1 atm or 14.7 psia				At 1,000 psia				At 10,000 psia			
	ρ	c_p	μ	k	ρ	c_p	μ	k	ρ	c_p	μ	k†
32	62.42	1.007	366	0.3286	62.62	0.999	365	0.3319	64.5	0.937	357	0.3508
40	62.42	1.004	323	0.334	62.62	0.997	323	0.337	64.5	0.945	315	0.356
50	62.42	1.002	272	0.3392	62.62	0.995	272	0.3425	64.5	0.951	267	0.3610
60	62.38	1.000	235	0.345	62.58	0.994	235	0.348	64.1	0.956	233	0.366
70	62.31	0.999	204	0.350	62.50	0.994	204	0.353	64.1	0.960	203	0.371
80	62.23	0.998	177	0.354	62.42	0.994	177	0.358	64.1	0.962	176	0.376
90	62.11	0.998	160	0.359	62.31	0.994	160	0.362	63.7	0.964	159	0.380
100	62.00	0.998	142	0.3633	62.19	0.994	142	0.3666	63.7	0.965	142	0.3841
110	61.88	0.999	126	0.367	62.03	0.994	126	0.371	63.7	0.966	126	0.388
120	61.73	0.999	114	0.371	61.88	0.995	114	0.374	63.3	0.967	114	0.391
130	61.54	0.999	105	0.374	61.73	0.995	105	0.378	63.3	0.968	105	0.395
140	61.39	0.999	96	0.378	61.58	0.996	96	0.381	63.3	0.969	98	0.398
150	61.20	1.000	89	0.3806	61.39	0.996	89	0.3837	63.0	0.970	91	0.4003
160	61.01	1.001	83	0.383	61.20	0.997	83	0.386	62.9	0.971	85	0.403
170	60.79	1.002	77	0.386	60.98	0.998	77	0.389	62.5	0.972	79	0.405
180	60.57	1.003	72	0.388	60.75	0.999	72	0.391	62.5	0.973	74	0.407
190	60.35	1.004	68	0.390	60.53	1.001	68	0.393	62.1	0.974	70	0.409
200	60.10	1.005	62.5	0.3916	60.31	1.002	62.9	0.3944	62.1	0.975	65.4	0.4106
250	boiling point 212°F				59.03	1.001	47.8	0.3994	60.6	0.981	50.6	0.4158
300					57.54	1.024	38.4	0.3993	59.5	0.988	41.3	0.4164
350					55.83	1.044	32.1	0.3944	58.1	0.999	35.1	0.4132
400					53.91	1.072	27.6	0.3849	56.5	1.011	30.6	0.4064
500					49.11	1.181	21.6	0.3508	52.9	1.051	24.8	0.3836
600					boiling point 544.58°F				48.3	1.118	21.0	0.3493

†At 7,500 psia.

*From: "1967 ASME Steam Tables", American Society of Mechanical Engineers, Tables 9, 10, and 11 and Figures 6, 7, 8, and 9.

The ASME compilation is a 330-page book of tables and charts, including a 2½ × 3½-ft Mollier chart. All values have been computed in accordance with the 1967 specifications of the International Formulation Committee (IFC) and are in conformity with the 1963 International Skeleton Tables. This standardization of tables began in 1921 and was extended through the International Conferences in London (1929), Berlin (1930), Washington (1934), Philadelphia (1954), London (1956), New York (1963) and Glasgow (1966). Based on these world-wide standard data, the 1967 ASME volume represents detailed computer output in both tabular and graphic form. Included are density and volume, enthalpy, entropy, specific heat, viscosity, thermal conductivity, Prandtl number, isentropic exponent, choking velocity, p-v product, etc., over the entire range (to 1500 psia 1500°F). English units are used, but all conversion factors are given.

TABLE B.2 Physical and Thermal Properties of Common Liquids

Part a. SI Units

(At 1.0 Atm Pressure (0.101 325 MN/m²), 300 K, except as noted.)

Common name	*Density, kg/m³*	*Specific heat, kJ/kg·K*	*Viscosity, N·s/m²*	*Thermal conductivity, W/m·K*	*Freezing point, K*	*Latent heat of fusion, kJ/kg*	*Boiling point, K*	*Latent heat of evaporation, kJ/kg*	*Coefficient of cubical expansion per K*
Acetic acid	1 049	2.18	.001 155	0.171	290	181	391	402	0.001 1
Acetone	784.6	2.15	.000 316	0.161	179.0	98.3	329	518	0.001 5
Alcohol, ethyl	785.1	2.44	.001 095	0.171	158.6	108	351.46	846	0.001 1
Alcohol, methyl	786.5	2.54	.000 56	0.202	175.5	98.8	337.8	1 100	0.001 4
Alcohol, propyl	800.0	2.37	.001 92	0.161	146	86.5	371	779	
Ammonia (aqua)	823.5	4.38		0.353					
Benzene	873.8	1.73	.000 601	0.144	278.68	126	353.3	390	0.001 3
Bromine		.473	.000 95		245.84	66.7	331.6	193	0.001 2
Carbon disulfide	1 261	.992	.000 36	0.161	161.2	57.6	319.40	351	0.001 3
Carbon tetrachloride	1 584	.866	.000 91	0.104	250.35	174	349.6	194	0.001 3
Castor oil	956.1	1.97	.650	0.180	263.2				
Chloroform	1 465	1.05	.000 53	0.118	209.6	77.0	334.4	247	0.001 3
Decane	726.3	2.21	.000 859	0.147	243.5	201	447.2	263	
Dodecane	754.6	2.21	.001 374	0.140	247.18	216	489.4	256	
Ether	713.5	2.21	.000 223	0.130	157	96.2	307.7	372	0.001 6
Ethylene glycol	1 097	2.36	.016 2	0.258	260.2	181	470	800	
Fluorine refrigerant R-11	1 476	.870[a]	.000 42	0.093[a]	162		297.0	180[b]	
Fluorine refrigerant R-12	1 311	.971[a]		0.071[a]	115	34.4	243.4	165[b]	
Fluorine refrigerant R-22	1 194	1.26[a]		0.086[a]	113	183	232.4	232[b]	
Glycerine	1 259	2.62	.950	0.287	264.8	200	563.4	974	0.000 54
Heptane	679.5	2.24	.000 376	0.128	182.54	140	371.5	318	
Hexane	654.8	2.26	.000 297	0.124	178.0	152	341.84	365	
Iodine		2.15			386.6	62.2	457.5	164	
Kerosene	820.1	2.09	.001 64	0.145				251	
Linseed oil	929.1	1.84	.033 1		253		560		
Mercury		.139	.001 53		234.3	11.6	630	295	0.000 18
Octane	698.6	2.15	.000 51	0.131	216.4	181	398	298	0.000 72
Phenol	1 072	1.43	.008 0	0.190	316.2	121	455		0.000 90
Propane	493.5	2.41[a]	.000 11		85.5	79.9	231.08	428[b]	
Propylene	514.4	2.85	.000 09		87.9	71.4	225.45	342	
Propylene glycol	965.3	2.50	.042		213		460	914	
Sea water	1 025	3.76–4.10			270.6				
Toluene	862.3	1.72	.000 550	0.133	178	71.8	383.6	363	
Turpentine	868.2	1.78	.001 375	0.121	214		433	293	0.000 99
Water	997.1	4.18	.000 89	0.609	273	333	373	2 260	0.000 20

[a]At 297 K, liquid.
[b]At .101 325 meganewtons, saturation temperature.

TABLE B.2 (continued) Physical and Thermal Properties of Common Liquids

Part b. English Units

(At 1.0 Atm Pressure 77°F (25°C), except as noted.)

For viscosity in N·s/m² (=kg m·s), multiply values in centipoises by 0.001. For surface tension in N/m, multiply values in dyne/cm by 0.001.

Common name	Density, lb/ft³	Specific gravity	Viscosity		Sound velocity, meters/sec	Dielectric constant	Refractive index
			$lb_m/ft\ sec \times 10^4$	cp			
Acetic acid	65.493	1.049	7.76	1.155	1584[50]	6.15	1.37
Acetone	48.98	.787	2.12	0.316	1174	20.7	1.36
Alcohol, ethyl	49.01	.787	7.36	1.095	1144	24.3	1.36
Alcohol, methyl	49.10	.789	3.76	0.56	1103	32.6	1.33
Alcohol, propyl	49.94	.802	12.9	1.92	1205	20.1	1.38
Ammonia (aqua)	51.411	.826	—	—	—	16.9	—
Benzene	54.55	.876	4.04	0.601	1298	2.2	1.50
Bromine	—	—	6.38	0.95	—	3.20	—
Carbon disulfide	78.72	1.265	2.42	0.36	1149	2.64	1.63
Carbon tetrachloride	98.91	1.59	6.11	0.91	924	2.23	1.46
Castor oil	59.69	0.960	—	650	1474	4.7	—
Chloroform	91.44	1.47	3.56	0.53	995	4.8	1.44
Decane	45.34	.728	5.77	0.859	—	2.0	1.41
Dodecane	47.11	—	9.23	1.374	—	—	1.41
Ether	44.54	0.715	1.50	0.223	985	4.3	1.35
Ethylene glycol	68.47	1.100	109	16.2	1644	37.7	1.43
Fluorine refrigerant R-11	92.14	1.480	2.82	0.42	—	2.0	1.37
Fluorine refrigerant R-12	81.84	1.315	—	—	—	2.0	1.29
Fluorine refrigerant R-22	74.53	1.197	—	—	—	2.0	1.26
Glycerine	78.62	1.263	6380	950	1909	40	1.47
Heptane	42.42	.681	2.53	0.376	1138	1.92	1.38
Hexane	40.88	.657	2.00	0.297	1203	—	1.37
Iodine	—	—	—	—	—	11	—
Kerosene	51.2	0.823	11.0	1.64	1320	—	—
Linseed oil	58.0	0.93	222	33.1	—	3.3	—
Mercury	—	13.633	10.3	1.53	1450	—	—
Octane	43.61	.701	3.43	0.51	1171	—	1.40
Phenol	66.94	1.071	54	8.0	1274[100]	9.8	—
Propane	30.81	.495	0.74	0.11	—	1.27	1.34
Propylene	32.11	.516	0.60	0.09	—	—	1.36
Propylene glycol	60.26	.968	—	42	—	—	1.43
Sea water	64.0	1.03	—	—	1535	—	—
Toluene	53.83	0.865	3.70	0.550	1275[30]	2.4	1.49
Turpentine	54.2	0.87	9.24	1.375	1240	—	1.47
Water	62.247	1.00	6.0	0.89	1498	78.54[a]	1.33

[a] The dielectric constant of water near the freezing point is 87.8; it decreases with increase in temperature to about 55.6 near the boiling point.

Appendix C. Properties of Solids

TABLE C.1 Properties of Common Solids*

Material	Specific gravity	Specific heat		Thermal conductivity	
		Btu/(lbm·deg R)	kJ/(kg·K)	Btu/(hr·ft·deg F)	W/(m·K)
Asbestos cement board	1.4	0.2	.837	0.35	0.607
Asbestos millboard	1.0	0.2	.837	0.08	0.14
Asphalt	1.1	0.4	1.67		
Beeswax	0.95	0.82	3.43		
Brick, common	1.75	0.22	.920	0.42	0.71
Brick, hard	2.0	0.24	1.00	0.75	1.3
Chalk	2.0	0.215	.900	0.48	0.84
Charcoal, wood	0.4	0.24	1.00	0.05	0.088
Coal, anthracite	1.5	0.3	1.26		
Coal, bituminous	1.2	0.33	1.38		
Concrete, light	1.4	0.23	.962	0.25	0.42
Concrete, stone	2.2	0.18	.753	1.0	1.7
Corkboard	0.2	0.45	1.88	0.025	0.04
Earth, dry	1.4	0.3	1.26	0.85	1.5
Fiberboard, light	0.24	0.6	2.51	0.035	0.058
Fiber hardboard	1.1	0.5	2.09	0.12	0.2
Firebrick	2.1	0.25	1.05	0.8	1.4
Glass, window	2.5	0.2	.837	0.55	0.96
Gypsum board	0.8	0.26	1.09	0.1	0.17
Hairfelt	0.1	0.5	2.09	0.03	0.050
Ice (32°)	0.9	0.5	2.09	1.25	2.2
Leather, dry	0.9	0.36	1.51	0.09	0.2
Limestone	2.5	0.217	.908	1.1	1.9
Magnesia (85%)	0.25	0.2	.837	0.04	0.071
Marble	2.6	0.21	.879	1.5	2.6
Mica	2.7	0.12	.502	0.4	0.71
Mineral wool blanket	0.1	0.2	.837	0.025	0.04
Paper	0.9	0.33	1.38	0.07	0.1
Paraffin wax	0.9	0.69	2.89	0.15	0.2
Plaster, light	0.7	0.24	1.00	0.15	0.2
Plaster, sand	1.8	0.22	.920	0.42	0.71
Plastics, foamed	0.2	0.3	1.26	0.02	0.03
Plastics, solid	1.2	0.4	1.67	0.11	0.19
Porcelain	2.5	0.22	.920	0.9	1.5
Sandstone	2.3	0.22	.920	1.0	1.7
Sawdust	0.15	0.21	.879	0.05	0.08
Silica aerogel	0.11	0.2	.837	0.015	0.02
Vermiculite	0.13	0.2	.837	0.035	0.058
Wood, balsa	0.16	0.7	2.93	0.03	0.050
Wood, oak	0.7	0.5	2.09	0.10	0.17
Wood, white pine	0.5	0.6	2.51	0.07	0.12
Wool, felt	0.3	0.33	1.38	0.04	0.071
Wool, loose	0.1	0.3	1.26	0.02	0.3

*Compiled from several sources.

TABLE C.2 Density of Various Solids:* Approximate Density of Solids at Ordinary Atmospheric Temperature

Substance	Grams per cu cm	Pounds per cu ft	Substance	Grams per cu cm	Pounds per cu ft	Substance	Grams per cu cm	Pounds per cu ft
Agate	2.5–2.7	156–168	Glass			Tallow		
Alabaster			Common	2.4–2.8	150–175	Beef	0.94	59
Carbonate	2.69–2.78	168–173	Flint	2.9–5.9	180–370	Mutton	0.94	59
Sulfate	2.26–2.32	141–145	Glue	1.27	79	Tar	1.02	66
Albite	2.62–2.65	163–165	Granite	2.64–2.76	165–172	Topaz	3.5–3.6	219–223
Amber	1.06–1.11	66–69	Graphite†	2.30–2.72	144–170	Tourmaline	3.0–3.2	190–200
Amphiboles	2.9–3.2	180–200	Gum arabic	1.3–1.4	81–87	Wax, sealing	1.8	112
Anorthite	2.74–2.76	171–172	Gypsum	2.31–2.33	144–145	Wood (seasoned)		
Asbestos	2.0–2.8	125–175	Hematite	4.9–5.3	306–330	Alder	0.42–0.68	26–42
Asbestos slate	1.8	112	Hornblende	3.0	187	Apple	0.66–0.84	41–52
Asphalt	1.1–1.5	69–94	Ice	0.917	57.2	Ash	0.65–0.85	40–53
Basalt	2.4–3.1	150–190	Ivory	1.83–1.92	114–120	Balsa	0.11–0.14	7–9
Beeswax	0.96–0.97	60–61	Leather, dry	0.86	54	Bamboo	0.31–0.40	19–25
Beryl	2.69–2.7	168–169	Lime, slaked	1.3–1.4	81–87	Basswood	0.32–0.59	20–37
Biotite	2.7–3.1	170–190	Limestone	2.68–2.76	167–171	Beech	0.70–0.90	32–56
Bone	1.7–2.0	106–125	Linoleum	1.18	74	Birch	0.51–0.77	32–48
Brick	1.4–2.2	87–137	Magnetite	4.9–5.2	306–324	Blue gum	1.00	62
Butter	0.86–0.87	53–54	Malachite	3.7–4.1	231–256	Box	0.95–1.16	59–72
Calamine	4.1–4.5	255–280	Marble	2.6–2.84	160–177	Butternut	0.38	24
Calcspar	2.6–2.8	162–175	Meerschaum	0.99–1.28	62–80	Cedar	0.49–0.57	30–35
Camphor	0.99	62	Mica	2.6–3.2	165–200	Cherry	0.70–0.90	43–56
Caoutchouc	0.92–0.99	57–62	Muscovite	2.76–3.00	172–187	Dogwood	0.76	47
Cardboard	0.69	43	Ochre	3.5	218	Ebony	1.11–1.33	69–83
Celluloid	1.4	87	Opal	2.2	137	Elm	0.54–0.60	34–37
Cement, set	2.7–3.0	170–190	Paper	0.7–1.15	44–72	Hickory	0.60–0.93	37–58
Chalk	1.9–2.8	118–175	Paraffin	0.87–0.91	54–57	Holly	0.76	47
Charcoal			Peat blocks	0.84	52	Juniper	0.56	35
Oak	0.57	35	Pitch	1.07	67	Larch	0.50–0.56	31–35
Pine	0.28–0.44	18–28	Porcelain	2.3–2.5	143–156	Lignum vitae	1.17–1.33	73–83
Cinnabar	8.12	507	Porphyry	2.6–2.9	162–181	Locust	0.67–0.71	42–44
Clay	1.8–2.6	112–162	Pressed wood			Logwood	0.91	57
Coal			pulp board	0.19	12	Mahogany		
Anthracite	1.4–1.8	87–112	Pyrite	4.95–5.1	309–318	Honduras	0.66	41
Bituminous	1.2–1.5	75–94	Quartz	2.65	165	Spanish	0.85	53
Cocoa butter	0.89–0.91	56–57	Resin	1.07	67	Maple	0.62–0.75	39–47
Coke	1.0–1.7	62–105	Rock salt	2.18	136	Oak	0.60–0.90	37–56
Copal	1.04–1.14	65–71	Rubber, hard	1.19	74	Pear	0.61–0.73	38–45
Cork	0.22–0.26	14–16	Rubber, soft			Pine		
Cork linoleum	0.54	34	Commercial	1.1	69	Pitch	0.83–0.85	52–53
Corundum	3.9–4.0	245–250	Pure gum	0.91–0.93	57–58	White	0.35–0.50	22–31
Diamond	3.01–3.52	188–220	Sandstone	2.14–2.36	134–147	Yellow	0.37–0.60	23–37
Dolomite	2.84	177	Serpentine	2.50–2.65	156–165	Plum	0.66–0.78	41–49
Ebonite	1.15	72	Silica			Poplar	0.35–0.5	22–31
Emery	4.0	250	Fused trans-			Satinwood	0.95	59
Epidote	3.25–3.50	203–218	parent	2.21	138	Spruce	0.48–0.70	30–44
Feldspar	2.55–2.75	159–172	Translucent	2.07	129	Sycamore	0.40–0.60	24–37
Flint	2.63	164	Slag	2.0–3.9	125–240	Teak		
Fluorite	3.18	198	Slate	2.6–3.3	162–205	Indian	0.66–0.88	41–55
Galena	7.3–7.6	460–470	Soapstone	2.6–2.8	162–175	African	0.98	61
Gamboge	1.2	75	Spermaceti	0.95	59	Walnut	0.64–0.70	40–43
Garnet	3.15–4.3	197–268	Starch	1.53	95	Water gum	1.00	62
Gas carbon	1.88	117	Sugar	1.59	99	Willow	0.40–0.60	24–37
Gelatin	1.27	79	Talc	2.7–2.8	168–174			

†Some values reported as low as 1.6
*Based largely on: "Smithsonian Physical Tables", 9th rev. ed., W.E. Forsythe, Ed., The Smithsonian Institution, 1956, p. 292.

Note: In the case of substances with voids, such as paper or leather, the bulk density is indicated rather than the density of the solid portion. For density in kg/m^3, multiply values in g/cm^3 by 1,000.

TABLE C.3 Specific Stiffness of Metals, Alloys, and Certain Non-Metallics*

Specific stiffness is usually expressed as the modulus of elasticity (in tension) per unit weight-density, i.e., E/ρ, in units of pounds and inches. While the stiffness of similar alloys varies considerably, there are definite ranges and groups to be recognized. Since the specific stiffness of steel is about 100 million, the values in the following table are also approximately the percentage stiffness, referred to steel.

Material	*Specific stiffness, millions*
Beryllium	650
Silicon carbide	600
Alumina ceramics	400
Mica	350
Titanium carbide cermet	250
Alumina cermet	200
Molybdenum and alloys; silica glass	130
Titanium and alloys; cobalt superalloys; soda-lime glass	110
Carbon and low-alloy steel; wrought iron	105
Stainless steel; nodular cast iron; magnesium and alloys; aluminum and alloys	100
Nickel and alloys; malleable iron	95
Iron silicon alloys (cast); iridium; vanadium	90
Monel alloys; tungsten	80
Gray cast iron; columbium alloys	70
Aluminum bronze; beryllium copper	65
Nickel silver; cupronickel; zirconium	55
Yellow brass; nickel cast iron; bronze; Muntz metal; antimony	50
Copper; red brass; tantalum	45
Silver and alloys; pewter; platinum and alloys; white gold	30
Tin; thorium	25
Gold	20
Tin-lead alloy	10
Lead	5

*Compiled from several sources.

TABLE C.4 Thermal Properties of Pure Metals—Metric Units

Metal	AT ATMOSPHERIC PRESSURE								LIQUID METAL			
				At 100°K		At 25°C (77°F)				Vapor pressure		
	Melting point, °C	Boiling point, °C	Latent heat of fusion, cal/g**	Thermal conductivity, watts/cm °C	Specific heat, cal/g °C**	Specific heat, cal/g °C**	Coeff. of linear expansion, (× 10⁶) (°C)⁻¹	Thermal conductivity, watts/cm °C	Specific heat (liquid) at 2000°K, cal/g °C**	10^{-3} atm	10^{-6} atm	10^{-9} atm
										Boiling point temperatures, °K		
Aluminum	660.	2441.	95	3.00*	.115	0.215	25	2.37	.26	1,782	1,333	1,063
Antimony	630.	1440.	38.5	—	.040	.050	9	.185	.062	1,007	741	612
Beryllium	1285.	2475.	324.	—	.049	.436	12	2.18	.78	1,793	1,347	1,085
Bismuth	271.4	1660.	12.4	—	.026	.030	13	.084	.036	1,155	851	677.
Cadmium	321.	767.	13.2	1.03	.047	.055	30	.93	.063	655	486	388
Chromium	1860.	2670.	79	1.58	.046	.110	6	.91	.224	1,992	1,530	1,247
Cobalt	1495.	2925.	66	—	.057	.10	12	.69	.164	2,167	1,652	1,345
Copper	1084.	2575.	49	4.83*	.061	.092	16.6	3.98	.118	1,862	1,391	1,120
Gold	1063.	2800.	15	3.45*	.026	.031	14.2	3.15	.0355	2,023	1,510	1,211
Iridium	2450.	4390.	33	—	.022	.031	6	1.47	.0434	3,253	2,515	2,062
Iron	1536.	2870.	65	1.32*	.052	.108	12	.803	.197	2,093	1,594	1,297
Lead	327.5	1750.	5.5	0.396	.028	.031	29	.346	.033	1,230	889	698
Magnesium	650.	1090.	88.0	1.69	.016	.243	25	1.59	.32	857	638	509
Manganese	1244.	2060.	64	—	.064	.114	22	—	.20	1,495	1,131	913
Mercury	−38.86	356.55	2.7	—	.029	.033	—	.0839	—	393	287	227
Molybdenum	2620.	4651.	69	1.79	.033	.060	5	1.4	.089	3,344	2,558	2,079
Nickel	1453.	2800.	71	1.58	.055	.106	13	.899	.175	2,156	1,646	1,343
Niobium (Columbium)	2470.	4740.	68	0.552	.045	.064	7	.52	.083	3,523	2,721	2,232
Osmium	3025.	4225.	34	—	—	.031	5	.61	.039	—	—	—
Platinum	1770.	3825.	24	0.79*	.024	.032	9	.73	.043	2,817	2,155	1,757
Plutonium	640.	3230.	3	—	.019	.032	54	.08	.041	2,200	1,596	1,252
Potassium	63.3	760.	14.5	—	.150	.180	83	.99	—	606	430	335
Rhodium	1965.	3700.	50	—	—	.058	8	1.50	.092	—	—	—
Selenium	217.	700.	16	—	—	.077	37	.005	—	—	—	—
Silicon	1411.	3280.	430	—	.062	.17	3	.835	.217	2,340	1,749	1,427
Silver	961.	2212.	26.5	4.50*	.045	.057	19	4.27	.068	1,582	1,179	952
Sodium	97.83	884.	27	—	.234	.293	70	1.34	—	701	504	394
Tantalum	2980.	5365.	41	0.592	.026	.034	6.5	.54	.040	3,959	3,052	2,495
Thorium	1750.	4800.	17	—	.024	.03	12	.41	.047	3,251	2,407	1,919
Tin	232.	2600.	14.1	0.85	.039	.054	20	.64	.058	1,857	1,366	1,080
Titanium	1670.	3290.	100	0.312	.072	.125	8.5	.2	.188	2,405	1,827	1,484
Tungsten	3400.	5550.	46	2.35*	.021	.032	4.5	1.78	.040	4,139	3,228	2,656
Uranium	1132.	4140.	12	—	.022	.028	13.4	.25	.048	2,861	2,128	1,699
Vanadium	1900.	3400.	98	—	.061	.116	8	.60	.207	2,525	1,948	1,591
Zinc	419.5	910.	27	1.32	.063	.093	35	1.15	—	752	559	449

* Temperatures of maximum thermal conductivity (conductivity values in watts/cm °C): Aluminum 13°K, cond. = 71.5; copper 10°K, cond. = 196; gold 10°K, cond. = 28.2; iron 20°K, cond. = 9.97; platinum 8°K, cond. = 12.9; silver 7°K, cond. = 193; tungsten 8°K, cond. = 85.3.

** To convert to SI units note that 1 cal = 4.186 J.

TABLE C.5 Mechanical Properties of Metals and Alloys:* Typical Composition, Properties, and Uses of Common Materials

For MN/m^2 multiply strength in thousands of psi by 6.895.

No.	*Material*	*Nominal composition*	*Form and condition*	*Typical mechanical properties*				*Comments*
				Yield strength (0.2% offset), 1000 lb/sq in.	*Tensile strength, 1000 lb/sq in.*	*Elongation, in 2 in., %*	*Hardness, Brinell*	
	FERROUS ALLOYS Ferrous alloys comprise the largest volume of metal alloys used in engineering. The actual range of mechanical properties in any particular grade of alloy steel depends on the particular history and heat treatment. The steels listed in this table are intended to give some idea of the range of properties readily obtainable. Many hundreds of steels are available. Cost is frequently an important criterion in the choice of material; in general the greater the percentage of alloying elements present in the alloy, the greater will be the cost.							
	IRON							
1	Ingot iron (Included for comparison)	Fe 99.9	Hot-rolled	29	45	26	90	
			Annealed	19	38	45	67	
	PLAIN CARBON STEELS							
2	AISI–SAE 1020	C 0.20 Mn 0.45 Si 0.25 Fe bal.	Hot-rolled	30	55	25	111	Bolts, crankshafts, gears, connecting rods; easily weldable
			Hardened (water-quenched, 1000°F-tempered)	62	90	25	179	
3	AISI 1025	C 0.25 Fe bal. Mn 0.45	Bar stock					
			Hot-rolled	32	58	25	116	
			Cold-drawn	54	64	15	126	
4	AISI–SAE 1035	C 0.35 Mn 0.75	Hot-rolled	39	72	18	143	Medium-strength, engineering steel
			Cold-rolled	67	80	12	163	
5	AISI–SAE 1045	C 0.45 Fe bal. Mn 0.75	Bar stock					
			Annealed	73	80	12	170	
			Hot-rolled	45	82	16	163	
			Cold-drawn	77	91	12	179	
6	AISI–SAE 1078	C 0.78 Fe bal. Mn 0.45	Bar stock					
			Hot-rolled; spheroidized	55	100	12	207	
			Annealed	72	94	10	192	
7	AISI–SAE 1095	C 0.95 Fe bal. Mn 0.40						
8	AISI–SAE 1120	C 0.2 Mn 0.8 S 0.1	Cold-drawn	58	69	—	137	Free-cutting, leaded, resulphurized steel; high-speed, automatic machining
	ALLOY STEELS							
9	ASTM A202/56	C 0.17 Mn 1.2 Cr 0.5 Si 0.75	Stress-relieved	45	75	18	—	Low alloy; boilers, pressure vessels

TABLE C.5 (continued) Mechanical Properties of Metals and Alloys:* Typical Composition, Properties, and Uses of Common Materials

No.	Material	Nominal composition	Form and condition	Typical mechanical properties: Yield strength (0.2% offset), 1000 lb/sq in.	Tensile strength, 1000 lb/sq in.	Elongation, in 2 in., %	Hardness, Brinell	Comments
10	AISI 4140	C 0.40, Si 0.3 Cr 1.0, Mo 0.2 Mn 0.9	Fully-tempered Optimum properties	95 132	108 150	22 18	240 —	High strength; gears, shafts
11	12% Manganese steel	12% Mn, C	Tempered 600°F Rolled and heat-treated stock	200 44	220 160	10 40	— 170	Machine tool parts; wear, abrasion-resistant
12	VASCO 300	Ni 18.5, Ti 0.6 Co 9.0, C 0.03 Mo 4.8	Solution treatment 1500°F; aged 900°F	110	150	18	—	Very high strength, maraging, good machining properties in annealed state
13	T1 (AISI)	W 18.0, V 1.0 Cr 4.0, C 0.7	Quenched; tempered				R(c)	High speed tool steel, cutting tools, punches, etc.
14	M2 (AISI)	W 6.5, Mo 5.0 Cr 4.0, C 0.85 V 2.0	Quenched; tempered				65–66	M-grade, cheaper, tougher
15	Stainless steel type 304	Ni 9.0, C 0.08 max Cr 19.0	Annealed; cold-rolled	35 to 160	85 to 185	60 8	160 to 400	General purpose, weldable; nonmagnetic austenitic steel
16	Stainless steel type 316	Cr 18.0, C 0.10 max Ni 11.0, Fe bal. Mo 2.5	Annealed	30 to 120	90 to 150	50 8	165 275	For severe corrosive media, under stress; nonmagnetic austenitic steel
17	Stainless steel type 431	Cr 16.0, Si 1.0 Ni 2.0, C 0.20 Mn 1.0, Fe bal.	Annealed Heat-treated	85 150	120 195	25 20	250 400	Heat-treated stainless steel, with good mechanical strength; magnetic
18	Stainless steel 17–4 PH	Cr 17.0, Co 0.35 Ni 4.0, C 0.07 Cu 4.0, Fe bal.	Annealed	110	150	10	363	Precipitation hardening; heat-resisting type; retains strength up to approx. 600°F

TABLE C.5 (continued) Mechanical Properties of Metals and Alloys:* Typical Composition, Properties, and Uses of Common Materials

No.	Material	Nominal composition	Form and condition	Typical mechanical properties				Comments
				Yield strength (0.2% offset), 1000 lb/sq in.	Tensile strength, 1000 lb/sq in.	Elongation, in 2 in., %	Hardness, Brinell	

CAST IRONS AND CAST STEELS

These alloys are used where large and/or intricate-shaped articles are required or where over-all dimensional tolerances are not critical. Thus the article can be produced with the fabrication and machining costs held to a minimum. Except for a few heat-treatable cast steels, this class of alloys does not demonstrate high-strength qualities.

No.	Material	Nominal composition	Form and condition	Yield strength (0.2% offset), 1000 lb/sq in.	Tensile strength, 1000 lb/sq in.	Elongation, in 2 in., %	Hardness, Brinell	Comments
	CAST IRONS							
19	Cast gray iron ASTM A48–48, Class 25	C 3.4, Si 1.8, Mn 0.5	Cast (as cast)	—	25 min	0.5 max	180	Engine blocks, fly-wheels, gears, machine-tool bases
20	White	C 3.4, Si 0.7, Mn 0.6	Cast	—	25	0	450	
21	Malleable iron ASTM A47	C 2.5, Si 1.0, Mn 0.55 max	Cast (annealed)	33	52	12	130	Automotives, axle bearings, track wheels, crankshafts
22	Ductile or nodular iron (Mg-containing) ASTM A339 ASTM A395	C 3.4, P 0.1 max, Mn 0.40, Mg 0.06, Ni 1%, Si 2.5, Fe bal.	Cast Cast (as cast) Cast (quenched, tempered)	53 68 108	70 90 135	18 7 5	170 235 310	Heavy-duty machines, gears, cams, crankshafts
23	Ni-hard type 2	C 2.7, Si 0.6, Mn 0.5, Ni 4.5, Cr 2.0, Fe bal.	Sand-cast Chill-cast (tempered)	— —	55 75	— —	550 625	Strength, with heat- and corrosion-resistance
24	Ni-resist type 2	C 3.0, Si 2.0, Mn 1.0, Ni 20.0, Cr 2.5, Fe bal.	Cast (as cast)	—	27	2	140	
	CAST STEELS							
25	ASTM A27–62 (60–30)	C 0.3, Mn 0.6, Si 0.8, Ni 0.5, Cr 0.4, Mo 0.2		30	60	24	—	Low alloy, medium strength, general application
26	ASTM A148–60 (105–85)			85	105	17	—	High strength; structural application

TABLE C.5 (continued) Mechanical Properties of Metals and Alloys:* Typical Composition, Properties, and Uses of Common Materials

No.	Material	Nominal composition	Form and condition	Typical mechanical properties				Comments
				Yield strength (0.2% offset), 1000 lb/sq in.	Tensile strength, 1000 lb/sq in.	Elongation, in 2 in., %	Hardness, Brinell	
27	Cast 12 Cr alloy (CA-15)	C 0.15 max Si 1.50 max Ni 1.00 max Mn 1.00 max Cr 11.5–14 Fe bal.	Air-cooled from 1800°F; tempered at 600°F Air-cooled from 1800°F; tempered at 1400°F	150 75	200 100	7 30	390 185	Stainless, corrosion-resistant to mildly corrosive alkalis and acids
28	Cast 29–9 alloy (CE-30) ASTM A296 63T	C 0.30 max Si 2.00 max Ni 8–11 Mn 1.50 max Cr 26–30 Fe bal.	As cast	60	95	15	170	Greater corrosion resistance, especially for oxidizing condition
29	Cast 28–7 alloy (HD) ASTM A297–63T	C 0.50 max Si 2.00 max Ni 4–7 Mn 1.50 max Cr 26–30 Fe bal.	As cast	48	85	16	190	Heat-resistant

SUPER ALLOYS

The advent of engineering applications requiring high temperature and high strength, as in jet engines and rocket motors, has lead to the development of a range of alloys collectively called super alloys. These alloys require excellent resistance to oxidation together with strength at high temperatures, typically 1800°F in existing engines. These alloys are continually being modified to develop better specific properties, and therefore entries in this group of alloys should be considered "fluid". Both wrought and casting-type alloys are represented. As the high temperature properties of cast materials improve, these alloys become more attractive, since great dimensional precision is now attainable in investment castings.

No.	Material	Nominal composition	Form and condition	Yield strength (0.2% offset), 1000 lb/sq in.	Tensile strength, 1000 lb/sq in.	Elongation, in 2 in., %	Hardness, Brinell	Comments
	NICKEL BASE							
30	Hastelloy X	Co 1.5 max Cr 22.0 W 0.6 C 0.20 max (cast) Fe 18.5 Mo 9.0 C 0.15 max (wrought) Ni bal.	Wrought sheet Mill-annealed As investment cast	52 — 46.5	113.2 67 —	43 17 —	194 172 —	
31	Hastelloy C	Cr 16.0 W 4.0 Mo 17.0 Fe 6.0 C 0.15 max Ni bal.	Sand-cast (annealed) Rolled (annealed) Investment cast	50 71 50	78 130 80	5 45 10	199 204 215	

TABLE C.5 (continued) Mechanical Properties of Metals and Alloys:* Typical Composition, Properties, and Uses of Common Materials

No.	Material	Nominal composition	Form and condition	Typical mechanical properties: Yield strength (0.2% offset), 1000 lb/sq in.	Tensile strength, 1000 lb/sq in.	Elongation, in 2 in., %	Hardness, Brinell	Comments
	NICKEL BASE (Cont.)							
32	Inconel 713C	Ni (+Co) bal. Mo 4.5 Al 6.0 Cr 13.0 Cb 2.0 Ti 0.6	Investment cast	102	120	6	—	
33	In 100	C 18.0 Mo 3.0 Al 55.0 V 1.0 Cr 10.0 Ti 4.7 Co 15.0	Cast					
34	Taz 8	C 125.0 Mo 4.0 W 4.0 Ta 8.0 Cr 6.0 Al 6.0 Zr 1.0 V 2.5	Cast					
35	Nimonic 90	Ni (+Co) 57.00 Mn 0.50 S 0.007 Cu 0.05 Al 1.65 Co 16.90 C 0.05 Fe 0.45 Si 0.20 Cr 20.55 Ti 2.60	Annealed; wrought	90	155	—	260	General elevated temperature applications
36	Inconel X	Ni (+Co) 72.85 Mn 0.65 S 0.007 Cu 0.05 Al 0.75 Cb (+Ta) 0.85 C 0.04 Fe 6.80 Si 0.30 Cr 15.0 Ti 2.50	Annealed Annealed; age-hardened	50 115	115 175	50 25	150 300	
37	Waspaloy	C 0.08 Mo 4.3 Co 13.5 Cr 19.5 Ti 3.0	Cold-rolled	270	275	8	Rc 51	
38	Rene 41	C 0.09 Mo 10.0 Al 1.5 Cr 19.0 Ti 3.1 Co 11.0	Wrought	100	145	—	—	

TABLE C.5 (continued) Mechanical Properties of Metals and Alloys:* Typical Composition, Properties, and Uses of Common Materials

No.	Material	Nominal composition	Form and condition	Typical mechanical properties: Yield strength (0.2% offset), 1000 lb/sq in.	Tensile strength, 1000 lb/sq in.	Elongation, in 2 in., %	Hardness, Brinell	Comments
39	Udimet 700	C 0.08 Cr 15.0 Mo 5.0 Ti 3.5 Al 4.3 Co 18.5	Cold-rolled	280	285	6	Rc 53	
40	T.D. Nickel	Ni 97.5 ThO_2 2.4	Extended and cold-worked	85	100	13	—	High temperature; jet engine parts
	COBALT BASE							
41	Haynes Stellite alloy 25 (L605)	C 0.15 max Cr 20.0 W 15.0 Ni 10.0 Co bal. Mn 1.5	Wrought sheet; mill annealed	63	140	60	244	Wrought products
42	Haynes Stellite alloy 21 AMS 5385 (cast)	C 0.25 Mo 5.5 Ni 2.5 Co bal. Cr 28.5	As investment cast	82	103	8	313 max	For castings

ALUMINUM ALLOYS

Although the strength of aluminum alloys is in general less than that attainable in ferrous alloys or copper-base alloys, their major advantage lies in their high strength-to-weight ratio due to the low density of aluminum. Aluminum alloys have good corrosion resistance for most applications except in alkaline solutions.

No.	Material	Nominal composition	Form and condition	Yield strength (0.2% offset), 1000 lb/sq in.	Tensile strength, 1000 lb/sq in.	Elongation, in 2 in., %	Hardness, Brinell	Comments
43	3003 ASTM B221	Cu 0.12 Al bal. Mn 1.2	Annealed-O Cold-rolled-H14 Cold-rolled-H18	6 21 27	16 22 29	40 16 10	28 40 55	Good formability, weldable, medium strength; chemical equipment
44	2017 ASTM B221	Mn 0.5 Mg 0.5 Cu 4.0 Al bal.	Annealed-O Heat-treated-T4	10 40	26 62	22 22	45 105	High strength; structural parts, aircraft, heavy forgings
45	2024 ASTM B211	Cu 4.5 Mg 1.5 Mn 0.6 Al bal.	Heat-treated-T4	47	68	19	120	
46	5052 ASTM B211	Cr 0.25 Al bal. Mg 2.5	Annealed-O Cold-rolled and stabilized-H34	13 31	28 38	30 14	47 68	Medium strength, good fatigue properties; street-light standards
47	ASTM B209		Cold-rolled and stabilized-H38	37	42	8	77	

TABLE C.5 (continued) Mechanical Properties of Metals and Alloys:* Typical Composition, Properties, and Uses of Common Materials

No.	Material	Nominal composition	Form and condition	Typical mechanical properties: Yield strength (0.2% offset), 1000 lb/sq in.	Tensile strength, 1000 lb/sq in.	Elongation, in 2 in., %	Hardness, Brinell	Comments
48	7075 ASTM B211	Cu 1.6 Mg 2.5 Cr 0.3 Al bal. Zn 5.6	Annealed-O	15	33	17	60	High strength, good corrosion resistance
			Heat-treated and artificially aged-T6	73	83	11	150	
49	380 ASTM SC84B	Si 9.0 Al bal. Cu 3.5	Die-cast	24	48	3	—	General purpose die casting
50	195 ASTM C4A	Si 0.8 Al bal. Cu 4.5	Sand-cast; heat-treated-T4	16	32	8.5	60	Structural elements, aircraft, and machines
			Sand-cast; heat-treated and artificially aged-T6	24	36	5	75	
51	214 ASTM G4A	Mg 3.8 Al bal.	Sand-cast-F	12	25	9	50	Chemical equipment, marine hardware, architectural
52	220 ASTM G10A	Mg 10.0 Al bal.	Sand-cast; heat-treated-T4	26	48	16	75	Strength with shock resistance; aircraft

COPPER ALLOYS

Because of their corrosion resistance and the fact that copper alloys have been used for many thousands of years, the number of copper alloys available is second only to the ferrous alloys. In general copper alloys do not have the high-strength qualities of the ferrous alloys, while their density is comparable. The cost per strength-weight ratio is high; however, they have the advantage of ease of joining by soldering, which is not shared by other metals that have reasonable corrosion resistance.

No.	Material	Nominal composition	Form and condition	Yield strength (0.2% offset), 1000 lb/sq in.	Tensile strength, 1000 lb/sq in.	Elongation, in 2 in., %	Hardness, Brinell	Comments
53	Copper ASTM B152 ASTM B124, B133 ASTM B1, B2, B3	Cu 99.9 plus	Annealed Cold-drawn Cold-rolled	10 40 40	32 45 46	45 15 5	42 90 100	Bus-bars, switches, architectural, roofing, screens
54	Gilding metal ASTM B36	Cu 95.0 Zn 5.0	Cold-rolled	50	56	5	114	Coinage, ammunition
55	Cartridge 70–30 brass ASTM B14 ASTM B19 ASTM B36 ASTM B134 ASTM B135	Cu 70.0 Zn 30.0	Cold-rolled	63	76	8	155	Good cold-working properties; radiator covers, hardware, electrical
56	Phosphor bronze 10% ASTM B103 ASTM B139 ASTM B159	Cu 90.0 Sn 10.0 P 0.25	Spring temper	—	122	4	241	Good spring qualities, high-fatigue strength

TABLE C.5 (continued) Mechanical Properties of Metals and Alloys:* Typical Composition, Properties, and Uses of Common Materials

No.	Material	Nominal composition		Form and condition	Typical mechanical properties				Comments
					Yield strength (0.2% offset), 1000 lb/sq in.	Tensile strength, 1000 lb/sq in.	Elongation, in 2 in., %	Hardness, Brinell	
57	Yellow brass (high brass) ASTM B36 ASTM B134 ASTM B135	Cu 65.0	Zn 35.0	Annealed Cold-drawn Cold-rolled (HT)	18 55 60	48 70 74	60 15 10	55 115 180	Good corrosion resistance; plumbing, architectural
58	Manganese bronze ASTM B138	Cu 58.5 Fe 1.0 Mn 0.3	Zn 39.2 Sn 1.0	Annealed Cold-drawn	30 50	60 80	30 20	95 180	Forgings
59	Naval brass ASTM B21	Cu 60.0 Sn 0.75	Zn 39.25	Annealed Cold-drawn	22 40	56 65	40 35	90 150	Condensor tubing; high resistance to salt-water corrosion
60	Muntz metal ASTM B111	Cu 60.0	Zn 40.0	Annealed	20	54	45	80	Condensor tubes; valve stress
61	Aluminum bronze ASTM B169, alloy A ASTM B124 ASTM B150	Cu 92.0	Al 8.0	Annealed Hard	25 65	70 105	60 7	80 210	
62	Beryllium copper 25 ASTM B194 ASTM B197 ASTM B196	Be 1.9 Co or Ni 0.25	Cu bal.	Annealed, solution-treated Cold-rolled Cold-rolled	32 104 70	70 110 190	45 5 3	B60 (Rockwell) B81 C40	Bellows, fuse clips, electrical relay parts, valves, pumps
63	Free-cutting brass	Cu 62.0 Pb 2.5	Zn 35.5	Cold-drawn	44	70	18	B80 (Rockwell)	Screws, nuts, gears, keys
64	Nickel silver 18% Alloy A (wrought) ASTM B122, No. 2	Cu 65.0 Ni 18.0	Zn 17.0	Annealed Cold-rolled Cold-drawn wire	25 70 —	58 85 105	40 4 —	70 170 —	Hardware, optical goods, camera parts
65	Nickel silver 13% (cast) 10A ASTM B149, No. 10A	Ni 12.5 Sn 2.0 Zn 20.0	Pb 9.0 Cu bal.	Cast	18	35	15	55	Ornamental castings, plumbing; good machining qualities
66	Cupronickel 10% ASTM B111 ASTM B171	Cu 88.35 Fe 1.25	Ni 10.0 Mn 0.4	Annealed Cold-drawn tube	22 57	44 60	45 15	— —	Condensor, salt-water piping

TABLE C.5 (continued) Mechanical Properties of Metals and Alloys:* Typical Composition, Properties, and Uses of Common Materials

No.	Material	Nominal composition	Form and condition	Typical mechanical properties: Yield strength (0.2% offset), 1000 lb/sq in.	Tensile strength, 1000 lb/sq in.	Elongation, in 2 in., %	Hardness, Brinell	Comments
67	Cupronickel	Cu 70.0 Ni 30.0	Wrought					Heat-exchanger process equipment, valves
68	Red brass (cast) ASTM B30, No. 4A	Cu 85.0 Zn 5.0 Pb 5.0 Sn 5.0	As-cast	17	35	25	60	
69	Silicon bronze ASTM B30, alloy 12A	Si 4.0 Fe 2.0 Zn 4.0 Al 1.0 Mn 1.0	Castings					Cheaper substitute for tin bronze
70	Tin bronze ASTM B30, alloy 1B	Sn 8% Zn 4.0	Castings					Bearings, high-pressure bushings, pump impellers
71	Navy bronze		Cast					
	TIN AND LEAD-BASE ALLOYS							
	Major uses for these alloys are as "white"-metal bearing alloys, extruded cable sheathing, and solders. Tin forms the basis of pewter used for culinary applications.							
72	Lead-base Babbitt ASTM B23, alloy 19	Pb 85.0 Sn 5.0 Sb 10.0 As 0.6 Cu 0.5	Chill cast	—	10	5	19	Bearings, light loads and low speeds
73	Arsenical-lead Babbitt ASTM B23, alloy 15	Pb 83.0 Sn 1.0 Sb 16.0 As 1.1 Cu 0.6	Chill cast	—	10.3	2	20	Bearings, high loads and speeds, diesel engines, steel mills
74	Chemical lead	Pb 99.9 Cu 0.06 Bi 0.005 max	Rolled 95%	1.9	2.5	50	5	
75	Antimonial lead (hard lead)	Pb 94.0 Sb 6.0	Chill cast Rolled 95%	— —	6.8 4.1	22 47	(500 kg) 9	Good corrosion resistance and strength
76	Calcium lead	Pb 99.9 Ca 0.025 Cu 0.10	Extruded and aged	—	4.5	25	—	Cable sheathing, creep-resistant pipe
77	Tin Babbitt alloy ASTM B23–61, grade 1	Sb 4.5 Sn bal. Cu 4.5	Chill cast	—	9.3	2	17	General bearings and die casting
78	Tin die-casting alloy ASTM B102–52	Sb 13.0 Sn bal. Cu 5.0	Die-cast	—	10	1	29	Die-casting alloy

TABLE C.5 (continued) Mechanical Properties of Metals and Alloys:* Typical Composition, Properties, and Uses of Common Materials

No.	Material	Nominal composition		Form and condition	Typical mechanical properties				Comments
					Yield strength (0.2% offset), 1000 lb/sq in.	Tensile strength, 1000 lb/sq in.	Elongation, in 2 in., %	Hardness, Brinell	
79	Pewter	Sn 91.0 Cu 2.0	Sb 7.0	Rolled sheet, annealed	—	8.6	40	9.5	Ornamental and household items
80	Solder 50–50	Sn 50.0	Pb 50.0	Cast	4.8	6.1	60	14	General-purpose solder
81	Solder	Sn 20.0	Pb 80.0	Cast	3.6	5.8	16	11	Coating and joining, filling seams on automobile bodies

MAGNESIUM ALLOYS

Because of their low density these alloys are attractive for use where weight is at a premium. The major drawback to the use of these alloys is their ability to ignite in air (this can be a problem in machining); they are also costly. Magnesium alloys are used in both the wrought and die-cast forms, the latter being the most frequently used form.

No.	Material	Nominal composition		Form and condition	Yield strength (0.2% offset), 1000 lb/sq in.	Tensile strength, 1000 lb/sq in.	Elongation, in 2 in., %	Hardness, Brinell	Comments
82	Magnesium alloy AZ31B	Zn 1.0 Al 3.0	Mn 0.20 min Mg bal.	Rolled-plate (strain-hardened, then partially annealed)	24	37	18	—	Structural applications of medium strength
				Rolled-sheet (strain-hardened, then partially annealed)	32	42	15	73	
				Annealed	22	37	21	56	
				Extruded	28	38	14	—	
83	Magnesium alloy AZ80A	Zn 0.5 Al 8.5	Mn 0.15 min Mg bal.	Extruded	36	49	11	60	General extruded and forged products
				Extruded (age-hardened)	39	53	6	82	
				Forged (age-hardened)	34	50	6	72	
84	Magnesium alloy AZ92A	Zn 2.0 Al 9.0	Mn 0.10 min Mg bal.	Sand-cast (as cast)	14	24	6	50	
				Sand-cast (solution heat-treated)	14	40	12	55	
				Sand-cast (solution heat-treated and aged)	19	40	5	83	
				Sand-cast (age-hardened)	16	30	18	—	
				Sand-cast and tempered	22	40	3	81	Pressure-tight sand and permanent mold castings; high UTS and good yield strength
85	Magnesium alloy ZK60A	Zn 5.7 Zr 0.55	Mg bal.	Extruded	43	52	12	82	

TABLE C.5 (continued) Mechanical Properties of Metals and Alloys:* Typical Composition, Properties, and Uses of Common Materials

No.	Material	Nominal composition	Form and condition	Typical mechanical properties: Yield strength (0.2% offset), 1000 lb/sq in.	Tensile strength, 1000 lb/sq in.	Elongation, in 2 in., %	Hardness, Brinell	Comments
86	Magnesium alloy AZ91A and AZ91B	Zn 0.6 Mn 0.13 min Al 9.0 Mg bal.	Die-cast (as cast)	22	33	3	67	General die-casting applications
	BERYLLIUM							
87	Beryllium		 Hot-pressed Cross-rolled	27 38 40 60	33 51 60 90	1–3 10–40	— —	Windows, X-ray tubes Moderator- and reflector-cladding nuclear reactors; heat-shield and structural-member missiles
	NICKEL ALLOYS							

Nickel and its alloys are expensive and used mainly either for their high-corrosion resistance in many environments or for high-temperature and strength applications. (See Super Alloys, above.)

No.	Material	Nominal composition	Form and condition	Yield strength (0.2% offset), 1000 lb/sq in.	Tensile strength, 1000 lb/sq in.	Elongation, in 2 in., %	Hardness, Brinell	Comments
88	Nickel (cast)	Ni 95.6 Cu 0.5 Fe 0.5 Mn 0.8 Si 1.5 C 0.8	As cast	25	57	22	110	Good corrosion-resistance applications
89	K Monel	Ni (+Co) 65.25 C 0.15 Mn 0.60 Fe 1.00 S 0.005 Si 0.15 Cu 29.60 Al 2.75 Ti 0.45	Annealed Annealed, age-hardened Spring Spring, age-hardened	45 100 140 160	100 155 150 185	40 25 5 10	155 270 300 335	High strength and corrosion resistance; aircraft parts, valve stems, pumps
90	A nickel ASTM B160 ASTM B161 ASTM B162	Ni (+Co) 99.40 C 0.06 Fe 0.15 Mn 0.25 Si 0.05 S 0.005 Cu 0.05	Annealed Hot-rolled Cold-drawn Cold-rolled	20 25 70 95	70 75 95 105	40 40 25 5	100 110 170 210	Chemical industry for resistance to strong alkalis, plating nickel
91	Duranickel	Ni (+Co) 93.90 C 0.15 Fe 0.15 Mn 0.25 Si 0.55 S 0.005 Al 4.50 Cu 0.05 Ti 0.45	Annealed Annealed, age-hardened Spring Spring, age-hardened	45 125 — —	100 170 175 205	40 25 5 10	160 330 320 370	High strength and corrosion resistance; pump rods, shafts, springs

TABLE C.5 (continued) Mechanical Properties of Metals and Alloys:* Typical Composition, Properties, and Uses of Common Materials

No.	Material	Nominal composition	Form and condition	Typical mechanical properties				Comments
				Yield strength (0.2% offset), 1000 lb/sq in.	Tensile strength, 1000 lb/sq in.	Elongation, in 2 in., %	Hardness, Brinell	
92	Cupronickel 55–45 (Constantan)	Cu 55.0 Ni 45.0	Annealed Cold-drawn Cold-rolled	30 50 65	60 65 85	45 30 20	— -- —	Electrical-resistance wire; low temperature coefficient, high resistivity
93	Nichrome	Ni 80.0 Cr 20.0						Heating elements for furnaces
94	"S" Monel	Ni 60.0 Cu 29.0 Fe 2.50 max Mn 1.5 max Si 4.0 Al 0.5 max	Sand-casting	80–115	110–145	2	270–350	High-strength casting alloy; good bearing properties for valve seats

TITANIUM ALLOYS

The main application for these alloys is in the aerospace industry. Because of the low density and high strength of titanium alloys, they present excellent strength-to-weight ratios.

No.	Material	Nominal composition	Form and condition	Yield strength (0.2% offset), 1000 lb/sq in.	Tensile strength, 1000 lb/sq in.	Elongation, in 2 in., %	Hardness, Brinell	Comments
95	Commercial titanium ASTM B265–58T	Ti 99.4	Annealed at 1100 to 1350°F (593 to 732°C)	70	80	20	—	Moderate strength, excellent fabricability; chemical industry pipes
96	Titanium alloy ASTM B265–58T–5 Ti–6 Al–4V		Water-quenched from 1750°F (954°C); aged at 1000°F (538°C) for 2 hr	160	170	13	—	High-temperature strength needed in gas-turbine compressor blades
97	Titanium alloy Ti–4 Al–4Mn		Water-quenched from 1450°F (788°C); aged at 900°F (482°C) for 8 hr	170	185	13	—	Aircraft forgings and compressor parts
98	Ti–Mn alloy ASTM B265–58T–7	Fe 0.5 Ti bal. Mn 7.0–8.0	Sheet	140	150	18	—	Good formability, moderate high-temperature strength; aircraft skin

ZINC ALLOYS

A major use for these alloys is for low-cost die-cast products, such as household fixtures, automotive parts, and trim.

No.	Material	Nominal composition	Form and condition	Yield strength (0.2% offset), 1000 lb/sq in.	Tensile strength, 1000 lb/sq in.	Elongation, in 2 in., %	Hardness, Brinell	Comments
99	Zinc ASTM B69	Cd 0.35 Zn bal. Pb 0.08	Hot-rolled	—	19.5	65	38	Battery cans, grommets, lithographer's sheet

TABLE C.5 (continued) Mechanical Properties of Metals and Alloys:* Typical Composition, Properties, and Uses of Common Materials

No.	Material	Nominal composition	Form and condition	Typical mechanical properties: Yield strength (0.2% offset), 1000 lb/sq in.	Tensile strength, 1000 lb/sq in.	Elongation, in 2 in., %	Hardness, Brinell	Comments
100	Zilloy-15	Cu 1.00 Zn bal. Mg 0.010	Hot-rolled Cold-rolled	— —	29 36	20 25	61 80	Corrugated roofs, articles with maximum stiffness
101	Zilloy-40	Cu 1.00 Zn bal.	Hot-rolled Cold-rolled	— —	24 31	50 40	52 60	Weatherstrip, spun articles
102	Zamac-5 ASTM 25	Zn (99.99% pure remainder) Al 3.5–4.3 Cu 0.75–1.25 Mg 0.03–0.08	Die-cast	—	47.6	7	91	Die casting for automobile parts, padlocks; used also for die material
	ZIRCONIUM ALLOYS These alloys have good corrosion resistance but are easily oxidized at elevated temperatures in air. The major application is for use in nuclear reactors.							
103	Zirconium, commercial	O_2 0.07 C 0.15 Hf 1.90 Zr bal.	Annealed	40	65	27	B80 (Rockwell)	
104	Zircaloy 2	Hf 0.02 Ni 0.05 Fe 0.15 Other 0.25 Sn 1.46 Zr bal.	Annealed	50	75	22	B90 (Rockwell)	Nuclear power-reactor cores at elevated temperatures

*Compiled from various sources.

TABLE C.6 Miscellaneous Properties of Metals and Alloys

Part a. Pure Metals

At Room Temperature

	PROPERTIES (TYPICAL ONLY)						
Common name	*Thermal conductivity, Btu/hr ft °F*	*Specific gravity*	*Coeff. of linear expansion, μ in./in. °F*	*Electrical resistivity, microhm-cm*	*Poisson's ratio*	*Modulus of elasticity, millions of psi*	*Approximate melting point, °F*
Aluminum	137	2.70	14	2.655	0.33	10.0	1220
Antimony	10.7	6.69	5	41.8		11.3	1170
Beryllium	126	1.85	6.7	4.0	0.024–.030	42	2345
Bismuth	4.9	9.75	7.2	115		4.6	521
Cadmium	54	8.65	17	7.4		8	610
Chromium	52	7.2	3.3	13		36	3380
Cobalt	40	8.9	6.7	9		30	2723
Copper	230	8.96	9.2	1.673	0.36	17	1983
Gold	182	19.32	7.9	2.35	0.42	10.8	1945
Iridium	85.0	22.42	3.3	5.3		75	4440
Iron	46.4	7.87	6.7	9.7		28.5	2797
Lead	20.0	11.35	16	20.6	0.40–.45	2.0	621
Magnesium	91.9	1.74	14	4.45	0.35	6.4	1200
Manganese		7.21–7.44	12	185		23	2271
Mercury	4.85	13.546		98.4			−38
Molybdenum	81	10.22	3.0	5.2	0.32	40	4750
Nickel	52.0	8.90	7.4	6.85	0.31	31	2647
Niobium (Columbium)	30	8.57	3.9	13		15	4473
Osmium	35	22.57	2.8	9		80	5477
Platinum	42	21.45	5	10.5	0.39	21.3	3220
Plutonium	4.6	19.84	30	141.4	0.15–.21	14	1180
Potassium	57.8	0.86	46	7.01			146
Rhodium	86.7	12.41	4.4	4.6		42	3569
Selenium	0.3	4.8	21	12.0		8.4	423
Silicon	48.3	2.33	2.8	1×10^5		16	2572
Silver	247	10.50	11	1.59	0.37	10.5	1760
Sodium	77.5	0.97	39	4.2			208
Tantalum	31	16.6	3.6	12.4	0.35	27	5400
Thorium	24	11.7	6.7	18	0.27	8.5	3180
Tin	37	7.31	11	11.0	0.33	6	450
Titanium	12	4.54	4.7	43	0.3	16	3040
Tungsten	103	19.3	2.5	5.65	0.28	50	6150
Uranium	14	18.8	7.4	30	0.21	24	2070
Vanadium	35	6.1	4.4	25		19	3450
Zinc	66.5	7	19	5.92	0.25	12	787

TABLE C.6 Miscellaneous Properties of Metals and Alloys

Part b. Commercial Metals and Alloys

CLASSIFICATION AND DESIGNATION		PROPERTIES (TYPICAL ONLY)					
Material No. (from Table 1-57)	*Common name and classification*	*Thermal conductivity, Btu/hr ft °F*	*Specific gravity*	*Coeff. of linear expansion, μ in./in. °F*	*Electrical resistivity, microhm-cm*	*Modulus of elasticity, millions of psi*	*Approximate melting point, °F*
1	Ingot iron (included for comparison)	42.	7.86	6.8	9.	30	2800
2	Plain carbon steel						
	AISI–SAE 1020	30.	7.86	6.7	10.	30	2760
15	Stainless steel type 304	10.	8.02	9.6	72.	28	2600
19	Cast gray iron						
	ASTM A48–48, Class 25	26.	7.2	6.7	67.	13	2150
21	Malleable iron						
	ASTM A47	—	7.32	6.6	30.	25	2250
22	Ductile cast iron						
	ASTM A339, A395	19	7.2	7.5	60.	25	2100
24	Ni-resist cast iron, type 2	23	7.3	9.6	170.	15.6	2250
29	Cast 28–7 alloy (HD)						
	ASTM A297–63T	1.5	7.6	9.2	41.	27	2700
31	Hastelloy C	5	3.94	6.3	139.	30	2350
36	Inconel X, annealed	9	8.25	6.7	122.	31	2550
41	Haynes Stellite alloy 25 (L605)	5.5	9.15	7.61	88.	34	2500
43	Aluminum alloy 3003, rolled						
	ASTM B221	90	2.73	12.9	4.	10	1200
44	Aluminum alloy 2017, annealed						
	ASTM B221	95	2.8	12.7	4.	10.5	1185
49	Aluminum alloy 380						
	ASTM SC84B	56	2.7	11.6	7.5	10.3	1050
53	Copper						
	ASTM B152, B124, B133, B1, B2, B3	225	8.91	9.3	1.7	17	1980
57	Yellow brass (high brass)						
	ASTM B36, B134, B135	69	8.47	10.5	7.	15	1710
61	Aluminum bronze						
	ASTM B169, alloy A; ASTM B124, B150	41	7.8	9.2	12.	17	1900
62	Beryllium copper 25						
	ASTM B194	7	8.25	9.3	—	19	1700
64	Nickel silver 18% alloy A (wrought)						
	ASTM B122, No. 2	19	8.8	9.0	29.	18	2030
67	Cupronickel 30%	17	8.95	8.5	35.	22	2240
68	Red brass (cast)						
	ASTM B30, No. 4A	42	8.7	10.	11.	13	1825
74	Chemical lead	20	11.35	16.4	21.	2	621
75	Antimonial lead (hard lead)	17	10.9	15.1	23.	3	554
80	Solder 50–50	26	8.89	13.1	15.	—	420
82	Magnesium alloy AZ31B	45	1.77	14.5	9.	6.5	1160
89	K Monel	11	8.47	7.4	58.	26	2430
90	Nickel						
	ASTM B160, B161, B162	35	8.89	6.6	10.	30	2625
92	Cupronickel 55–45 (Constantan)	13	8.9	8.1	49.	24	2300
95	Commercial titanium	10	5.	4.9	80.	16.5	3300
99	Zinc						
	ASTM B69	62	7.14	18	6.	—	785
103	Zirconium, commercial	10	6.5	2.9	41.	12	3350

*Compiled from several sources.

TABLE C.7 Composition and Melting Points of Binary Eutectic Alloys:* Binary Alloys and Solid Solutions of Metallic Components

This table represents most of the common binary combinations of metals. For many pairs no eutectic exists; for many others, the informations is uncertain or unavailable. In a fair number of cases, there is complete mutual solubility in all proportions; hence, there is a smooth temperature vs. composition curve, with no point of inflection from the melting point of one constituent to that of the other. For purposes of comparison, all values must be considered approximate in view of the experimental difficulties and the many sources of data.

Those pairs for which the liquidus curve exhibits more than one cusp are designated by a superscript *a*. In a few cases, the cusp selected for this table does not represent the lowest possible melting point for the binary mixture.

Constituents		*Composition*		*Melting point*	
A	*B*	*Mol % B*	*Weight % B*	*K*	*deg F*
Ag	Al	57	25	835	1 044
Ag	As	24	18	813	1 004
Ag	Ca^a	37	18	820	1 017
Ag	Ce^a	80	84	798	977
Ag	Cu	40	28	1 050	1 431
Ag	Ge	25	18	924	1 204
Ag	La^a	72	77	791	964
Ag	Li	99	89	418	293
Ag	Mg^a	83	52	745	882
Ag	Pb	95.3	97.5	577	579
Ag	Pd	25.9	25.6	924	1 204
Ag	Sb	41	44	758	905
Ag	Si	10.5	2.96	1 110	1 539
Ag	Sr^a	77	73	709	817
Ag	Te	65	69	623	662
Ag	Th	7.6	15	1 167	1 641
Ag	Zr	97	93	1 100	1 521
Al	Au^a	59.5	90.0	842	1 056
Al	Ca^a	65	73	818	1 013
Al	Cd	81	90	1 650	2 511
Al	Ce	69	92	928	1 211
Al	Cu^a	17.3	33.0	821	1 018
Al	Fe	32	49.34	1 426	2 107
Al	Ge	29	55	700	801
Al	In	5	18	910	1 179
Al	Mg	70	67.0	710	819
Al	Ni^a	76	87	1 658	2 525
Al	Pt^a	57	90	1 533	2 300
Al	Si	13	13	850	1 071
Al	Th	80	97	1 510	2 259
Al	Zn	88.7	95.0	655	720
As	Co	75	70	1 189	1 681
As	Cu^a	81.6	78.0	958	1 265
As	Fe	75	69	1 103	2 017
As	In	13	18	1 004	1 348
As	Mn	57	49	1 143	1 598
As	Ni^a	63	57	1 077	1 479
As	Sb	80	87	878	1 121
As	Sn^a	40	51	852	1 074
As	Zn^a	20	18	996	1 333
Au	Bi	86.8	85	514	466
Au	Cd	70	57.1	773	932
Au	Ce^a	86	81	793	968
Au	Ge	27	12	629	673
Au	La^a	83	78	834	1 042
Au	Mg	93	62	848	1 067
Au	Mn^a	32	12	1 233	1 760
Au	Na	17	2.3	1 149	1 609
Au	Pb	84	85	488	419
Au	Sb	34.8	24.8	633	680
Au	Si	18.6	3.15	636	685
Au	Sn^a	29.3	19.9	551	532
Au	Te	88	83	689	781
Au	Tl	72	73	404	268
Au	U	14	16	1 128	1 571
B	Hf	13	71	2 130	3 375
B	Ni	57	88	1 263	1 814
B	Ti	7	25	1 700	2 601
B	Zr	88	98	1 920	2 997
Ba	Mg	97	87	891	1 144
Be	Ni	33	76	1 468	2 183
Be	Pu	97	99	910	1 179
Be	Si	33	61	1 363	1 994
Be	Ti	75	94	1 300	2 061
Be	Y	61	94	1 390	2 043
Be	Zr	65	95	1 250	1 791
Bi	Ca	88	58.5	1 059	1 447
Bi	Cd	56	40	420	297
Bi	In^a	78	66	340	153
Bi	K	50	16	615	648
Bi	Mg	85	40	820	1 017
Bi	Na	22	3.0	500	441
Bi	Pb	44	44	397	255
Bi	Sn	57	43	415	288
Bi	Te	90	84	686	775
Bi	Tl^a	53	52	465	378
C	Cr	87	96	1 775	2 736
C	Hf	35	88	3 450	5 751
C	Mo	17	45	2 480	4 005
C	Nb	40	84	3 580	5 985

*Compiled from several sources.

TABLE C.7 (continued) Composition and Melting Points of Binary Eutectic Alloys:* Binary Alloys and Solid Solutions of Metallic Components

Constituents		*Composition*		*Melting point*	
A	*B*	*Mol % B*	*Weight % B*	*K*	*deg F*
C	Ti	36	69	3 050	5 031
C	V	84	96	1 900	2 961
C	W	59	96	2 980	4 905
Ca	Cu	51	62	833	1 040
Ca	Mg[a]	32	22	718	833
Ca	Na	22	14	983	1 310
Ca	Ni	16	22	878	1 121
Ca	Sn	19	41	1 010	1 359
Cd	Cu	52	38	810	999
Cd	In	74	74	400	261
Cd	Pb	71	82	540	513
Cd	Pu	40	59	1 170	1 647
Cd	Sb	7.4	8	563	554
Cd	Sn	68	69	450	351
Cd	Tl	73	83	475	396
Cd	Zn	27	18	540	513
Ce	Cu[a]	28	15	688	779
Ce	Ru	33	26	923	1 202
Co	Gd	65	83	913	1 184
Co	Mo	27	38	1 610	2 439
Co	Nb	15	22	1 500	2 241
Co	Si[a]	71	54	1 486	2 215
Co	Sn	21	35	1 380	2 025
Co	Ti[a]	22	19	1 430	2 115
Co	V	41	38	1 521	2 278
Cr	Mo	14	23	1 973	3 092
Cr	Ni	46	47	1 610	2 439
Cr	Ta	13	34	1 950	3 051
Cr	Ti	86	85	950	1 251
Cr	V	33	32	2 050	3 231
Cs	K	50	23	235	−36
Cs	Na	20.9	4.37	241	−26
Cs	Rb	50	39	282	48
Cu	Ge	34	37	913	1 184
Cu	Mg[a]	85.5	69.3	758	905
Cu	Mn	37	34	1 143	1 598
Cu	Pb	15	36	1 230	1 755
Cu	Pr[a]	69	83	745	882
Cu	Sb[a]	63	76	800	981
Cu	Si	30	16	1 075	1 476
Cu	Te	69	82	617	207
Cu	Ti[a]	27	22	1 133	1 580
Cu	Tl	14.5	35.3	1 357	1 983
Cu	U	8.2	25	1 213	1 724
Cu	Zr	9.4	13	1 253	1 796
Fe	Gd	69	86	1 123	1 562
Fe	Mo	21	31	1 725	2 646
Fe	Nb	12	18.49	1 643	2 498
Fe	Sb	88	94.10	1 021	1 378
Fe	Si[a]	35	21	1 475	2 196
Fe	Sn	31	49	1 400	2 061
Fe	Y	65	75	1 173	1 652
Fe	Zr[a]	11	17	1 600	2 421
Ga	Mg[a]	80	58	698	797
Ga	Ni	70	66	1 477	2 199
Gd	Ni[a]	32	15	943	1 238
Ge	Mg	38	17	953	1 256
Ge	Mn[a]	48	41	970	1 287
Hf	Ta	24	24	1 300	1 881
In	Ni	30	17.97	1 143	1 598
In	Sb	68	69	780	945
In	Sn	47	48	390	243
Ir	Mo	68	52	2 350	3 771
Ir	Nb	55	23	2 110	3 339
Ir	W	22	12	2 590	4 203
K	Na	32	21.67	260	−8.6
K	Rb	70	84	307	93
K	Sb[a]	68	84	680	765
K	Tl	84	96	440	333
La	Mg[a]	38	9.7	970	1 287
La	Pb[a]	11	15	1 049	1 429
La	Sn[a]	10	9	993	1 328
La	Tl	16	22	913	1 184
Mg	Ni	11	22.98	780	945
Mg	Pr	4.9	23	858	1 085
Mg	Pu	15	63	815	1 008
Mg	Sb[a]	86	97	855	1 080
Mg	Si	53	57	1 223	1 742
Mg	Sr[a]	70	89	699	799
Mg	Th	7	42	855	1 080
Mg	Zn	30	53	615	648
Mn	Ni	40	42	1 300	1 881
Mn	Pd	26	41	1 398	2 057
Mn	Sb	82	91	843	1 058
Mn	Ti[a]	9	7.9	1 460	2 169
Mn	U[a]	75	93	988	1 319
Mn	Y[a]	65	75	1 163	1 634
Mo	Nb	66	65	2 570	4 167
Mo	Ni	64	52	1 590	2 403
Mo	Os	21	34	2 650	4 311
Mo	Pd	54	57	2 020	3 177
Mo	Re	48	64	2 780	4 545
Mo	Ru	41	42	2 200	3 501
Mo	Si[a]	17	5.7	2 350	3 771
Na	Rb	82.1	94.5	269	25
Na	Sb	60	89	678	761
Na	Sn	37	75	718	833
Na	Te	55	87	592	606
Nb	Ni	58	47	1 450	2 151
Nb	Pt	54	71	1 970	3 087
Nb	Rh	45	31	1 770	2 727
Nb	Ru[a]	64	49	2 050	3 231
Nb	Zr	77	77	2 010	3 159
Ni	Sb	22	36.90	1 375	2 016
Ni	Sn	19	32.16	1 403	2 066
Ni	Th[a]	35	68	1 303	1 886
Ni	Ti[a]	39	34	1 390	2 043
Ni	V	52	48	1 473	2 192
Ni	W	20.7	45	1 773	2 732
Ni	Zn	69	71	1 148	1 607

TABLE C.7 (continued) Composition and Melting Points of Binary Eutectic Alloys:* Binary Alloys and Solid Solutions of Metallic Components

Constituents		*Composition*		*Melting point*		*Constituents*		*Composition*		*Melting point*	
A	*B*	*Mol % B*	*Weight % B*	*K*	*deg F*	*A*	*B*	*Mol % B*	*Weight % B*	*K*	*deg F*
Pb	Pr	40	31	1 315	1 908	Si	Th[a]	88	98	1 710	2 619
Pb	Pt	5.3	5.0	563	554	Si	Ti[a]	86	91	1 600	2 421
Pb	Sb	18	11	520	477	Si	Zr[a]	9	24	1 570	2 367
Pb	Sn	73	61	460	369	Sn	Te	84	85	678	761
Pb	Te	85	78	680	765	Sn	Tl	31	44	440	333
Pb	Ti	92	74	998	1 337	Sn	Zn	16	9.5	465	378
Pd	Sb	89	90	868	1 103	Te	Tl	30	41	483	410
Pt	Sn	40	29	1 345	1 962	Th	Ti	40	12	1 463	2 174
Pu	Zn	73	42	1 100	1 521	Th	Zn[a]	49	21	1 220	1 737
Re	W	26	26	3 100	5 121	Ti	U	17	51	933	1 220
Sb	Tl	70	80	468	383	Ti	Y	6.8	12	1 593	2 408
Sb	Zn	33	21	780	945	Ti	Zr	50	66	790	963
Sb	Zr	82	77	1 700	2 601	U	Zr	70	47	879	1 123
Se	Sn	39	49	913	1 184						
Se	Tl	26	48	424	304						

REFERENCES

"Selected Values of Thermodynamic Properties of Metals and Alloys," R. Hultgren, R.L. Orr, P.D. Anderson, K.K. Kelley, John Wiley & Sons, inc., 1963; a supplement to this publication has been issued periodically by the University of California, 1964-1971.

"Constitution of Binary Alloys," 2nd ed., M. Hansen, McGraw-Hill Book Company, 1958.

"Metals Reference Book," 4th ed., C.J. Smithells, Vol. 2, Butterworth & co., London, 1967.

"Handbook of Binary Metallic Systems," 2 volumes; translated from Russian, Israel Program for Scientific Translations, JJerusalem. Available from Clearinghouse for Federal Scientific and Technical Information, Springfield, Virginia 22151.

See also *Trans. AIME, J. Inst. Metals,* and *Z. Metallkunde,* by indexes.

TABLE C.8 Melting Points of Mixtures of Metals**

Melting Points, °C

Metals		*Percentage of metal in second column*										
		0%*	10%	20%	30%	40%	50%	60%	70%	80%	90%	100%
Pb.	Sn.	326	295	276	262	240	220	190	185	200	216	232
	Bi.	322	290	...	...	179	145	126	168	205	...	268
	Te.	322	710	790	880	917	760	600	480	410	425	446
	Ag.	328	460	545	590	620	650	705	775	840	905	959
	Na.	...	360	420	400	370	330	290	250	200	130	96
	Cu.	326	870	920	925	945	950	955	985	1005	1020	1084
	Sb.	326	250	275	330	395	440	490	525	560	600	632
Al.	Sb.	650	750	840	925	945	950	970	1000	1040	1010	632
	Cu.	650	630	600	560	540	580	610	755	930	1055	1084
	Au.	655	675	740	800	855	915	970	1025	1055	675	1062
	Ag.	650	625	615	600	590	580	575	570	650	750	954
	Zn.	654	640	620	600	580	560	530	510	475	425	419
	Fe.	653	860	1015	1110	1145	1145	1220	1315	1425	1500	1515
	Sn.	650	645	635	625	620	605	590	570	560	540	232
Sb.	Bi.	632	610	590	575	555	540	520	470	405	330	268
	Ag.	630	595	570	545	520	500	505	545	680	850	959
	Sn.	622	600	570	525	480	430	395	350	310	255	232
	Zn.	632	555	510	540	570	565	540	525	510	470	419

Metals		*Percentage of metal in second column*										
		0%*	10%	20%	30%	40%	50%	60%	70%	80%	90%	100%
Ni.	Sn.	1455	1380	1290	1200	1235	1290	1305	1230	1060	800	232
Na.	Bi.	96	425	520	590	645	690	720	730	715	570	268
	Cd.	96	125	185	245	285	325	330	340	360	390	322
Cd.	Ag.	322	420	520	610	700	760	805	850	895	940	954
	Tl.	321	300	285	270	262	258	245	230	210	235	302
	Zn.	322	280	270	295	313	327	340	355	370	390	419
Au.	Cu.	1063	910	890	895	905	925	975	1000	1025	1060	1084
	Ag.	1064	1062	1061	1058	1054	1049	1039	1025	1006	982	963
	Pt.	1075	1125	1190	1250	1320	1380	1455	1530	1610	1685	1775
K.	Na.	62	17.5	−10	−3.5	5	11	26	41	58	77	97.5
	Hg.	...	...	...	...	...	90	110	135	162	265	...
	Tl.	62.5	133	165	188	205	215	220	240	280	305	301
Cu.	Ni.	1080	1180	1240	1290	1320	1355	1380	1410	1430	1440	1455
	Ag.	1082	1035	990	945	910	870	830	788	814	875	960
	Sn.	1084	1005	890	755	725	680	630	580	530	440	232
	Zn.	1084	1040	995	930	900	880	820	780	700	580	419
Ag.	Zn.	959	850	755	705	690	660	630	610	570	505	419
	Sn.	959	870	750	630	550	495	450	420	375	300	232
Na.	Hg.	96.5	90	80	70	60	45	22	55	95	215	...

*The data in this table are compiled from various sources—hence, the variations in the melting point of the metals as shown in this column.

**Based largely on: "Smithsonian Physical Tables," 9th rev. ed., W.E. Forsythe, Ed., The Smithsonian Institution, 1956.

TABLE C.9 Trade Names, Composition, and Manufacturers of Various Plastics

Trade name	Composition	Manufacturer
Abson	Acrylonitrile-butadiene, ABS polymers	B. F. Goodrich Chemical Co.
Alathon	Polyethylene resins	E. I. du Pont de Nemours & Co., Inc.
Alkor	Furane resin cement	Atlas Minerals & Chemicals Div., The Electric Storage Battery Co.
Amres	Phenolics, urea, and melamine resins	American Marietta Co., Pacific Resins & Chemicals, Inc.
Araldite	Epoxy resins	CIBA Products Co., Div. CIBA Corp.
Atlac	Polyester resins	Atlas Chemical Industries, Inc.
Bakelite	Acrylics, epoxies, phenolics, polyethylenes, copolymers	Union Carbide Corp., Chemicals and Plastics Div.
Bavick-11	Methyl methacrylate and methylstyrene copolymer	J. T. Baker Chemical Co.
Boltaflex	Supported and unsupported flexible vinyl sheeting	The General Tire & Rubber Co.
Boltaron	Rigid polyvinyl chloride sheet	The General Tire & Rubber Co., Chemical & Plastics Div.
Butacite	Polyvinyl butyral resins	E. I. du Pont de Nemours & Co., Inc.
Conolite	Polyester resins and laminates	Shellmar-Betner, Div. Continental Can Co. Woodall Industries, Inc., Conolite Div.
Corvel	Epoxies, vinyls Fusion-bond finishes	The Polymer Corp., Export-Polypenco Div.
Cumar	Paracoumarone-indene resins	Allied Chemical Corp., Plastics Div.
Cycolac	ABS polymers, acrylonitrile-butadiene-styrene copolymers	Marbon Chemical Div., Borg-Warner Corporation
Dacovin	Polyvinyl chlorides	Diamond Shamrock Corp.
Dapon	Diallyl phthalate resins	FMC Corp., Organic Div.
Delrin	Acetal resin and pipe	E. I. du Pont de Nemours & Co., Inc.
Dylan	Polyethylene	Sinclair-Koppers Co.
Dylene	Polystyrene	Sinclair-Koppers Co.
Dylite	Expandable polystyrene	Sinclair-Koppers Co.
Epi-Rez	Epoxy resins	Celanese Coatings Co., Celanese Resin Div.
Epolene	Low molecular-weight polyethylene resins	Eastman Chemical Products, Inc., Sub. Eastman Kodak Company
Epoxical	Epoxy resins	United States Gypsum Co.
Epon	Epoxy resins and curing agents	The Shell Chemical Company, Plastics and Resins Div.
Escon	Polypropylene resins	Enjay Chemical Co., Div. Humble Oil & Refining Company
Estane	Polyurethane materials	B. F. Goodrich Chemical Company
Fluorogreen	Teflon with glass and ceramic fibers, fluorocarbons	John L. Dore Co.
Fluororay	Ceramic-filled fluorocarbons	Raybestos-Manhattan, Inc., Plastic Products Div.
Formica	Melamines	Formica Corp. of American Cyanamid
Forticel	Cellulose propionate sheet films, molding powders	Celanese Plastics Co.
Fortiflex	Polyethylene resins	Celanese Plastics Co.
Fosta-Tuf-Flex	Polystyrene, high-impact	Foster-Grant, Inc.
Furnane	Furanes	Atlas Minerals & Chemicals Div., The Electric Storage Battery Co.
GenEpoxy	Epoxy resins for adhesives, coatings, etc.	General Mills, Inc., Chemical Div.
Genetron	Fluorinated hydrocarbons, monomers, and polymers	Allied Chemical Corp., General Chemical Div.
Geon	Polyvinyl chloride materials	B. F. Goodrich Chemical Co.
Grex	High-density polyethylenes	Allied Chemical Corp., Plastics Div.
Halon	Fluorohalocarbon resins	Allied Chemical Corp.
Hetron	Fire-retardant polyester resin	Hooker Chemical Corp., Durez Plastics Div.
Isothane	Polyurethane foam, ester, and ether	Bernel Foam Products Co., Inc.
Kel-F	Chlorotrifluoroethylene, molding resins, and dispersions	3M Company
Kralac	High-styrene resins, styrene-butadiene copolymers	Uniroyal Chemical, Div. of Uniroyal Inc.
Kralastic	ABS polymers, copolymers	Uniroyal Chemical, Div. of Uniroyal Inc.
Kynar	Polyvinylidene fluoride	Pennsalt Chemical Corp.
Lexan	Polycarbonate resin, film, and sheet	General Electric Company, Plastics Dept.
Lucite	Acrylic resin and syrup	E. I. du Pont de Nemours & Co., Inc.
Lustran	ABS polymers	Monsanto Co.
Lustrex	Styrene molding and extrusion resins	Monsanto Co.
Lytron	Styrene molding and extrusion resins	Monsanto Co.
Madurit	Melamine resins and compounds	Cassella Farbwerke Mainkur, A.G.
Maraglas	Epoxy-casting resin	The Marblette Corporation, Div. of Allied Products
Marlex	Polyethylenes, polypropylenes, copolymers	Phillips Petroleum Co.
Marvinol	Vinyl chloride resins and compounds	Uniroyal Chemical, Div. of Uniroyal Inc.
Merlon	Polycarbonate resins	Mobay Chemical Co.
Micarta	Melamines, phenolics, polyesters	Westinghouse Electric Co., Industrial Micarta Div.
Microthene	Polyethylenes, polyolefins	U.S. Industrial Chemicals Co.
Multrathane	Urethane elastomers	Mobay Chemical Company
Nopcofoam	Polyurethane plastics	Nopco Chemical Co., Plastics Div.
Novodur	Polyacrylonitrile-butadiene-styrene	Farbenfabriken Bayer, A. G.
Opalon	Vinyl chloride resins and compounds	Monsanto Co.
Paraplex	Polyester resins, acrylic-modified polyester resins	Rohm & Haas Company

TABLE C.9 (continued) Trade Names, Composition, and Manufacturers of Various Plastics

Trade name	Composition	Manufacturer
Permelite	Melamines	Melamine Plastics, Inc., Div. of Fiberlite Corp.
Petrothene	Polyethylene resins, polypropylene resins	U.S. Industrial Chemicals Co.
Piccoflex	Styrene-copolymer resins	Pennsylvania Industrial Chemical Corp.
Piccolastic	Styrene-polymer resins	Pennsylvania Industrial Chemical Corp.
Plaskon	Nylons, melamines, phenolics, polyesters	Allied Chemical Corp.
Pleogen	Alkyds, polyesters, copolymers	Mol-Rez Div., American Petrochemical Corp.
Plexiglas	Acrylics	Rohm & Haas Company
Pliovic	Polyvinyl chlorides	The Goodyear Tire & Rubber Co., Chemical Div.
Plyophen	Phenolic resins	Reichhold Chemicals, Inc.
Poly-Eth	Polyethylene resins	Gulf Oil Corp., U.S. Div. of Gulf Oil Corp.
Polylite	Polyester resins	Reichhold Chemicals, Inc.
Polypenco	Acrylics, chlorinated polyethers, fluorocarbons, nylons, polycarbonates	Polymer Corp.
Resimene	Urea and melamine resins	Monsanto Co.
Resinox	Phenolic resins and compounds	Monsanto Co.
Rhonite	Urea resins	Rohm & Haas Company
Roylar	Polyurethanes	Uniroyal Chemical, Div. of Uniroyal Inc.
Ryertex	Laminated phenolics and rigid polyvinyl chloride extrusions	Joseph T. Ryerson & Son, Inc., Industrial Plastics and Bearings Sales Div.
Super Dylan	Polyethylene	Sinclair-Koppers Co.
Supreme	Polyethylenes	Johns-Manville Company
Sylplast	Urea-formaldehyde compounds	FMC Corp., Organic Chemicals Div.
Teflon	Fluorocarbon resins	E. I. du Pont de Nemours & Co., Inc.
Tenite	Cellulose acetate, cellulose-acetate-polyethylene, polypropylenes, urethane elastomers, copolymers	Eastman Chemical Products, Inc., Sub. Eastman Kodak Co.
Tetran	Fluorocarbons	Pennsalt Chemicals Corp.
Texin	Urethane elastomers	Mobay Chemical Company
Thioment	Polyisoprenes	Atlas Minerals & Chemicals Div., The Electric Storage Battery Co.
Ultrapas	Melamine resins	Dynamit Nobel, A. G.
Ultrathene	Ethylene-vinyl acetates	U.S. Industrial Chemicals Co.
Ultron	Polyvinyl chlorides	Monsanto Co.
Vibrathane	Urethane elastomers	Uniroyal Chemical, Div. of Uniroyal Inc.
Vibrin	Polyester resins	Uniroyal Chemical, Div. of Uniroyal Inc.
Vitel	Polyesters	The Goodyear Tire & Rubber Co., Chemical Div.
Viton	Synthetic rubbers	E. I. du Pont de Nemours & Co., Inc.
Vitroplast	Polyester cements	Atlas Minerals & Chemicals Div., The Electric Storage Battery Co.
Vyron	Polyvinyl chlorides	Industrial Vinyls, Inc.

TABLE C.10 Properties of Commercial Nylon Resins*

Property	Type 6/6	Type 6	Type 6/10	Type 11	Glass-reinforced Type 6/6, 40%	MoS_2-filled, 2½%	Direct polymerized, castable
Mechanical							
Tensile strength, psi	11,800	11,800	8200	8500	30,000	10,000 to 14,000	11,000 to 14,000
Elongation, %	60	200	240	120	1.9	5 to 150	10 to 50
Tensile yield stress, psi	11,800	11,800	8500		30,000		
Flexural modulus, psi	410,000	395,000	280,000	151,000	1,800,000	450,000	
Tensile modulus, psi	420,000	380,000	280,000	178,000		450,000 to 600,000	350,000 to 450,000
Hardness, Rockwell	118R	119R	111R	55A	75E–80E	110R–125R	112R–120R
Impact strength, tensile, ft-lb/sq in.	76		160			50–180	80–100
Impact strength, Izod, ft-lb/in. of notch	0.9	1.0	1.2	3.3	3.7**	0.6	0.9
Deformation under load, 2000 psi, 122°F, %	1.4	1.8	4.2	2.02†	0.4§	0.5 to 2.5	0.5 to 1
Thermal							
Heat-deflection temp, °F							
At 66 psi	360	365	300	154	509	400 to 490	400 to 425
At 264 psi	150	152	135	118	502	200 to 470	300 to 425
Coefficient of thermal expansion, per °F	4.5×10^{-5}	4.6×10^{-5}	5×10^{-5}	10×10^{-5}	0.9×10^{-5}	3.5×10^{-5}	5.0×10^{-5}
Coefficient of thermal conductivity, Btu in./hr ft² °F	1.7	1.7	1.5				
Specific heat	0.3–0.5	0.4	0.3–0.5	0.58			
Brittleness temp, °F	−112		−166				
Electrical							
Dielectric strength, short time, v/mil	385	420	470	425	480	300 to 400	500 to 600‡
Dielectric constant							
At 60 hz	4.0	3.8	3.9		4.45		3.7
At 10^3 hz	3.9	3.7	3.6	3.3	4.40		3.7
At 10^6 hz	3.6	3.4	3.5		4.10		3.7
Power factor							
At 60 hz	0.014	0.010	0.04	0.03	0.009		0.02
At 10^3 hz	0.02	0.016	0.04	0.03	0.011		0.02
At 10^6 hz	0.04	0.020	0.03	0.02	0.018		0.02
Volume resistivity, ohm-cm	10^{14} to 10^{15}	3×10^{15}	10^{14} to 10^{15}	2×10^{13}	2.6×10^{15}	2.5×10^{13}	
General							
Water absorption, 24 hr, %	1.5	1.6	0.4	0.4	0.6	0.5 to 1.4	0.9
Specific gravity	1.13 to 1.15	1.13	1.07 to 1.09	1.04	1.52	1.14 to 1.18	1.15 to 1.17
Melting point, °F	482 to 500	420 to 435	405 to 430	367	480 to 490	496±9	430±10
Flammability	self ext	self ext	self ext	self ext	self ext	self ext	self ext
Chemical resistance to							
Strong acids	Poor	Poor	Poor	Poor	Poor	Poor	Poor
Strong bases	Good	Good	Good	Fair	Good	Good	Good
Hydrocarbons	Excellent	Excellent	Excellent	Good	Excellent	Excellent	Excellent
Chlorinated hydrocarbons	Good	Good	Good	Fair	Good	Good	Good
Aromatic alcohols	Good	Good	Good	Good	Good	Good	Good
Aliphatic alcohols	Good	Good	Good	Fair	Good	Good	Good

Notes:

Most nylon resins listed in this table are used for injection molding, and test values are determined from standard injection-molded specimens. In these cases, a single typical value is listed. Exceptions are MoS2-filled nylon and direct-polymerized (castable) nylon, which are sold principally in semifinished stock shapes. Ranges of values listed are based on tests on various forms and sizes produced under varying conditions.

Because single values apply only to standard molded specimens, and properties vary in finished parts of different sizes and forms produced by various processes, these values should be used for comparison and preliminary design considerations only. For final design purposes the manufacturer should be consulted for test experience with the form being considered. Limited values should not be used for specification purposes.

† 2000 psi, 73°F.

‡0.040-in. thick.

**½x¼-in. bar

§4000 psi, 122°F.

*From: "Nylons," D.D. Carswell, *Machine Design*, 40(29):62, Dec. 12, 1968.

For Conversion factors see Table C.10.

TABLE C.11 Properties of Silicate Glasses*

Most of the commercially produced glass is for windows, bottles, and inexpensive containers; it is a soda-lime-silica glass of fairly uniform composition, similar to glass No. 0080 in the table below. The following tables on glasses deal largely with that one-tenth of the glass output for which special properties are required. All data are subject to normal manufacturing variations.

Silica glass is inherently high in viscosity and melting point. These are reduced by fluxes such as Na_2O, K2O, and B2O. Soda and potash glasses have a high expansion coefficient (column 7), while that of fused silica is very low. Because the borosilicate glasses are intermediate, and their thermal shock resistance is very high (e.g., Corning Code 7740 glasses), they are widely used for laboratory and kitchen glassware. Aluminosilicate glasses are hard, heat-resisting, and of high chemical durability. Glass hardness (indentation) correlates closely with elastic modulus (column 14). Lead oxide is laso used with flux, with a result of reduced softening point and high refractive index: hence, its uses for optical glass and art glass.

Sealing of glass with metal calls for close control of the coefficient of expansion (column 7).

EXPLANATION OF COLUMNS:

Column 5:

B—blown ware
P—pressed ware
S—plate glass
M—multiform
R—rolled sheet
T—tubing and rod
U—panels
LC—large castings

Column 6:

[2]Since weathering is determined primarily by clouding, which changes transmission, a rating for the opal glasses is omitted.

[3]These borosilicate glasses may rate differently if subjected to excessive heat treatment.

Column 8:

Normal service: No breakage from excessive thermal shock is assumed.

Extreme limits: Glass will be very vulnerable to thermal shock. Recommendations in this range are based on mechanical stability considerations only. Tests should be made before adopting final designs. These data are approximate only.

Column 9:

Based on plunging sample into cold water after oven heating. Resistance of 100°C means no breakage if heated to 110°C and plunged into water at 10°C. Tempered samples have over twice the resistance of annealed glass.

these data are approximate only.

Column 10:

[4]These data are estimated.

Resistance in °C is the temperature differential between the two surfaces of a tube or a constrained plate that will cause a tensile stress of 1000 psi on the cooler surface.

Column 11:

Viscosity is given in poises. At the strain point the stresses are significantly reduced in a matter of hours, while at the annealing point there is adequate stress reduction in minutes.

Column 12:

Data show relative resistance to sandblasting.

Column 15:

Data at 25°C are extrapolated from high temperature readings and are approximate only.

*From: "Properties of Selected Commercial Glasses," Publications B-83, Corning Glass Works.

TABLE C.11 (continued) Properties of Silicate Glasses*

1	2	3	4	5	6			7		8				9		
					Corrosion Resistance			Thermal Expansion 10^{-7}in./in./°C.		Upper Working Temperatures (Mechanical Considerations Only)				Thermal Shock Res. Plates 6" × 6"		
										Annealed		Tempered		Annealed		
Glass Code†	Type	Color	Principal Use	Forms Usually Available	Weath-ering	Water	Acid	0-300°C 32 572°F	Room Temp.-Setting Point	Normal Service °C.	Extreme Limit °C.	Normal Service °C.	Extreme Limit °C.	⅛" Thk. °C.	¼" Thk. °C.	½" Thk. °C.
0010	Potash Soda Lead	Clear	Lamp Tubing	T	2	2	2	93	100	110	380	—	—	65	50	35
0080	Soda Lime	Clear	Lamp Bulbs	B M T	3	2	2	92	103	110	460	220	250	65	50	35
0120	Potash Soda Lead	Clear	Lamp Tubing	T M	2	2	2	89	98	110	380	—	—	65	50	35
1720	Aluminosilicate	Clear	Ignition Tube	B T	1	1	3	42	52	200	650	400	450	135	115	75
1723	Aluminosilicate	Clear	Electron Tube	B T	1	1	3	46	54	200	650	400	450	125	100	70
1990	Potash Soda Lead	Clear	Iron Sealing	—	3	3	4	124	136	100	310	—	—	45	35	25
2405	Borosilicate	Red	General	B P U		—	—	43	51	200	480	—	—	135	115	75
2475	Soda Zinc	Red	Neon Signs	T	3	2	2	93	—	110	440	—	—	65	50	35
3320	Borosilicate	Canary	Tungsten Sealing	—	[3]1	[3]1	[3]2	40	43	200	480	—	—	145	110	80
6720	Soda Zinc	Opal	General	P	[2]—	1	2	80	92	110	480	220	275	70	60	40
6750	Soda Barium	Opal	Lighting Ware	B P R	[2]—	2	2	88	—	110	420	220	220	65	50	35
6810	Soda Zinc	Opal	Lighting Ware	B P R	[2]—	1	2	69	—	120	470	240	270	85	70	45
7040	Borosilicate	Clear	Kovar Sealing	B T	[3]3	[3]3	[3]4	48	54	200	430	—	—	—	—	—
7050	Borosilicate	Clear	Series Sealing	T	[3]3	[3]3	[3]4	46	51	200	440	235	235	125	100	70
7052	Borosilicate	Clear	Kovar Sealing	B M P T	[3]2	[3]2	[3]4	46	53	200	420	210	210	125	100	70
7056	Borosilicate	Clear	Kovar Sealing	B T P	2	2	4	51	57	200	460	—	—	—	—	—
7070	Borosilicate	Clear	Low Loss Electrical	B M P T	[3]2	[3]2	[3]2	32	39	230	430	230	230	180	150	100
7250	Borosilicate	Clear	Seal Beam Lamps	P	[3]1	[3]2	[3]2	36	38	230	460	260	260	160	130	90
7570	High Lead	Clear	Solder Sealing	—	1	1	4	84	92	100	300	—	—	—	—	—
7720	Borosilicate	Clear	Tungsten Sealing	B P T	[3]2	[3]2	[3]2	36	43	230	460	260	260	160	130	90
7740	Borosilicate	Clear	General	B P S T U	[3]1	[3]1	[3]1	33	35	230	490	260	290	180	150	100
7760	Borosilicate	Clear	General	B P	2	2	2	34	37	230	450	250	250	160	130	90
7900[1]	96% Silica	Clear	High Temp.	B P T U M	1	1	1	8	7	800	1100	—	—	1250	1000	750
7913	96% Silica	Clear	High Temp.	B P R S T	1	1	1	8	7	900	1200	—	—	—	—	—
7940	Fused Silica	Clear	Ultrasonic	U	1	1	1	5.5	7	900	1100	—	—	1250	1000	750
8160	Potash Soda Lead	Clear	Electron Tubes	P T	2	2	3	91	100	110	380	—	—	65	50	35
8161	Potash Lead	Clear	Electron Tubes	P T	2	1	4	90	97	110	390	—	—	—	—	—
8363	High Lead	Clear	Radiation Shielding	L C	3	1	4	104	112	100	200	—	—	—	—	—
8871	Potash Lead	Clear	Capacitors	—	2	1	4	102	113	125	300	—	—	55	45	35
9010	Potash Soda Barium	Grey	TV Bulbs	P	2	2	2	89	102	110	380	—				—
9700	Borosilicate	Clear	u v Trans-mission	T U	[3]1	[3]1	[3]2	39	39	220	500	—	—	150	120	80
9741	Borosilicate	Clear	u v Trans-mission	B U T	[3]3	[3]3	[3]4	39	49	200	390	—	—	150	120	80

† Corning Glass Works code numbers are used in this table.

TABLE C.11 (continued) Properties of Silicate Glasses*

10	11				12	13	14			15			16			17	18
Thermal Stress Resistance °C.	Viscosity Data†				Impact Abrasion Resistance	Density grams per C.C.	Young's Modulus		Poisson's Ratio	Log_{10} of Volume Resistivity			Dielectric Properties at 1 Mc and 20°C			Pefractive Index Sod. D Line (.5893 Microns)	Glass Code
	Strain Point °C.	Annealing Point °C.	Softening Point °C.	Working Point °C.			(10^6lb / sq. in)	(10^6kg / cm²)		25°C. 77°F	250°C. 482°F	350°C. 662°F	Power Factor	Dielectric Const.	Loss Factor		
19	395	435	625	985	0.8	2.86	8.9	0.63	.21	17.+	8.9	7.0	.16%	6.7	1.%	1.539	0010
17	470	510	695	1005	1.2	2.47	10.0	0.70	.24	12.4	6.4	5.1	.9	7.2	6.5	1.512	0080
20	395	435	630	980	0.8	3.05	8.6	0.60	.22	17.+	10.1	8.0	.12	6.7	.8	1.560	0120
28	670	715	915	1190	2.0	2.52	12.7	0.89	0.25	—	11.4	9.5	.38	7.2	2.7	1.530	1720
25	670	710	910	1175	2.0	2.64	12.5	0.88	0.25	—	13.5	11.3	.16%	6.3	1.0%	1.547	1723
14	330	360	500	755	—	3.47	8.4	0.59	.25	—	10.1	7.7	.04	8.3	.33	—	1990
'37	500	530	770	1085	—	2.50	9.9	0.70	0.21	—	—	—	—	—	—	1.507	2405
'17	440	480	690	1040	—	2.59	10.0	0.70	—	—	7.8	6.2	—	—	—	1.511	2475
'40	500	540	780	1155	—	2.27	9.4	0.66	0.19	—	8.6	7.1	.30	4.9	1.5	1.481	3320
19	510	550	775	1010	—	2.58	10.2	0.72	.21	—	—	—	—	—	—	1.507	6720
'18	445	485	670	1040	—	2.59	—	—	—	—	—	—	—	—	—	1.513	6750
'23	490	530	770	1010	—	2.65	—	—	—	—	—	—	—	—	—	1.508	6810
37	450	490	700	1080	—	2.24	8.6	0.60	.23	—	9.6	7.8	.20	4.8	1.0	1.480	7040
39	460	500	705	1025	—	2.24	8.7	0.61	.22	16.	8.8	7.2	.33	4.9	1.6	1.479	7050
41	435	480	710	1115	—	2.28	8.2	0.58	.22	17.	9.2	7.4	.26	4.9	1.3	1.484	7052
34	470	510	720	1045	—	2.29	9.2	0.65	.21	—	10.2	8.3	.27	5.7	1.5	1.487	7056
66	455	495	—	1070	4.1	2.13	7.4	0.52	.22	17.+	11.2	9.1	.06	4.1	.25	1.469	7070
48	490	540	780	1190	3.2	2.24	9.2	0.65	.20	15.	8.2	6.7	.27	4.7	1.3	1.475	7250
21	340	365	440	560	—	5.42	8.0	0.56	.28	—	10.6	8.7	.22	15.	3.3	—	7570
49	485	525	755	1140	3.2	2.35	9.1	0.64	.20	16.	8.8	7.2	.27	4.7	1.3	1.487	7720
53	515	565	820	1245	3.1	2.23	9.1	0.64	.20	15.	8.1	6.6	.50	4.6	2.6	1.474	7740
52	480	525	780	1210	—	2.23	9.1	0.64	—	17.	9.4	7.7	.18	4.5	.79	1.473	7760
202	820	910	1500	—	3.5	2.18	10.0	0.70	.19	17.	9.7	8.1	.05	3.8	.19	1.458	7900'
211	820	910	1500	—	3.5	2.18	9.6	0.67	.19	—	9.7	8.1	.04	3.8	0.15	1.458	7913
290	990	1050	1580	—	3.6	2.20	10.5	0.74	.16	—	11.8	10.2	.001	3.8	.0038	1.459	7940
'18	395	435	630	975	—	2.98	—	—	—	—	10.6	8.4	.09	7.0	.63	1.553	8160
22	400	435	600	860	—	4.00	7.8	0.55	.24	—	12.0	9.9	.06	8.3	0.50	1.659	8161
19	300	315	380	460	—	6.22	7.4	0.52	.27	—	9.2	7.5	.19	17.0	3.2	1.97	8363
17	350	385	525	785	—	3.84	8.4	0.59	.26	—	11.1	8.8	.05	8.4	.42	—	8871
18	405	445	650	1010	—	2.64	9.8	0.69	.21	—	8.9	7.0	.17	6.3	1.1	1.507	9010
45	520	565	805	1200	—	2.26	9.6	0.67	.20	15.	8.0	6.5	—	—	—	1.478	9700
55	410	450	705	—	—	2.16	7.2	0.51	.23	17.+	9.4	7.6	—	—	—	1.468	9741

†Viscosities at these four temperatures are approximately as follows: $10^{14.5}$ poises at the strain point, 10^{13} poises at the annealing point, $10^{7.8}$ poises at the softening point, at 10^4 poises at the working point.

TABLE C.12 Properties of Window Glass*: Transmittance of Sheet and Plate Glass

Type or tint	Nominal thickness, in.	Weight, lb/ft²	Transmittance	
			Total visible daylight, %	Direct 90° solar energy, %
Sheet	$\frac{1}{16}$	0.81	91	89
Sheet	$\frac{5}{64}$	1.00	91	88
Sheet	$\frac{3}{32}$	1.22	90	87
Sheet	$\frac{1}{8}$	1.64	90	86
Sheet	$\frac{3}{16}$	2.47	89	84
Sheet	$\frac{7}{32}$	2.85	89	82
Plate or float	$\frac{1}{8}$	1.64	90	86
Plate or float	$\frac{1}{4}$	3.28	88	79
Plate or float	$\frac{5}{16}$	4.09	88	77
Plate or float	$\frac{3}{8}$	4.91	87	74
Plate or float	$\frac{1}{2}$	6.55	86	70
Plate or float	$\frac{5}{8}$	8.18	85	65
Plate or float	$\frac{3}{4}$	9.83	83	60
Plate or float	$\frac{7}{8}$	11.45	81	55
Plate or float	1	13.13	79	49
Gray[a]	$\frac{1}{4}$	3.28	43	46
Bronze[a]	$\frac{1}{4}$	3.28	49	45
Green[a]	$\frac{1}{4}$	3.28	75	46
Double[b]	$\frac{1}{4}$ each	6.56	78	—

Note: Many types of glass are available, including tempered heat-strengthened glass, laminated shatter-proof glass, conductive-coated glass, and reflective-coated glass. Several double-pane combinations are offered.

Direct 90° transmittance of solar ultraviolet radiation through non-tinted window glass is about 85 percent az high as the values for toal solar energy transmittance. Ultraviolet transmittance of gray or bronze glass is lower.

Infrared transmittance is considerably lower than visual transmittance. This is significant in view of the large percentage of infrared radiation from most sources.

Approximate shading coefficients, ASHRAE, 1/4-in. glass only: clear, 0.93; gray, 0.67; bronze, 0.65; green, 0.67.

Overall heat transfer coefficient of window area (air to air) is usually assumed to be 1.0 Btu/ft² hr, but it is lower if there is no wind.

[a]Transmittance of tinted glass depends on depth of tint..
[b]Two 1/2-in. panes with 1/2-in. air space, sealed.

*Tables compiled from several sources.

TABLE C.13 Properties and Uses of American Woods*

Species	Specific gravity		Characteristics	Uses	Weight		
	Green	Dry			lb/cu ft, green	lb/cu ft, air-dry 12%	lb/1000 board ft, air-dry 12%
Alder, red	0.37	0.41	Low shrinkage; moderate in strength, shock resistance, hardness, and weight†	Furniture; sash; doors; millwork	46	28	2330
Ash, black	0.45	0.49	Light in weight†	Cabinets; veneer; cooperage, containers	52	34	2830
Ash, Oregon	0.50	0.55	Similar to but lighter than white ash†	Similar to white ash	46	38	3160
Ash, white	0.54	0.58	Heavy; hard; stiff; strong; high shock resistance†	Handles; ladder rungs; baseball bats; farm implements; car parts	48	41	3420
Bald cypress (Southern cypress)			Moderate in strength, weight, hardness, and shrinkage**	Building construction; beams; posts; ties; tanks; ships; paneling	51	32	2670
Beech, American	0.56	0.64	Heavy; high strength, shock resistance, and shrinkage; uniform texture†	Flooring; furniture; handles; kitchenwear; ties (treated)	54	45	3750
Birch	0.57	0.63	Heavy; high strength, shock resistance, and shrinkage; uniform texture†	Interior finish; dowels; ties (treated); veneer; musical instruments	57	44	3670
Cottonwood	0.37	0.40	Uniform texture; does not split readily; moderate in weight, strength, hardness, and shrinkage	Crates; trunks; car parts; farm implements	49	28	2330
Douglas fir	0.41	0.44	Moderate in strength, weight, shock resistance, and shrinkage‡	Building and construction; poles; veneer; plywood; ships; furniture; boxes	38	34	2830
Elm	0.57	0.63	Moderate in strength, weight, and hardness; high in shock resistance and shrinkage; good in bending†	Cooperage; baskets; crates; veneer; vehicle parts	54	34	2920
Hemlock, Eastern	0.38	0.40	Moderate in weight, strength, and hardness†	Building and construction; boxes	50	28	2330
Hemlock, Western	0.38	0.42	Moderate in weight, strength, and hardness†	Sash; doors; posts; piles; building and construction	41	29	2420
Hickory, true	0.65	0.73	High toughness, hardness, shock resistance, strength, and shrinkage†	Dowels; spokes; poles; shafts; gymnasium equipment	63	51	4250
Incense cedar	0.35		Uniform texture; easy to season; low shrinkage; shock resistance, weight, and stiffness**	Lumber; fence posts; ties; poles; shingles	45		
Larch, Western	0.48	0.52	Moderate in strength, weight, shock resistance, hardness, and shrinkage‡	Doors; sash; posts; pilings; building and construction	48	36	3000

TABLE C.13 (continued) Properties and Uses of American Woods*

Species	Specific gravity		Characteristics	Uses	Weight		
	Green	Dry			lb/cu ft, green	lb/cu ft, air-dry 12%	lb/1000 board ft, air-dry 12%
Locust, black	0.66	0.69	High in shock resistance, weight, and hardness; very high strength; moderate shrinkage**	Mine timbers; posts; poles; ties	58	48	4000
Maple	0.44	0.48	High in hardness, weight, strength, shock resistance, and shrinkage; uniform texture†	Flooring; furniture; trim; spools; farm implements	54	40	3330
Oak, red and white	0.57	0.63	High in hardness, weight, strength, shock resistance, and shrinkage; red†, white‡	Trim; ships; flooring; ties; furniture; cooperage; piles	64	44	3670
Pine, jack			Coarse texture; low strength, stiffness, shock resistance, and shrinkage	Box lumber; fuel; mine timber; ties; poles; posts			
Pine, lodgepole	0.38	0.41	Moderate in weight, hardness, strength, shock resistance, and shrinkage; easy to work‡	Poles; mine timber; ties; construction	39	29	2420
Pine			High shrinkage; moderate strength, stiffness, hardness, and shock resistance	General construction; ties; poles; posts			
Pine, Ponderosa	0.38	0.40	Moderate in weight, shock resistance, shrinkage, and hardness; easy to work†	Building; paneling; sash; frames	45	28	2330
Pine, S. yellow	0.47	0.51	Moderate in shock resistance, shrinkage, and hardness; high in strength‡	Building and construction; poles; pilings; boxes	55	41	3420
Pine, sugar	0.35	0.36	Low shock resistance; easy to work; moderate strength†	Sash; counters; blinds; patterns	52	25	2080
Pine, Western white	0.36	0.38	Moderate in strength, shock resistance, shrinkage, and hardness; easy to work†	Building and construction; patterns; boxes	35	27	2250
Red cedar, Eastern and Western	0.44	0.47	High shock resistance; low stiffness and shrinkage; moderate in strength and hardness**	Fence posts; closet liners; chests; flooring	37	37	2750
Redwood	0.38	0.40	Low shrinkage; medium in weight, strength, hardness, and shock resistance**	Posts; doors; interiors; cooling towers	50	28	2330
Spruce, Eastern	0.38	0.40	Moderate in hardness, shock resistance, weight, shrinkage, and strength†	Building; millwork; boxes; ladders	34	28	2330

TABLE C.13 (continued) Properties and Uses of American Woods*

Species	Specific gravity		Characteristics	Uses	Weight		
	Green	Dry			lb/cu ft, green	lb/cu ft, air-dry 12%	lb/1000 board ft, air-dry 12%
Spruce, Engelmann	0.31	0.33	Generally straight grained; light in weight; low strength as a beam or post; low shock resistance; moderate shrinkage	Mine timber; ties; poles; flooring; studding; paper	39	23	1920
Spruce, Sitka	0.37	0.40	Moderate in weight, hardness, strength, shock resistance, and shrinkage†	Important in boat and plane construction; sash; doors; boxes; siding	33	28	2330
Sycamore	0.46	0.49	High shrinkage; moderate in weight, strength, hardness, and shock resistance†	Boxes; ties; posts; veneer; flooring; butcher blocks	52	34	2830
Tamarack	0.49	0.53	Coarse texture; moderate in strength, hardness, shrinkage, and shock resistance	Ties; mine timber; posts; poles; tanks; scaffolding	47	37	3080
Tupelo			Uniform texture; moderate in strength, hardness, shock resistance; high shrinkage; interlocked grain makes splitting difficult†	Flooring; planking; crates; furniture			
Walnut, black	0.51	0.55	Moderate shrinkage; high weight, strength, hardness, and shock resistance; easily worked and glued**	Gun stocks; cabinets; plywood; furniture; veneer	58	38	3170
White cedar	0.31	0.32	Low shrinkage, weight, shock resistance, and strength; soft; easily worked**	Poles; posts; ties; tanks; ships	24	23	1920
Willow, black			High strength and shock resistance; low beam strength and weight; interlocked grain	Lumber; veneer; charcoal; furniture; sub-flooring; studding			

†Decay resistance low.
‡Decay resistance medium.
**Decay resistance high.
*From: "Materials Data Book", E.R. Parker, McGraw–Hill Book Company, 1967, pp. 252–255.

Note: For weight-density in kg/m^3, multiply value in lb/ft^3 by 16.02.

TABLE C.14 Properties of Natural Fibers*

Because there are great variations within a given fiber class, average properties may be misleading. The following typical values are only a rough comparative guide.

Name	*Specific gravity*	*Tenacity, g/denier*	*Tensile strength, 10^3 psi*	*Elongation at break (dry), %*	*Standard regain, % of dry*[b]	*Fiber diameter, microns*	*Fiber length, in.*	*Fiber shape and kind*	*Resistant to*
ANIMAL ORIGIN									
Wool	1.32	1.0–1.7	17–29	23–35	15–18	17–40	1.5–5	Oval, crimped, scales	Age, weak acids, solvents
Silk	1.25	3.5–5	90	20–25	10	10–13		Flexible, soft, smooth	Heat, solvents, weak acids, wear
Cashmere						15–16	1–4	Round, scales, soft	
Mohair	1.32	1.2–1.5		30	13	24–50	6–12	Round, silky	Wear, age, solvents, weak acids
Camel hair	1.32	1.8		40	13	10–40	1–6	Oval, striated	Age, solvents
VEGETABLE ORIGIN									
Cotton	1.54	2–5	30–120	5–11	7.5–8.5	10–20	0.5–2	Flat, convoluted, ribbon	Age, heat, washing, wear, solvents, alkalies, insects
Jute (bast)	1.5		50	1–1.5	14	15–20		Woody, rough, polygon	
Sisal (leaf)	1.49	2.2	75	2–2.5	13	10–30	Strand 30–40	Stiff, straight	
Flax (bast)	1.52	4–7		2–3	12	15–18	Strand 40–50	Soft, fine	Age, solvents, washing, insects, weak acids, and alkalies
Kenaf (bast)			45			15–30		Polygon or oval	
Hemp (bast)	1.48			2		18–25	Strand 30–70	Polygon or oval, irregular	
Henequen (leaf)			60				Strand 30–60	Finer than sisal	
Abaca (leaf) (Manila)	1.48	2.3–2.9	100	2–3	13		Strand 30–120		
MINERAL ORIGIN									
Asbestos	2.5		40–200			Various	0.5–10	Smooth, straight	Heat to 400 deg C, acids, chemicals, organisms
Glass[a]	2.5	7–12	200–500	3–4.5	0	Various		Circular, smooth	Chemicals, insects
Silicate[a] (Ca, Al, Mg)	2.85				0				Heat to 900 deg C, most chemicals, insects, rot

Note: Wide variations may be expected, especially for different grades of cotton. Wet strength is lower (for rayon, very much lower), but it depends on the duration of soaking. The strength of yarn is only a fraction of the cumulative strength of all individual fibers.

Most fibers exhibit relaxation of stress at constant strain and also increase in elongation at constant load (creep). The stress-strain curve is greatly affected by the rate of extension. When the stress is removed, there is a quick elastic recovery, a delayed recovery, and a permanent set. Hence the elastic behavior of any fiber depends on its stress-strain history. The elastic recoveries of nylon and wool are high; those of cotton, flax, and rayon are much lower.

The heat capacity (specific heat) of most fibers is about one-third that of water.

Other fibers: Fur hair is slightly coarser than silk fibers. Camel and llama hairs are almost as coarse as wool but only about one-third the size of human hair. Horse hair is over 100 microns; hog bristles, over 200 microns. Jute, sisal, and hemp are intermediate between cotton and wool. These are rough average sizes, and many natural fibers range 50% above or below such averages.

[a]Here classified as natural fibers for convenience, although they are man-made by processing.
[b]Expected equilibrium moisture regain of dry fiber, in percent of dry weight, when exposed in air at 70 deg F, 65% relative humidity.

*Compiled from several sources.

TABLE C.15 Properties of Manufactured Fibers*

Chemical class; common name (sources)	*Specific gravity*	*Tenacity, g/denier*	*Tensile strength, 10^3 psi*	*Elongation at break, %*	*Regain (standard)*	*Softening point, deg C*	*Melting point, deg C*	*Flammability*	*Brittleness temp, deg C*
CELLULOSE FIBERS (NATURAL)									
Acetate	1.30	1.-1.3	18–25	20–30	6.5	140	230	Melts and burns	
Triacetate	1.32	1.2–1.4	20–28	25–30	3–4.5	225	300	Melts and burns	
Viscose rayon	1.51	2–2.6	30–46	17–25	13.		200[a]	Burns readily	
High-tenacity viscose	1.53	3–5	60–80	10–12	10		200[a]	Burns readily	< −114
Polynosic viscose	1.53	3–5	60–80	8–20	7		200[a]	Burns readily	
Cuprammonium rayon (cupro)	11.52	1.7–2.3	30–45	10–17	12.5		250[a]	Burns readily	
PROTEIN FIBERS (NATURAL)									
Animal: casein (milk)	1.3	1.0	15	60–70	14	100	150	Slow	
Vegetable—seed: soybeans, peanuts, corn	1.3	0.7–0.9	11–14	40–60	11–15	150	250	Slow	
Vegetable—latex: rubber (vulcanized)	1.0	0.4–0.6	4–7	700–900	0	300		Burns	−60
SYNTHETIC FIBERS									
Polyacrylonitrile (acrylic)	1.17	2–5	50–75	25–40	2	190	260	Burns	
Polyamide (nylon)	1.14	4–9	70–120	20–40	4	200	215–250	Slow	< −100
Polyester (PET dacron)	1.38	4–8	70–120	10–50	0.4	225	250–290	Low	
Polyethylene (olefin, low density)	0.92	3–6	40–70	25–40	0.15	90–120	120	Slow	−114
Polyethylene (olefin, high density)	0.95	5–7	60–80	10–20	0.01	120–130	140	Slow	−114
Polypropylene (olefin)	0.91	4.5–8	45–80	15–30	0–0.5	145	160–170	Self-ext. low	−70
Polyurethane (spandex)	1.1	0.5–1.0	7–16	500–700	1.0	190	250	Burns	
Polyvinyl chloride (PVC)	1.38	0.7–2	12–17	100–125	0.1	70	140[a]	No; chars	< −100
Polyvinyl alcohol (PVA)	1.3	3–7	60–90	15–28	5	230	240	Slow	
Polyvinylidene chloride (saran)	1.7	2	40	20–30	0.1	115–135	170	No	
Polytetrafluoroethylene (PTFE)	2.1	1.2–1.4	33	15–30	0	225	300[a]	No	

Note: Mechanical properties are for room temperature and humidity and based on unstressed cross section.

[a]Decomposition; does not melt.

*Compiled from several sources.

TABLE C.16 Properties of Rubbers and Elastomers*

Elastomers cannot be classified in any brief and simple manner, nor are they well characterized by the usual mechanical tests. The terms *rubber* and *synthetic rubber* are loosely applied to a great variety of elastic materials, from pure gum natural rubber and pure synthetics to cured, compounded, filled, and even reinforced products.

ASTM designations (D1418) by chemical polymer description are used in the following table; yet within each class the properties can vary widely, depending on the exact composition, heat treatment service temperature, and application. Typical uses, such as rubber springs and cushioning, permit an almost unlimited number of combinations of design variables.

Mechanically, rubbers may be expected to lose strength rapidly with increase in temperature, to show a large hysteresis in stress-strain behavior, to exhibit marked creep and set, and to be greatly affected by rates of load application or frequency of repeated stress. "Heat build-up", i.e., increase in temperature in service, as well as deterioration from environment (sunlight, oils, ozone, etc.) will reduce the valuable properties of many rubbers, both natural and synthetic.

The following data apply to typical samples of commercial elastomers for common uses.

KEY:

A—Acetone
B—Benzene
C—Carbon tetrachloride
D—Carbon disulfide
E—Phenol
F—Sulfur compounds
G—Glycerol or glycol
H—Hexane
I—Acids
J—Alkalies
K—Ketones
L—Alcohols
M—Ammonia
N—Turpentine
O—Coal derivatives; bitumens
P—Petroleum products
R—Aromatics
S—Salts
T—Heat or high temperature
U—Ultraviolet
V—Vegetable oils
W—Weathering
X—Oxidation
Y—Aging
Z—Ozone

Chemical name	*Polyisoprene*	*Butadiene*	*Styrene-butadiene*	*Acrylonitrile butadiene*
Other names	*Natural (or synthetic) rubber NR (IR)*	*BR Cis 4*	*Buna S Styrene SBR, GR-S*	*Nitrile, Buna N Hycar NBR, GR-A*
CHEMICAL AND PHYSICAL				
Specific gravity	0.93	1.0	1.0	1.0
Specific heat	0.40	0.45	0.40	0.47
Thermal conductivity				
W/cm·K	0.001 7	0.002 5	0.002 6	0.002 5
Btu/hr·ft·deg F	0.10	0.14	0.15	0.14
Service temperature, deg C				
min	−25	−40	−20	−20
max	90	90	75	110
Solvents, softeners	D,K,P,V	D,H,N,P	K,P,R,V	C,K,O,R
Resistant to	A,I,J,L	G,I,J,W,Y	G,I,L,S,X	G,I,K,L,P,S, T,V,W
Swelled by	D,P,V	A,P,V	P,V	A,E,N
MECHANICAL AND ELECTRICAL				
Tensile strength				
kg/cm^2 (max)	300.	210.	210.	295.
kpsi (max)	4.3	3.0	3.0	4.2
Elongation at break, %	600.	700.	600.	600.
Vol. resistivity, ohm-cm	10^{15}	10^{15}	10^{14}	10^{10}
Dielectric strength				
kV/cm	235		235	185
V/mil	600.		600.	475.
Dielectric constant	3.0	2.3	2.8	3.0
Power factor (50–100 Hz)	0.003	0.005	0.005	0.007
Rebound	Good	Good	Fair	Good
COMPARATIVE RATINGS—RESISTANCE TO				
Abrasion	Good	Excellent	Good	Excellent
Cold flow (set)	Excellent		Good	Good
Tearing	Good		Poor	Fair
Air permeability	Fair	Good	Fair	Excellent
Oxidation	Fair	Fair	Fair	Fair
Flame	Poor		Poor	Poor

*Compiled from several sources.

TABLE C.16 (continued) Properties of Rubbers and Elastomers*

Chemical name	*Polychloroprene*	*Isobutylene-isoprene*	*Polysulfide*	*Polymethane*
Other names	*Neoprene*[a] *CR, GR-M*	*Butyl* *IIR, GR-I*	*Thiokol*[a] *PS, GR-P*	*Adiprene*[a] *PU*
CHEMICAL AND PHYSICAL				
Specific gravity	1.25	0.95	1.4	1.2
Specific heat	0.5	0.45	0.31	0.45
Thermal conductivity				
W/cm·K	0.002 1	0.001 3	0.003	0.001 3
Btu/hr·ft·deg F	0.12	0.075	0.17	0.075
Service temperature, deg C				
min	−20	−40	−15	−35
max	100	120	90	120
Solvents, softeners	A,B,C,D,I,N,R	D,P	C	
Resistant to	G,L,P,S,T,U,V, W,Y,Z	E,G,J,S,U,V, W,X,Y,Z	L,P,U,Z	P,V,X,Z
Swelled by	C,D,N,R	D,H,P	C,R	B,C,K,R
MECHANICAL AND ELECTRICAL				
Tensile strength				
kg/cm^2 (max)	240.	175.	90.	350.
kpsi (max)	3.5	2.5	1.3	5.0
Elongation at break, %	800.	700.	500.	550.
Vol. resistivity, ohm-cm	10^{11}	10^{17}	10^{8}	10^{11}
Dielectric strength				
kV/cm	195	295	125	195
V/mil	500	750	325	500
Dielectric constant	7.	2.4	8.	7.
Power factor (50–100 Hz)	.04	0.004	0.02	0.04
Rebound	Good	Poor	Poor	
COMPARATIVE RATINGS—RESISTANCE TO				
Abrasion	Excellent	Fair	Poor	Excellent
Cold flow (set)	Excellent	Fair	Poor	Poor
Tearing	Good	Good	Poor	Excellent
Air permeability	Good	Excellent	Good	Excellent
Oxidation	Good	Good	Good	Good
Flame	Excellent	Poor	Poor	Poor

[a]Proprietary.

Appendix D. Gases and Vapors

TABLE D.1 SI Units — Definitions, Abbreviations and Prefixes

BASIC UNITS—MKS

Length	meter	m	Electric current	ampere	A
Mass	kilogram	kg	Thermodynamic temperature	kelvin	K
Time	second	s	Luminous intensity	candela	cd

DERIVED UNITS

Property	*Units*†	*Abbreviations and dimensions*	
Acceleration	meter per second squared	m/s^2	
Activity (of radioactive source)	1 per second	s^{-1}	
Angular acceleration	radian per second squared	rad/s^{-1}	
Angular velocity	radian per second	rad/s	
Area	square meter	m^2	
Density	kilogram per cubic meter	kg/m^3	
Dynamic viscosity	newton-second per sq meter	$N\cdot s/m^2$	
Electric capacitance	farad	F	$(A\cdot s/V)$
Electric charge	coulomb	C	$(A\cdot s)$
Electric field strength	volt per meter	V/m	
Electric resistance	ohm		(V/A)
Entropy	joule per kelvin	J/K	
Force	newton	N	$(kg\cdot m/s^2)$
Frequency	hertz	hz	(s^{-1})
Illumination	lux	lx	(lm/m^2)
Inductance	henry	H	$(V\cdot s/A)$
Kinematic viscosity	sq meter per second	m^2/s	
Luminance	candela per sq meter	cd/m^2	
Luminous flux	lumen	lm	$(cd\cdot sr)$
Magnetomotive force	ampere	A	
Magnetic field strength	ampere per meter	A/m	
Magnetic flux	weber	Wb	$(V\cdot s)$
Magnetic flux density	tesla	T	(Wb/m^2)
Power	watt	W	(J/s)
Pressure	newton per square meter	N/m^2	
Radiant intensity	watt per steradian	W/sr	
Specific heat	joule per kilogram kelvin	J/kg K	
Thermal conductivity	watt per meter kelvin	W/m K	
Velocity	meter per second	m/s	
Volume	cubic meter	m^3	
Voltage, potential difference, electromotive force	volt	V	(W/A)
Wave number	1 per meter	m^{-1}	
Work, energy, quantity of heat	joule	J	$(N\cdot m)$

PREFIX NAMES OF MULTIPLES AND SUBMULTIPLES OF UNITS

Decimal equivalent	*Prefix*	*Pronunciation*	*Symbol*	*Exponential expression*
1,000,000,000,000	tera	tĕr'á	T	10^{+12}
1,000,000,000	giga	jĭ'gá	G	10^{+9}
1,000,000	mega	mĕg'á	M	10^{+6}
1,000	kilo	kĭl'ō	k	10^{+3}
100	hecto	hĕk'tō	h	10^{+2}
10	deka	dĕk'á	da	10
0.1	deci	dĕs'ĭ	d	10^{-1}
0.01	centi	sĕn'tĭ	c	10^{-2}
0.001	milli	mĭl'ĭ	m	10^{-3}
0.000 001	micro	mī'krō	μ	10^{-6}
0.000 000 001	nano	năn'ō	n	10^{-9}
0.000 000 000 001	pico	pē'kō	p	10^{-12}
0.000 000 000 000 001	femto	fĕm'tō	f	10^{-15}
0.000 000 000 000 000 001	atto	ăt'tō	a	10^{-18}

Appendix E. Miscellaneous

TABLE E.1 Sizes and Allowable Unit Stresses for Softwood Lumber

American Softwood Lumber Standard. A voluntary standard for softwood lumber has been developing since 1922. Five editions of Simplified Practice Recommendation R16 were issued from 1924–53 by the Department of Commerce; the present NBS voluntary Product Standard PS 20-70, "American Softwood Lumber Standard," was issued in 1970. It was supported by the American Lumber Standards Committee, which functions through a widely representative National Grading Rule Committee.

Part a. Nominal and Minimum-Dressed Sizes of Lumber*

Item	*Thicknesses*			*Face widths*		
	Nominal	*Minimum-dressed*		*Nominal*	*Minimum-dressed*	
		Dry,[a] inches	*Green, inches*		*Dry,[a] inches*	*Green, inches*
Boards[b]				2	$1\frac{1}{2}$	$1\frac{9}{16}$
				3	$2\frac{1}{2}$	$2\frac{9}{16}$
				4	$3\frac{1}{2}$	$3\frac{9}{16}$
				5	$4\frac{1}{2}$	$4\frac{5}{8}$
	1	$\frac{3}{4}$	$\frac{25}{32}$	6	$5\frac{1}{2}$	$5\frac{5}{8}$
				7	$6\frac{1}{2}$	$6\frac{5}{8}$
	$1\frac{1}{4}$	1	$1\frac{1}{32}$	8	$7\frac{1}{4}$	$7\frac{1}{2}$
				9	$8\frac{1}{4}$	$8\frac{1}{2}$
	$1\frac{1}{2}$	$1\frac{1}{4}$	$1\frac{9}{32}$	10	$9\frac{1}{4}$	$9\frac{1}{2}$
				11	$10\frac{1}{4}$	$10\frac{1}{2}$
				12	$11\frac{1}{4}$	$11\frac{1}{2}$
				14	$13\frac{1}{4}$	$13\frac{1}{2}$
				16	$15\frac{1}{4}$	$15\frac{1}{2}$
Dimension				2	$1\frac{1}{2}$	$1\frac{9}{16}$
				3	$2\frac{1}{2}$	$2\frac{9}{16}$
				4	$3\frac{1}{2}$	$3\frac{9}{16}$
	2	$1\frac{1}{2}$	$1\frac{9}{16}$	5	$4\frac{1}{2}$	$4\frac{5}{8}$
	$2\frac{1}{2}$	2	$2\frac{1}{16}$	6	$5\frac{1}{2}$	$5\frac{5}{8}$
	3	$2\frac{1}{2}$	$2\frac{9}{16}$	8	$7\frac{1}{4}$	$7\frac{1}{2}$
	$3\frac{1}{2}$	3	$3\frac{1}{16}$	10	$9\frac{1}{4}$	$9\frac{1}{2}$
				12	$11\frac{1}{4}$	$11\frac{1}{2}$
				14	$13\frac{1}{4}$	$13\frac{1}{2}$
				16	$15\frac{1}{4}$	$15\frac{1}{2}$
Dimension				2	$1\frac{1}{2}$	$1\frac{9}{16}$
				3	$2\frac{1}{2}$	$2\frac{9}{16}$
				4	$3\frac{1}{2}$	$3\frac{9}{16}$
				5	$4\frac{1}{2}$	$4\frac{5}{8}$
	4	$3\frac{1}{2}$	$3\frac{9}{16}$	6	$5\frac{1}{2}$	$5\frac{5}{8}$
	$4\frac{1}{2}$	4	$4\frac{1}{16}$	8	$7\frac{1}{4}$	$7\frac{1}{2}$
				10	$9\frac{1}{4}$	$9\frac{1}{2}$
				12	$11\frac{1}{4}$	$11\frac{1}{2}$
				14		$13\frac{1}{2}$
				16		$15\frac{1}{2}$
Timbers	5 and thicker		$\frac{1}{2}$ off	5 and wider		$\frac{1}{2}$ off

[a]Maximum moisture content of 19% or less.

[b]Boards less than the minimum thickness for 1 in. nominal but 5/8-in. or greater thickness dry (11/16-in. green) may be regarded as American Standard Lumber, but such boards shall be marked to show the size and condition of seasoning at the time of dressing. †hey shall also be distinguished from 1-in. boards on invoices and certificates.

*Reprinted from: "american Softwood Lumber Standard," NBS PS 20-70, National Bureau of Standards, 1970; available from Superintendent of documents.

Note: This table applies to boards, dimensional lumber, and timbers. The thicknesses apply to all widths and all widths to all thicknesses.

TABLE E.1 (continued) Sizes and Allowable Unit Stresses for Softwood Lumber

The "American Softwood Lumber Standard", PS 20-70, gives the size and grade provisions for American Standard lumber and describes the organization and procedures for compliance enforcement and review. It lists commercial name classifications and complete definitions of terms and abbreviations.

Eleven softwood species are listed in PS 20-70, viz., cedar, cypress, fir, hemlock, juniper, larch, pine, redwood, spruce, tamarack, and yew. Five dimensional tables show the standard dressed (surface planed) sizes for almost all types of lumber, including matched tongue-and-grooved and shiplapped flooring, decking, siding, etc. Dry or seasoned lumber must have 19% or less moisture content, with an allowance for shrinkage of 0.7–1.0% for each four points of moisture content below the maximum. Green lumber has more than 19% moisture. Table A illustrates the relation between nominal size and dressed or green sizes.

National Design Specification. Part b is condensed from the 1971 edition of "National Design Specification for Stress-Grade Lumber and Its Fastenings," as recommended and published by the National Forest Products Association, Washington, D.C. This specification was first issued by the National Lumber Manufacturers Association in 1944; subsequent editions have been issued as recommended by the Technical Advisory Committee. The 1971 edition is a 65-page bulletin with a 20-page supplement giving "Allowable Unit Stresses, Structural Lumber," from which Part b has been condensed. The data on working stresses in this Supplement have been determined in accordance with the corresponding ASTM Standards, D245-70 and D2555-70.

Part b. Species, Sizes, Allowable Stresses, and Modulus of Elasticity of Lumber

Normal loading conditions: Moisture content not over 19%, No. 1 grade, visual grading. To convert psi to N/m^2, multiply by 6 895.

Species[a]	*Sizes, nominal*	*Typical grading agency, 1971*[b]	*Allowable unit stresses, psi*[d]				*Modulus of elasticity, psi*
			Extreme fiber in bending[c]	*Tension parallel to grain*	*Compression perpendicular*	*Compression parallel*	
CEDAR							
Northern white	2 × 4	NL, NH	1 100	600	205	675	800 000
	2 or 4 × 6+	NL, NH	1 000	575	205	675	800 000
Western	2 × 4	NC	1 450	725	285	975	1 100 000
	2 or 4 × 6+	NC, WW	1 250	725	285	975	1 100 000
FIR							
Balsam	2 × 4	NL, NH	1 300	675	170	825	1 200 000
	2 or 4 × 6+	NL, NH	1 150	650	170	825	1 200 000
Douglas (larch)	2 × 4	WC, NC	2 400	1 200	385	1 250	1 800 000
	2 or 4 × 6+	WC, NC	1 750	1 000	385	1 250	1 800 000
HEMLOCK							
Eastern (tamarack)	2 × 4	NL, NH	1 750	900	365	1 050	1 300 000
	2 or 4 × 6+	NL, NH	1 500	875	365	1 050	1 300 000
Hem-fir	2 × 4	WC, NC	1 600	825	245	1 000	1 500 000
	2 or 4 × 6+	WC, NC	1 400	800	245	1 000	1 500 000
Mountain	2 × 4	WC, WW	1 700	850	370	1 000	1 300 000
	2 or 4 × 6+	WC, WW	1 450	850	370	1 000	1 300 000
PINE							
Idaho white	2 × 4	WW	1 400	725	240	925	1 400 000
	2 or 4 × 6+	WW	1 200	700	240	925	1 400 000
Lodgepole	2 × 4	WW	1 500	750	250	900	1 300 000
	2 or 4 × 6+	WW	1 300	750	250	900	1 300 000

TABLE E.1 (continued) Sizes and Allowable Unit Stresses for Softwood Lumber

Species[a]	Sizes, nominal	Typical grading agency, 1971[b]	Allowable unit stresses, psi[d]				Modulus of elasticity, psi
			Extreme fiber in bending[c]	Tension parallel to grain	Compression perpendicular	Compression parallel	
PINE *(continued)*							
Northern	2×4	NL, NH	1 600	825	280	975	1 400 000
	2 or 4×6+	NL, NH	1 400	800	280	975	1 400 000
Ponderosa (sugar)	2×4	WW, NC	1 400	700	250	850	1 200 000
	2 or 4×6+	WW, NC	1 200	700	250	850	1 200 000
Red	2×4	NC	1 350	700	280	825	1 300 000
	2 or 4×6+	NC	1 150	675	280	825	1 300 000
Southern	2×4	SP	2 000	1 000	405	1 250	1 800 000
	2 or 4×6+	SP	1 750	1 000	405	1 250	1 800 000
REDWOOD							
California	2 or 4×2 or 4	RI	1 950	1 000	425	1 250	1 400 000
	2 or 4×6 to 12	RI	1 700	1 000	425	1 250	1 400 000
SPRUCE							
Eastern	2×4	NL, NH	1 500	750	255	900	1 400 000
	2 or 4×6+	NL, NH	1 250	750	255	900	1 400 000
Engelmann	2×4	WW	1 300	675	195	725	1 200 000
	2 or 4×6+	WW	1 150	650	195	725	1 200 000
Sitka	2×4	WC	1 550	775	280	925	1 500 000
	2 or 4×6+	WC	1 300	775	280	925	1 500 000

Note: Allowable unit stresses in horizontal shear are in the range of 60–100 psi for No. 1 grade.

[a] Grade designations are not entirely uniform. Values in the table apply approximately to "No. 1." There is seldom more than one better grade than No. 1, and this may be designated as select, select structural, dense, or heavy. In addition to lower grades 2 and 3, there may be other lower grades, designated as construction, standard, stud, and utility. In bending and tension the allowable unit stresses in the lowest recognized grade (utility) are of the order of $\frac{1}{8}$ to $\frac{1}{6}$ of the allowable stresses for grade No. 1. The tabular values for allowable bending stress are for the extreme fiber in "repetitive member uses," and edgewise use. The original tables give correction factors, which are less than unity for moist locations and for short-time loading; they are greater than unity if the moisture content of the wood in service is 15% or less. In general, all data apply to uses within covered structures. From the extensive tables, only the No. 1 grade in nominal 2×4 size and 2-in. or 4-in. planks, 6 in., and wider have been selected for illustration.

In a few cases the allowable stresses specified for the Canadian products will vary slightly from those given here for the same species by the U.S. agencies.

[b] Grading agencies represented by letters in this column are as follows:

NC = National Lumber Grades Authority (a Canadian agency)
NH = Northern Hardwood and Pine Manufacturers Association
NL = Northern Lumber Manufacturers Association
RI = Redwood Inspection Service
SP = Southern Pine Inspection Bureau
WC = West Coast Lumber Inspection Bureau
WW = Western Wood Products Association

[c] It is assumed that all members are so framed, anchored, tied, and braced that they have the necessary rigidity.

[d] For short term loads, these values may be increased: add 15% for 2-month snow load; add 33% for wind or earthquake; add 100% for impact load.

REFERENCES

"Wood Handbook," Handbook No. 72, U.S. Department of Agriculture, 1955.

"Timber Construction Manual," American Institute of Timber Construction, John Wiley & Sons, Inc., 1966.

"National Design Specification for Stress-Grade Lumber and its Fastenings," national Forest Products Association, Washington D.C., 1971.

TABLE E.2 Standard Grades of Bolts

Part a: SAE Grades for Steel Bolts

SAE grade no.	Size range incl.	Proof strength,† kpsi	Tensile strength,† kpsi	Material	Head marking
1	$\frac{1}{4}$–$1\frac{1}{2}$			Low- or medium-carbon steel	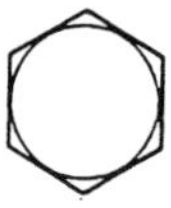
2	$\frac{1}{4}$–$\frac{3}{4}$	55	74		
	$\frac{7}{8}$–$1\frac{1}{2}$	33	60		
5	$\frac{1}{4}$–1	85	120	Medium-carbon steel, Q & T	
	$1\frac{1}{8}$–$1\frac{1}{2}$	74	105		
5.2	$\frac{1}{4}$–1	85	120	Low-carbon martensite steel, Q & T	
7	$\frac{1}{4}$–$1\frac{1}{2}$	105	133	Medium-carbon alloy steel, Q & T‡	
8	$\frac{1}{4}$–$1\frac{1}{2}$	120	150	Medium-carbon alloy steel, Q & T	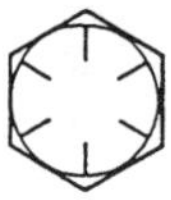
8.2	$\frac{1}{4}$–1	120	150	Low-carbon martensite steel, Q & T	

†Minimum values.

‡Roll threaded after heat treatment.

SOURCES: See "Helpful Hints," by Russell, Burdsall & Ward Corp., Mentor, Ohio 44060; and Chap. 23.

TABLE E.2 (continued) Standard Grades of Bolts

Part b: ASTM Grades for Steel Bolts

ASTM designation	Size range incl.	Proof strength,† kpsi	Tensile strength,† kpsi	Material	Head marking
A307	$\frac{1}{4}$ to 4			Low-carbon steel	
A325 type 1	$\frac{1}{2}$ to 1 $1\frac{1}{8}$ to $1\frac{1}{2}$	85 74	120 105	Medium-carbon steel, Q & T	 A325
A325 type 2	$\frac{1}{2}$ to 1 $1\frac{1}{8}$ to $1\frac{1}{2}$	85 74	120 105	Low-carbon martensite steel, Q & T	 A325
A325 type 3	$\frac{1}{2}$ to 1 $1\frac{1}{8}$ to $1\frac{1}{2}$	85 74	120 105	Weathering steel, Q & T	 A325
A354 grade BC				Alloy steel, Q & T	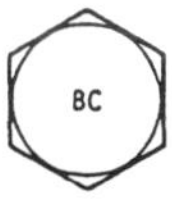 BC
A354 grade BD	$\frac{1}{4}$ to 4	120	150	Alloy steel, Q & T	
A449	$\frac{1}{4}$ to 1 $1\frac{1}{8}$ to $1\frac{1}{2}$ $1\frac{3}{4}$ to 3	85 74 55	120 105 90	Medium-carbon steel, Q & T	
A490type	$\frac{1}{2}$ to $1\frac{1}{2}$	120	150	Alloy steel, Q & T	 A490
A490type 3				Weathering steel, Q & T	 A490

†Minimum value.

Sources: See "Helpful Hints," by Russell, Burdsall & Ward Corp., Mentor, Ohio 44060; and Chapter 23.

TABLE E.2 (continued) Standard Grades of Bolts

Part c: Metric Mechanical Property Classes for Steel Bolts, Screws, and Studs

Property class	Size range incl.	Proof strength, MPa	Tensile strength, MPa	Material	Head marking
4.6	M5–M36	225	400	Low- or medium-carbon steel	4.6
4.8	M1.6–M16	310	420	Low- or medium-carbon steel	4.8
5.8	M5–M24	380	520	Low- or medium-carbon steel	5.8
8.8	M16–M36	600	830	Medium-carbon steel, Q & T	8.8
9.8	M1.6–M16	650	900	Medium-carbon steel, Q & T	9.8
10.9	M5–M36	830	1040	Low-carbon martensite steel, Q & T	10.9
12.9	M1.6–M36	970	1220	Alloy steel, Q & T	12.9

Sources: "Helpful Hints," by Russell, Burdsall & Ward Corp., Mentor, Ohio 44060; see also Chapter 23 and SAE standard J1199, and ASTM standard F569.

TABLE E.3 Steel Pipe Sizes

Nominal Pipe Size, in.	Outside Diameter, in.	Schedule Number or Weight	Wall Thickness, in.	Inside Diameter, in.	Surface Area: Outside, ft²/ft	Surface Area: Inside, ft²/ft	Areas and Weights Cross-sectional: Metal Area, in.²	Areas and Weights Cross-sectional: Flow Area, in.²	Weight: Pipe lb/ft
¾	1.05	40	0.113	0.824	0.275	0.216	0.333	0.533	1.131
		80	0.154	0.742	0.275	0.194	0.434	0.432	1.474
1	1.315	40	0.133	1.049	0.344	0.275	0.494	0.864	1.679
		80	0.179	0.957	0.344	0.250	0.639	0.719	2.172
1¼	1.660	40	0.140	1.38	0.434	0.361	0.668	1.496	2.273
		80	0.191	1.278	0.434	0.334	0.881	1.283	2.997
1½	1.900	40	0.145	1.61	0.497	0.421	0.799	2.036	2.718
		80	0.200	1.50	0.497	0.393	1.068	1.767	3.632
2	2.375	40	0.154	2.067	0.622	0.541	1.074	3.356	3.653
		80	0.218	1.939	0.622	0.508	1.477	2.953	5.022
2½	2.875	40	0.203	2.469	0.753	0.646	1.704	4.79	5.794
		80	0.276	2.323	0.753	0.608	2.254	4.24	7.662
3	3.5	40	0.216	3.068	0.916	0.803	2.228	7.30	7.58
		80	0.300	2.900	0.916	0.759	3.016	6.60	10.25
3½	4.0	40	0.226	3.548	1.047	0.929	2.680	9.89	9.11
		80	0.318	3.364	1.047	0.881	3.678	8.89	12.51
4	4.5	40	0.237	4.026	1.178	1.054	3.17	12.73	10.79
		80	0.337	3.826	1.178	1.002	4.41	11.50	14.99
5	5.563	10 S	0.134	5.295	1.456	1.386	2.29	22.02	7.77
		40	0.258	5.047	1.456	1.321	4.30	20.01	14.62
		80	0.375	4.813	1.456	1.260	6.11	18.19	20.78
6	6.625	10 S	0.134	6.357	1.734	1.664	2.73	31.7	9.29
		40	0.280	6.065	1.734	1.588	5.58	28.9	18.98
		80	0.432	5.761	1.734	1.508	8.40	26.1	28.58
8	8.625	10 S	0.148	8.329	2.258	2.180	3.94	54.5	13.40
		30	0.277	8.071	2.258	2.113	7.26	51.2	24.7
		80	0.500	7.625	2.258	1.996	12.76	45.7	43.4
10	10.75	10 S	0.165	10.420	2.81	2.73	5.49	85.3	18.7
		30	0.279	10.192	2.81	2.67	9.18	81.6	31.2
		Extra heavy	0.500	9.750	2.81	2.55	16.10	74.7	54.7
12	12.75	10 S	0.180	12.390	3.34	3.24	7.11	120.6	24.2
		30	0.330	12.09	3.34	3.17	12.88	114.8	43.8
		Extra heavy	0.500	11.75	3.34	3.08	19.24	108.4	65.4
14	14.0	10	0.250	13.5	3.67	3.53	10.80	143.1	36.7
		Standard	0.375	13.25	3.67	3.47	16.05	137.9	54.6
		extra heavy	0.500	13.00	3.67	3.40	21.21	132.7	72.1
16	16.0	10	0.250	15.50	4.19	4.06	12.37	188.7	42.1
		Standard	0.375	15.25	4.19	3.99	18.41	182.7	62.6
		extra heavy	0.500	15.00	4.19	3.93	24.35	176.7	82.8
18	18.0	10 S	0.188	17.624	4.71	4.61	10.52	243.9	35.8
		Standard	0.375	17.25	4.71	4.52	20.76	233.7	70.6
		extra heavy	0.500	17.00	4.71	4.45	27.49	227.0	93.5
20	20.0	10 S	0.218	19.564	5.24	5.12	13.55	300.6	46.1
		Standard	0.375	19.25	5.24	5.04	23.12	291	78.6
		extra heavy	0.500	19.00	5.24	4.97	30.6	283.5	104.1
22	22.0	10	0.250	21.50	5.76	5.63	17.1	363	58.1
		Standard	0.375	21.25	5.76	5.56	25.5	355	86.6
		extra heavy	0.500	21.00	5.76	5.50	33.8	346	114.8
24	24.0	10	0.250	23.50	6.28	6.15	18.7	434	63.4
		Standard	0.375	23.25	6.28	6.09	27.8	425	94.6
		extra heavy	0.500	23.00	6.28	6.02	36.9	415	125.5
26	26.0	Standard	0.375	25.25	6.81	6.61	30.2	501	102.6
		extra heavy	0.500	25.00	6.81	6.54	40.1	491	136.2
30	30.0	10	0.312	29.376	7.85	7.69	29.1	678	98.9
		Standard	0.375	29.250	7.85	7.66	34.9	672	118.7
		extra heavy	0.500	29.00	7.85	7.59	46.3	661	157.6
34	34.0	Standard	0.375	33.250	8.90	8.70	39.6	868	134.7
		extra heavy	0.500	33.00	8.90	8.64	52.6	855	178.9
36	36.0	Standard	0.375	35.25	9.42	9.23	42.0	976	142.7
		extra heavy	0.500	35.00	9.42	9.16	55.8	962	189.6
42	42.0	Standard	0.375	41.25	11.0	10.8	49.0	1336	166.7
		extra heavy	0.500	41.00	11.0	10.73	65.2	1320	221.6

TABLE E.4 Commercial Copper Tubing*

The following table gives dimensional data and weights of copper tubing used for automotive, plumbing, refrigeration, and heat exchanger services. For additional data see the standards handbooks of the Copper Development Association, Inc., the ASTM standards, and the "SAE Handbook."

Dimensions in this table are actual specified measurements, subject to accepted tolerances. Trade size designations are usually by actual OD, except for water and drainage tube (plumbing), which measures 1/8-in. larger OD. A 1/2-in. plumbing tube, for example, measures 5/8-in. OD, and 2-in. plumbing tube measures 2 1/8-in. OD.

KEY TO GAGE SIZES

Standard-gage wall thicknesses are listed by numerical designation (14 to 21), BWG or Stubs gage. These gage sizes are standard for tubular heat exchangers. The letter *A* designates SAE tubing sizes for automotive service. Letter designations *K* and *L* are the common sizes for plumbing services, soft or hard temper.

OTHER MATERIALS

These same dimensional sizes are also common for much of the commercial tubing available in aluminum, mild steel, brass, bronze, and other alloys. Tube weights in this table are based on copper at 0.323 lb/in³. For other materials the weights should be multiplied by the following approximate factors:

aluminum	0.30	monel	0.96
mild steel	0.87	stainless steel	0.89
brass	0.95		

Size, OD		Wall Thickness			Flow Area		Metal Area,	Surface Area		Weight,
in.	mm	in.	mm	gage	in.²	mm²	in.²	Inside, ft²/ft	Outside, ft²/ft	lb/ft
1/8	3.2	.030	0.76	A	0.003	1.9	0.012	0.017	0.033	0.035
3/16	4.76	.030	0.76	A	0.013	8.4	0.017	0.034	0.049	0.058
1/4	6.4	.030	0.76	A	0.028	18.1	0.021	0.050	0.066	0.080
1/4	6.4	.049	1.24	18	0.018	11.6	0.031	0.038	0.066	0.120
5/16	7.94	.032	0.81	21A	0.048	31.0	0.028	0.065	0.082	0.109
3/8	9.53	.032	0.81	21A	0.076	49.0	0.033	0.081	0.098	0.134
3/8	9.53	.049	1.24	18	0.060	38.7	0.050	0.072	0.098	0.195
1/2	12.7	.032	0.81	21A	0.149	96.1	0.047	0.114	0.131	0.182
1/2	12.7	.035	0.89	20L	0.145	93.6	0.051	0.113	0.131	0.198
1/2	12.7	.049	1.24	18K	0.127	81.9	0.069	0.105	0.131	0.269
1/2	12.7	.065	1.65	16	0.108	69.7	0.089	0.97	0.131	0.344
5/8	15.9	.035	0.89	20A	0.242	156	0.065	0.145	0.164	0.251
5/8	15.9	.040	1.02	L	0.233	150	0.074	0.143	0.164	0.285
5/8	15.9	.049	1.24	18K	0.215	139	0.089	0.138	0.164	0.344
3/4	19.1	.035	0.89	20A	0.363	234	0.079	0.178	0.196	0.305
3/4	19.1	.042	1.07	L	0.348	224	0.103	0.174	0.196	0.362
3/4	19.1	.049	1.24	18K	0.334	215	0.108	0.171	0.196	0.418
3/4	19.1	.065	1.65	16	0.302	195	0.140	0.162	0.196	0.542
3/4	19.1	.083	2.11	14	0.268	173	0.174	0.151	0.196	0.674
7/8	22.2	.045	1.14	L	0.484	312	0.117	0.206	0.229	0.455
7/8	22.2	.065	1.65	16K	0.436	281	0.165	0.195	0.229	0.641
7/8	22.2	.083	2.11	14	0.395	255	0.206	0.186	0.229	0.800
1	25.4	.065	1.65	16	0.594	383	0.181	0.228	0.262	0.740
1	25.4	.083	2.11	14	0.546	352	0.239	0.218	0.262	0.927
1 1/8	28.6	.050	1.27	L	0.825	532	0.176	0.268	0.294	0.655

*Compiled and computed.

TABLE E.4 (continued) Commercial Copper Tubing*

Size, OD		Wall Thickness			Flow Area		Metal Area,	Surface Area		Weight,
								Inside,	Outside,	
in.	*mm*	*in.*	*mm*	*gage*	*in.*^{2}	*mm*^{2}	*in.*^{2}	*ft*^{2}/*ft*	*ft*^{2}/*ft*	*lb/ft*
1 1/8	28.6	.065	1.65	16K	0.778	502	0.216	0.261	0.294	0.839
1 1/4	31.8	.065	1.65	16	0.985	636	0.242	0.293	0.327	0.938
1 1/4	31.8	.083	2.11	14	0.923	596	0.304	0.284	0.327	1.18
1 3/8	34.9	.055	1.40	L	1.257	811	0.228	0.331	0.360	0.884
1 3/8	34.9	.065	1.65	16K	1.217	785	0.267	0.326	0.360	1.04
1 1/2	38.1	.065	1.65	16	1.474	951	0.294	0.359	0.393	1.14
1 1/2	38.1	.083	2.11	14	1.398	902	0.370	0.349	0.393	1.43
1 5/8	41.3	.060	1.52	L	L.779	1148	0.295	0.394	0.425	1.14
1 5/8	41.3	.072	1.83	K	1.722	1111	0.351	0.388	0.425	1.36
2	50.8	.083	2.11	14	2.642	1705	0.500	0.480	0.628	1.94
2	50.8	.109	2.76	12	2.494	1609	0.620	0.466	0.628	2.51
2 1/8	54.0	.070	1.78	L	3.095	1997	0.449	0.520	0.556	1.75
2 1/8	54.0	.083	2.11	14K	3.016	1946	0.529	0.513	0.556	2.06
2 5/8	66.7	.080	2.03	L	4.77	3078	0.645	0.645	0.687	2.48
2 5/8	66.7	.095	2.41	13K	4.66	3007	0.760	0.637	0.687	2.93
3 1/8	79.4	.090	2.29	L	6.81	4394	0.950	0.771	0.818	3.33
3 1/8	79.4	.109	2.77	12K	6.64	4284	1.034	0.761	0.818	4.00
3 5/8	92.1	.100	2.54	L	9.21	5942	1.154	0.897	0.949	4.29
3 5/8	92.1	.120	3.05	11K	9.00	5807	1.341	0.886	0.949	5.12
4 1/8	104.8	.110	2.79	L	11.92	7691	1.387	1.022	1.080	5.38
4 1/8	104.8	.134	3.40	10K	11.61	7491	1.682	1.009	1.080	6.51

TABLE E.5 Standard Gages for Wire, Sheet, and Twist Drills

Gage	(1) Mfrs. steel sheet	(2) USS steel sheet (old)	(3) Birmingham or Stub	(4) W & M or Roebling steel wire	(5) AWG or B & S non-ferrous wire or sheet	Numbered twist drills	Copper wire (AWG) Circular mils	Ohms/1000 ft, 77°F	Lb/1000 ft	Sheet steel Lb/sq ft
0000000		0.500		0.4900						20.00
000000		0.469		0.4615	0.580					18.75
00000		0.438		0.4305	0.516					17.50
0000		0.406	.454	0.3938	0.460		212,000	0.0500	641.0	16.25
000		0.375	.425	0.3625	0.410		168,000	0.0630	508.0	15
00		0.344	.380	0.3310	0.365		133,000	0.0795	403.0	13.75
0		0.313	.340	0.3065	0.325		106,000	0.100	319.0	12.50
1		0.281	.300	0.2830	0.289	0.2280	83,700	0.126	253.0	11.25
2		0.266	.284	0.2625	0.258	0.2210	66,400	0.159	201.0	10.625
3	.2391	0.250	.259	0.2437	0.229	0.2130	52,600	0.201	159.0	10
4	.2242	0.234	.238	0.2253	0.204	0.2090	41,700	0.253	126.0	9.375
5	.2092	0.219	.220	0.2070	0.182	0.2055	33,100	0.319	100.0	8.75
6	.1943	0.203	.203	0.1920	0.162	0.2040	26,300	0.403	79.5	8.125
7	.1793	0.188	.180	0.1770	0.144	0.2010	20,800	0.508	63.0	7.5
8	.1644	0.172	.165	0.1620	0.128	0.1990	16,500	0.641	50.0	6.875
9	.1495	0.156	.148	0.1483	0.114	0.1960	13,100	0.808	39.6	6.25
10	.1345	0.141	.134	0.1350	0.102	0.1935	10,400	1.02	31.4	5.625
11	.1196	0.125	.120	0.1205	0.0907	0.1910	8,230	1.28	24.9	5
12	.1046	0.109	.109	0.1055	0.0808	0.1890	6,530	1.62	19.8	4.375
13	.0897	0.0937	.095	0.0915	0.0720	0.1850	5,180	2.04	15.7	3.75
14	.0747	0.0781	.083	0.0800	0.0641	0.1820	4,110	2.58	12.4	3.125
15	.0673	0.0703	.072	0.0720	0.0571	0.1800	3,260	3.25	9.86	2.813
16	.0598	0.0625	.065	0.0625	0.0508	0.1770	2,580	4.09	7.82	2.5
17	.0538	0.0562	.058	0.0540	0.0453	0.1730	2,050	5.16	6.20	2.25
18	.0478	0.0500	.049	0.0475	0.0403	0.1695	1,620	6.51	4.92	2
19	.0418	0.0437	.042	0.0410	0.0359	0.1660	1,290	8.21	3.90	1.75
20	.0359	0.0375	.035	0.0348	0.0320	0.1610	1,020	10.4	3.09	1.50
21	.0329	0.0344	.032	0.0318	0.0285	0.1590	810	13.1	2.45	1.375
22	.0299	0.0312	.028	0.0286	0.0253	0.1570	642	16.5	1.94	1.25
23	.0269	0.0281	.025	0.0258	0.0226	0.1540	509	20.8	1.54	1.125
24	.0239	0.0250	.022	0.0230	0.0201	0.1520	404	26.2	1.22	1
25	.0209	0.0219	.020	0.0204	0.0179	0.1495	320	33.0	0.970	0.875
26	.0179	0.0187	.018	0.0181	0.0159	0.1470	254	41.6	0.769	0.75
27	.0164	0.0172	.016	0.0173	0.0142	0.1440	202	52.5	0.610	0.6875
28	.0149	0.0156	.014	0.0162	0.0126	0.1405	160	66.2	0.484	0.625
29	.0135	0.0141	.013	0.0150	0.0113	0.1360	127	83.4	0.384	0.5625
30	.0120	0.0125	.012	0.0140	0.0100	0.1285	101	105	0.304	0.5
31	.0105	0.0109	.010	0.0132	0.0089	0.1200	79.7	133	0.241	0.4375
32	.0097	0.0102	.009	0.0128	0.0080	0.1160	63.2	167	0.191	0.4063
33	.0090	0.0094	.008	0.0118	0.0071	0.1130	50.1	211	0.152	0.375
34	.0082	0.0086	.007	0.0104	0.0063	0.1110	39.8	266	0.120	0.3438
35	.0075	0.0078	.005	0.0095	0.0056	0.1100	31.5	335	0.0954	0.3125
36	.0067	0.0070	.004	0.0090	0.0050	0.1065	25.0	423	0.0757	0.2813
37	.0064	0.0066		0.0085	0.0045	0.1040	19.8	533	0.0600	0.2656
38	.0060	0.0062		0.0080	0.0040	0.1015	15.7	673	0.0476	0.25
39				0.0075	0.0035	0.0995	12.5	848	0.0377	
40				0.0070	0.0031	0.0980	9.9	1070	0.0200	
41				0.0066	0.0028	0.0960				
42				0.0062	0.0025	0.0935				
43				0.0060	0.0022	0.0890				
44				0.0058	.0020	0.0860				
45				0.0055	.0018	0.0820				
46				0.0052	.0016	0.0810				
47				0.0050	.0014	0.0785				
48				0.0048	.0012	0.0760				
49				0.0046	.0011	0.0730				
50				0.0044	.0010	0.0700				

Note: The present trend, especially for sheet and strip, is to quote thickness as decimal or fraction of an inch rather than gage number. ANSI Standard preferred thicknesses have been adopted. These preferred sizes for thickness of uncoated sheet, strip, and plate under 0.25 in. are as follows: .224, .220, .180, .160, .140, .125, .112, .100, .090, .080, .071, .063, .056, .050, .045, .040, .036, .032, .028, .025, .022, .020, .018, .016, .014, .012, .011, .010, .009, .008, .007, .006, .005, .004.

KEY: (1) Manufacturer's standard for hot- and cold-rolled uncoated carbon steel sheet and most alloy steel sheet.
(2) U.S. Standard for cold-rolled steel strip and stainless and nickel alloy sheet.
(3) Birmingham or Stub for hot-rolled carbon and alloy steel strip and tubing.
(4) Washburn and Moen, Roebling, or U.S. Steel for steel wire.
(5) American wire gage or Brown and Sharpe for non-ferrous wire, sheet, and strip.

Dimensions in approximate decimals of an inch.

TABLE E.6 Properties of Typical Gaseous and Liquid Commercial Fuels*

m

Gaseous fuels	*Composition, percent by volume*								*Mol wt of fuel*	*Theor. air/fuel ratio by wt*	*Higher heating value, Btu/lb$_m$*	*Density, lb$_m$/ft^3*
	H_2	N_2	O_2	CH_4	CO	CO_2	C_2H_4	C_6H_6				
Blast furnace gas	1.0	60.0	—	—	27.5	11.5	—	—	29.6	0.667	1,170	.075 5[a]
Blue water gas	47.3	8.3	0.7	1.3	37.0	5.4	—	—	16.4	3.759	6,550	.042 2[a]
Carb. water gas	40.5	2.9	0.5	10.2	34.0	3.0	6.1	2.8	18.3	7.299	11,350	.046 6[a]
Coal gas	54.5	4.4	0.2	24.2	10.9	3.0	1.5	1.3	12.1	10.87	16,500	.031 1[a]
Coke-oven gas	46.5	8.1	0.8	32.1	6.3	2.2	3.5	0.5	13.7	17.24	17,000	.032 6[a]
Natural gas (15.8% C_2H_6)	—	0.8	—	83.4	—	—	—	—	18.3	17.24	24,100	.045 1[a]
Producer gas	14.0	50.9	0.6	3.0	27.0	4.5	—	—	24.7	14.29	2,470	.063 6[a]

Liquid commercial fuels	*Vapor*		*Gravity, API, 60°F*	*Distillation*			*Flash point, °F*	*Viscosity, centi-stokes, 100°F*	*Mol wt of fuel*	*Theor. air/fuel ratio by wt*	*Higher heating value, Btu/lb$_m$*	*Density, lb$_m$/ft^3*
	c_p, 60°F	*c_p/c_v, 60°F*		*10%, °F*	*90%, °F*	*End point, °F*						
	(approximately)											
Gasoline	0.4	1.05	63	121	320	397	0	—	113	14.93	20,460	43.8[b]
Gasoline	0.4	1.05	63	118	330	410	0	—	126[c]	14.97	20,260	46.1[b]
Kerosene	0.4	1.05	41.9	370	510	546	130	—	154[c]	14.99	19,750	51.5[b]
Diesel oil (1-D)	0.4	1.05	42	—	550	—	100	1.4–2.5	170	15.02	19,240	54.6[b]
Diesel oil (2-D)	0.4	1.05	36	—	540–576	—	125	2.0–5.8	184	15.06	19,110	57.4[b]
Diesel oil (4-D)	0.4	1.05	—	—	—	—	130	5.8–26.4	198	14.93	18,830	59.9[b]

[a]Based on dry air at 25°C and 760 mm Hg.
[b]Based on H_2O at 60°F, 1 atm (ρ = 62.367 lb$_m$/ft^3).
[c]Estimated.

*Abridged from: "Engineering Experimentation", G.L. Tuve and L.C. Domholdt, McGraw-Hill Book Company, 1966; and "The Internal Combustion Engine", 2nd ed., C.F. Taylor and E.S. Taylor, International Textbook Co., 1961.

Note: For heating value in J/kg, multiply the value in Btu/lb$_m$ by 2324. For density in kg/m^3, multiply the value in lb/ft^3 by 16.02.

TABLE E.7 Combustion Data for Hydrocarbons*

Hydrocarbon	*Formula*	*Higher heating value (vapor), Btu/lb$_m$*	*Theor. air/fuel ratio, by mass*	*Max flame speed, ft/sec*	*Adiabatic flame temp (in air), °F*	*Ignition temp (in air), °F*	*Flash point, °F*	*Flammability limits (in air), % by volume*	
PARAFFINS OR ALKANES									
Methane	CH_4	23875	17.195	1.1	3484	1301	gas	5.0	15.0
Ethane	C_2H_6	22323	15.899	1.3	3540	968–1166	gas	3.0	12.5
Propane	C_3H_8	21669	15.246	1.3	3573	871	gas	2.1	10.1
n-Butane	C_4H_{10}	21321	14.984	1.2	3583	761	−76	1.86	8.41
iso-Butane	C_4H_{10}	21271	14.984	1.2	3583	864	−117	1.80	8.44
n-Pentane	C_5H_{12}	21095	15.323	1.3	4050	588	< −40	1.40	7.80
iso-Pentane	C_5H_{12}	21047	15.323	1.2	4055	788	< −60	1.32	9.16
Neopentane	C_5H_{12}	20978	15.323	1.1	4060	842	gas	1.38	7.22
n-Hexane	C_6H_{14}	20966	15.238	1.3	4030	478	−7	1.25	7.0
Neohexane	C_6H_{14}	20931	15.238	1.2	4055	797	−54	1.19	7.58
n-Heptane	C_7H_{16}	20854	15.141	1.3	3985	433	25	1.00	6.00
Triptane	C_7H_{16}	20824	15.141	1.2	4035	849	—	1.08	6.69
n-Octane	C_8H_{18}	20796	15.093	—	—	428	56	0.95	3.20
iso-Octane	C_8H_{18}	20770	15.093	1.1	—	837	10	0.79	5.94
OLEFINS OR ALKENES									
Ethylene	C_2H_4	21636	14.807	2.2	4250	914	gas	2.75	28.6
Propylene	C_3H_6	21048	14.807	1.4	4090	856	gas	2.00	11.1
Butylene	C_4H_8	20854	14.807	1.4	4030	829	gas	1.98	9.65
iso-Butene	C_4H_8	20737	14.807	1.2	—	869	gas	1.8	9.0
n-Pentene	C_5H_{10}	20720	14.807	1.4	4165	569	—	1.65	7.70
AROMATICS									
Benzene	C_6H_6	18184	13.297	1.3	4110	1044	12	1.35	6.65
Toluene	C_7H_8	18501	13.503	1.2	4050	997	40	1.27	6.75
p-Xylene	C_8H_{10}	18663	13.663	—	4010	867	63	1.00	6.00
OTHER HYDROCARBONS									
Acetylene	C_2H_2	21502	13.297	4.6	4770	763–824	gas	2.50	81
Naphthalene	$C_{10}H_8$	17303	12.932	—	4100	959	174	0.90	5.9

*Based largely on: "Gas Engineers' Handbook", American Gas Association, Inc., Industrial Press, 1967.

REFERENCES

"American Institute of Physics Handbook", 2nd ed., D.E. Gray, Ed., McGraw-Hill Book Company, 1963.

"Chemical Engineers' Handbook", 4th ed., R.H. Perry, C.H. Chilton, and S.D. Kirkpatrick, Eds., McGraw-Hill Book Company, 1963.

"Handbook of Chemistry and Physics", 53rd ed., R.C. Weast, Ed., The Chemical Rubber Company, 1972; gives the heat of combustion of 500 organic compounds.

"Handbook of Laboratory Safety", 2nd ed., N.V. Steere, Ed., The Chemical Rubber Company, 1971.

"Physical Measurements in Gas Dynamics and Combustion", Princeton University Press, 1954.

Note: For heating value in J/kg, multiply the value in Btu/lb$_m$ by 2324. For flame speed in m/s, multiply the value in ft/s by 0.3048.

Index

A

Abrasion
- three-body, 137
- two-body, 137

Abrasive wear, 137
Absolute pressure, 2
Absolute viscosity, 19
- centrifugal, 13

Acceleration, Coriolis, 13, 204
Acoustic velocity, 57
Addition of variable, 33
Additive, 144
Adhesion, 209, 212
Adhesive wear, 136
Adiabatic flow, 90
Air-assist, 194
Airblast, 194 *see also* twin fluid/air blast
- prefilming, 193

Airfoils, 83
Air/vacuum valves, 58
Air valves, 58
Algebra, 30
Algebraic procedure, 30, 31
Analysis of two-phase flows, 106
Anemometer
- laser Doppler, 205
- thermal, 205

Angle
- Mach, 101
- spray, 191
- wave, 100
- wedge, 100

Angle of attack, 102
Apparent body forces, 13
Apparent viscosity, 121, 122
Archard wear equation, 137
Archimedes, principles of, 6
Area Mach number relation, 96
Area-velocity relation, 94
Arithmetic split coordinate plot, 39
Asymptotes (suitable), 38
Atomic force microscope, 210
Atomizers, 187, 189
Axial fans, 183
Axial flow, 180

B

Babbitts, 162
Backwater solution curve, 72
Ball bearings, 165
Bearing cages, 168
Bearing design, journal, 149
Bearing materials
- oil film, 162
- rolling element, 165

Bearings, 135
- ball, 165
- dry, 141
- flat-land, 157
- fluid film, 141 to 143
- journal, 149
- plastic, 163, 164
- porous metal, 163 to 165
- roller, 165
- rolling element, 141, 165
- semilubricated, 141, 163
- spherical roller, 167
- step, 157
- tapered-land, 158
- tapered roller, 168
- thrust, 155, 157

Bernoulli equation, 23, 76
Blowing, 84
Bluff bodies, 81
Bodies
- bluff, 81
- stability for partially submerged, 6, 7
- stability of submerged and floating, 6

Body forces, 17
- apparent, 13
- conservative, 17

Boronizing, 139
Boundary conditions, fluid flow, 22
Boundary layer, 76
- control, 83
- disturbance thickness, 76
- laminar, 77
- turbulent 77

Boundary lubrication, 217
Breakup, 191
Broad-crested weirs, 69
Bubbly flow, 113
Buckingham plastic, 122
Bulk viscosity, 19
Buoyancy, 6
- center of, 6
- laws of, 6

Buoyant
- force, 35
- neutrally, 6

Butterfly valve, 54

C

Calorically perfect gas, 89
Camber, 102
Capillary viscometry, 206
Carbonitriding, 139
Carburizing, 139
Cavitation, 55, 182
- choking, 55
- damage, 55

Center of buoyancy, 6
Center of pressure, 3
Centrifugal acceleration, 13
Centrifugal fans, 183
Centrifugal pumps, 180 to 181
Centroidal moments of inertia, 5
Characteristic Mach number, 96
Check valves, 57
Chemical vapor deposition, 140
- plasma-assisted, 139

Chézy coefficient, 66
Choking cavitation, 55
Choking limit, 118
Chord length, 102
Chord line, 83
Churchill-Usagi equation, 38
Coatings, 187
Coefficients
- drag, 80, 102
- flow, 199
- lift, 80, 103
- minor loss, 52
- pressure, 101
- skin friction, 79
- wear, 137

Coefficient of drag on particle, 116
Coefficient of friction, 136
Colebrook equation, 43
Combination valves, 58
Compressibility, 87
Compressible flow, 87
Compression, 18, 87
Computation fluid dynamics, 85
Cone
- full, 189
- hollow, 189
- solid, sprays, 189

Conformal mapping, 28
Conservation of energy, 14
Conservation of mass, 12
Conservation of momentum, 13
Conservative body force, 17
Continuity equation, 47
Controllability, 54

Control structure, 70
Control volume, 92
  analyses for fluid flow problems, 13
  integral relations for, 11
Convection, forced, 34
Convergent-divergent nozzle, 95
Coriolis acceleration, 204
Correlating equations, 38, 44
Correlations, 37
  drop size, 193
Correlations for limiting velocities in pneumatic conveying, 119
Correlations for void fraction and frictional pressure drop, 113
Corrosive wear, 136
Critical depth, 70
Critical flow, 66
Critical Reynolds number, 81
Crop spraying, 187
Curved surfaces, forces on, 5

D

Dam, 69
Darcy friction factor, 66, 126
Darcy-Weisbach equation, 48
de Laval nozzle, 95
Depth
  critical, 70
  normal, 70
Deviatoric stress tensor, 19
Diagrams
  Mollier, 93
  Moody, 66
Diesel injector, 192
Diffuser, 97
Dilatant, 122
Dimensional analysis, 29
Direct acting, 180
Discharge coefficient, 190, 198
Dispersed drop, 112
Displacement thickness, 76
Dissipation, rate of, 21
Disturbance thickness, boundary layer, 76
Doublet impingement, 192
Drag, 80
  coefficient, 80, 102, 103
  drop, 192
  friction, 80
  induced, 83
  pressure, 81
Drift-flux model, 112
Drop drag, 192
Drop size correlations, 193
Dry bearings, 141, 163
Dual-orifice nozzle, 194
Duplex nozzles, 194
Dynamic loading rating, 169
Dynamic viscosity, 19, 122

E

Effervescent atomizers, 194
Einstein summation convention, 19
Elastohydrodynamic lubrication, 170
Elbows, 50
Electric arc spraying, 139
Electroless nickel, 139
Electromagnetic, 180
Electromagnetic flow meters, 203
Electron beam hardening, 139
Electroplating, 139
Electrostatic atomizers, 195
Energy
  conservation, 14, 20
  equation, 48
  internal, 14
  specific, 14, 66
Enthalpy, specific, 21
Entropy, 21, 91
Equations
  Archard wear, 137
  Bernoulli, 23, 76
  Churchill-Usagi, 38
  Colebrook, 43
  continuity, 47
  correlating, 38, 44
  Darcy-Weisbach, 48
  energy, 48, 89
  integral, moment of momentum, 14
  Kozeny–Carman, 41
  Laplace, 25
  mass conservation-continuity, 16
  mechanical, 20
  modified power law, 124
  momentum, 47
  momentum integral, 78
  Navier-Stokes, 20, 23
  power law constitutive, 124
  stream function, 24
  system loss, 54
  thermal energy, 20 to 22
  vorticity, 23
Equilibrium model, homogeneous, 108
Equipotential lines, 27
Erosion, 136, 138
Eulerian viewpoint, 11
Evaluation of n, 38
Expansion waves, 98
Extensive quantity, 11
External loads, 53
External-mixing designs, 194

F

Fanning friction factor, 33, 42, 129
Fans, 182
Fan spray, 193
Feasibility study, 62
First law of thermodynamics, 14
Fixed (or deformable) volume controls, 11
Flame hardening, 139
Flaps, 84
Flashing, 194
Float meters, 203
Flow
  adiabatic, 90
  axial, 180
  characteristics in horizontal pneumatic conveying, 117
  characteristics in vertical pneumatic conveying, 117
  choked, 94
  coefficient, 199
  compressible, 87
  critical, 66
  curve, 122
  gradually varied, 70, 71
  hypersonic, 88
  incompressible, 13, 88
  index, 126
  irrotational, 23
  maps, parameters of, 109
  meters, 203
    electromagnetic, 203
    positive-displacement, 198
    restriction, 198
    turbine, 203
    ultrasonic, 203
  mixed, 180
  non-Newtonian, 121
  nozzle, 200 to 201
  one-dimensional, 89
  open channel, 65
  pattern map for horizontal flow, 111
  pattern map for vertical flow, 112
  pattern maps, 109
  patterns in gas-liquid horizontal flow, 110
  patterns in gas-liquid vertical flow, 110
  patterns for vertical up flow, 115
  quality, 106
  quasi-one-dimensional, 89
  radial, 180
  regimes, 109, 112
  steady, 13, 16
  subcritical, 66
  subsonic, 88
  supercritical, 66
  supersonic, 88
  through packed bed of spheres, 41
  transonic, 88
  uniform, 66
Fluid dynamics, computational, 85
Fluid film bearings, 141
Fluid flow
  boundary conditions for, 22
  problems, control volume analyses for, 13
Fluid motion, differential relations of, 16
Fluids
  friction, 48
  layered, hydrostatic forces in, 5
  Newtonian, 19
  non-Newtonian, 19

- purely viscous time-independent, 122
- time-dependent, 122
- viscoelastic, 19, 122, 131

Fluid seals, 173
Fluid statics, 1
Forced convection, 34
Forces
- body, 17
- hydrostatic in layered fluids, 5
- hydrostatic on submerged objects, 2
- potential, 17
- on curved surfaces, 5
- surface, 17

Forces and rate of deformation relationship between, 18
Fourier heat conduction law, 21
Frame
- inertial coordinate, 13
- noninertial, 13

Free overfall, 69
Fretting corrosion, 138
Friction, 136, 210, 212
- fluid, 48
- local losses and, 50

Frictional pressure drop, correlations for void fraction and, 113
Friction drag, 80
Friction factor, 41, 126
- Darcy, 66, 126
- Fanning, 33, 42
- Moody chart, 49

Friction loss, 48
Frontal area, 80
Froude number, 67, 80
Full cone, 189
Fully developed laminar pressure drops, 125
Fully developed turbulent flow pressure drops, 129, 131

G

Gas, calorically perfect, 89
Gas lift, 180
Gas-liquid two-phase flow, 109
Gas-solid/liquid-solid two-phase flow, 114
Gauge pressure, 2
Gradually varied flow, 70, 71
Graphical correlations, 37
Grashof number, 36
Greases, 147
Guidelines, pressure class, 52

H

Hardening
- electron beam, 139
- flame, 139
- induction, 139
- laser, 139
- thermal transformation, 139

Heat conduction, 20
Heat flux, 20
Helmholtz's theorem, 24
Hershel-Buckley, 122
High-velocity oxyfuel, 140
Hollow cone, 189
Homogeneous equilibrium model, 108
Horizontal flow, flow patterns in gas-liquid, 110
Hydraulic diameter, 48
Hydraulic jump, 68
Hydraulic radius, 66
Hydraulic ram, 180
Hydraulics, 47
Hydrostatic forces in layered fluids, 5
Hydrostatic forces on submerged objects, 2
Hydrostatic pressure, 1
Hydrostatic thrust bearings, 158
Hypersonic flow, 88

I

Impeller, 180
Incompressible flow, 23, 88
Indentation, 214
Induced drag, 83
Induction hardening, 139
Inertia, centroidal moments of, 5
Inertial coordinate frame, 13
Integral moment of momentum equation, 14
Integral relations for a control volume, 11
Intensive value, 11
Intermediate regimes, 40
Internal design pressure, 53
Internal energy, 14
Internal-mixing nozzles, 194
Ion implantation, 140
Isentropic compressibility, 87
Isentropic process, 88

J

Jerk pump, 192
Jet breakup length, 191, 192
Jet pumps, 181
Jets, 180
Journal bearing design, 149
Journal bearings, 149

K

Kelvin's theorem, 24
Kinematic viscosity, 24
Kinetic coefficient of friction, 136
Kozeny–Carman equation, 41

L

Lagrangian viewpoint, 11
Laminar boundary layers, 77
Laminar flow elements, 202
Laminar flow pressure drops, fully developed, 132
Laminar region, 43
Laplace equation, 25
Laser Doppler anemometers, 205
Laser hardening, 139
Law, Fourier heat conduction, 21
Laws of buoyancy, 6
Lift, 83
- coefficient, 80, 102
- gas, 180

Limiting velocities, 53
Lines, Mach, 101
Liquid atomization, 187
Liquid sheets, 189, 193
Local losses, 50
Lubricant properties, 143
Lubricants
- greases, 147
- mineral oils, 144
- solid, 148
- synthetic oils, 146

Lubricant supply methods, 171
Lubrication, 209
- boundary, 217
- effect of on friction and wear, 140
- elastohydrodynamic, 170
- rolling bearing, 170
- wick, 172

M

Mach angle, 101
Mach lines, 101
Mach number, 80, 88
- area relation, 96
- characteristic, 96

Macrotribology, 209
Manning's formula, 67
Manning's roughness parameter, 66
Manometry, 2
Mass, conservation of, 12
Mass conservation-continuity equation, 16
Material derivative, 12
Mechanical equation, 20
Mechanical face seals, 175
Metacenter, 6
Methods
- combination, 31
- inspection, 31

Micro/nanotribology, 209
Mineral oils, 144
Minimum fluidization, 114, 115
Minor loss, 50, 51
Mixed flow, 180
Modified power law, 125
Mollier diagram, 93
Momentum, 13, 16
- equation, 47
- integral equation, 78
- thickness, 77

Moody chart, 49, 67
Moody diagram, 66

Motion
  equations of, 11
  fluid, pressure variation in rigid-body, 7
Moving surfaces, 84
Multiphase flow, 105
Multiple rotor pump, 180

N

Nappe, 70
Navier-Stokes equations, 20, 23
Nebulizers, 195
Net positive suction head, 182
Neutrally buoyant, 6
Newtonian fluid, 19
Nitriding, 139
Nitrocarburizing, 139
Noise generation fans, 185
Nomenclature, fluid mechanics, 218 to 221
Noncircular ducts in turbulent fully developed flow, 130
Noninertial frame, 13
Non-Newtonian fluids, 19, 122
Normal depth, 70
Normal shock wave, 90
No-slip condition, 22
Nozzle
  convergent-divergent, 95
  de Laval, 95
  dual-orifice, 194
  flow, 200 to 201
  internal-mixing, 194
  plain hole, 190
  simplex, 194
  spill-return, 194

O

Oblique shock, 98
Oil film bearing materials, 162
Oil rings, 172
One-dimensional flow, 89
Open channel flow, 65
Orifice plates, 200
Orifices
  eccentric, 200
  plain, 194
Oxyfuel, 139, 140

P

Packed bed of spheres, 41
Parallel pump operation, 60
Parameters
  flow maps, 109
  Manning's roughness, 66
  roughness, 67
  shear rate, 125
Penetration distance, 192
Peripheral, 180
Physical vapor deposition, 140
Pipe design, 51
Pipes, 48
Piston, 180
Plain hole nozzle, 190
Plain orifice, 193
Planform area, 83
Plasma arc spraying, 139
Plasma-assisted chemical vapor deposition, 140
Plastic bearings, 163, 164
Plunger, 180
Pneumatic conveying, 118
Poiseuille's law, 43
Porous metal bearings, 163 to 165
Positive-displacement flow meters, 198
Positive displacement pump, 180, 181
Potential flow, 25
Power, 180
Power law, 124, 129
  equation, 124
  fluids, 129
  modified, 125
Prefilming air blast, 193
Pressure, 2
  absolute, 2
  center of, 3
  gauge, 2
  hydrostatic, 1
  internal design, 53
  stagnation, 76
  transient, 53
Pressure atomizers, 190
Pressure class guidelines, 53
Pressure coefficient, 102
Pressure difference at curved interface, 22
Pressure drag, 81
Pressure drop, fully developed turbulent flow, 129, 131
Pressure drop in flow through a packed bed of spheres, 41
Pressure gradient, 41, 78
Pressure swirl, 193
Pressure variation in rigid-body motion of a fluid, 7
Primary breakup, 191
Principal axes, 18
Principle, Archimedes, 6
Processes
  compression, 87
  electroplating, 139
  isentropic, 88
  thermal spray, 139
  weld hardfacing, 140
Properties of commercial nylon resins, C-62
Properties of dry air at atmospheric pressure, A-2
Properties of gases, A-18
Properties of liquid, B-35
Properties of manufactured fibers, C-70
Properties of moist air, A-10
Properties of natural fibers, C-70
Properties of rubbers and elastomers, C-72
Properties of saturated water and steam, A-12
Properties of silicate glasses, C-63
Properties of solids, C-38
Properties of superheated steam, A-15
Properties of typical gaseous and liquid commercial fuels, E-85
Properties and uses of American woods, C-67
Properties of window glass, C-66
Property measurements, rheological, 118
Pseudoplastic, 122
Pumps, 14, 180
  centrifugal, 180, 181
  curves, 60
  jerk, 192
  jet, 181
  multiple rotor, 180
  operation, parallel, 60
  positive displacement, 180, 181
  power, 14
  reciprocating, 180, 181
  rotary, 181
  selection, 59
  self-priming, 181
  single rotor, 180
  special effect, 180
  velocity head, 181
Pure extension, 18
Purely viscous time-independent fluids, 122
Pure metals, thermal properties of, C-41

Q

Quasi-one-dimensional flow, 89

R

Radial flow, 180
Radial-flow fans, 184
Radial lip seals, 175
Rate of deformation and forces, 18
Rate of dissipation, 21
Rate-of-strain, 18
Rayleigh curve, 94
Rayleigh number, 37
Reciprocating pumps, 180, 181
Release valves, 58
Restriction flow meters, 198
Reynolds number, 33, 80, 125
  critical, 81
Reynolds transport theorem, 11, 14
Rheological property measurements, 125
Rheology, 19
Rheopectic, 122
Ridge column theory, 56
Rigid-body motion of a fluid, pressure variation in, 7
Rigid-body rotation, 8, 18
Roller bearings, 165, 166
Rolling bearing lubrication, 170
Rolling element bearings, 141, 165, 168

Rotary atomizer, 193
Rotary pump, 180
Rotational viscometer, 207
Roughness parameters, 67
Round tube, 34

S

Sauter mean diameter, 189
Scratching, 214
Seals
  contact, 173
  fluid, 173
  mechanical face, 175
  radial lip, 175
Secondary breakup, 192
Second viscosity, 19
Self-priming pumps, 181
Semilubricated bearings, 141
Separated-phases model, 114
Separation, 75
Shaping-air, 194
Sharp-crested weirs, 69
Shear rate, 122
Shear thickening, 122
Shock, oblique, 98
Shock wave, normal, 90
Similar velocity profiles, 77
Similitude, 29
Simplex nozzles, 194
Single rotor pump, 180
Sizes and allowable unit stresses for softwood lumber, E-75
Skin friction coefficient, 79
Slats, 84
Sliding wear, 136
Solid cone sprays, 189
Solid lubricants, 148
Special effect pump, 180
Specific energy, 66
Specific enthalpy, 21
Specific heats, 88
Specific weight, 2
Speed of sound, 88, 89
Spherical roller bearings, 167
Spill-return nozzle, 194
Spray, fan, 193
Spraying, 187
Spray-painting, 195
Stability of submerged and floating bodies, 6, 7
Stagnation pressure, 76, 89
Stagnation speed of sound, 89
Stall, 83
Steady flow, 13, 16
Step bearings, 157
Stick-slip, 136
Stokes assumption, 20
Storage, 62
Stream function, 24
Streamlining, 80, 82
Stress tensor, 17, 18
Subsonic flow, 88
Suction, 84
Suitable asymptotes, 38
Supercritical flow, 66
Superficial velocity, 106
Superheated steam, properties of, A-15
Supersonic flow, 88, 98
Surface engineering, 138
Surface fatigue wear, 137
Surface fracture, 137
Surface forces, 17
Surface roughness, 212
Surface tension, 190
Surge, 53, 56
Synthetic oils, 146
System curve, 60
System loss equation, 54

T

Tapered-land bearings, 158
Tapered roller bearings, 166, 168
Terminal slip velocity, 116
Theorem
  Helmholtz's, 24
  Kelvin's, 24
  Reynolds transport, 11, 14
Thermal anemometers, 205
Thermal conductivity, 21
Thermal diffusion, 139
Thermal energy equation, 20
Thermal spray processes, 139
Thermal transformation hardening, 139
Thermodynamics , first law of, 14
Thickness, 102
  boundary layer disturbance, 76
  displacement, 76
  momentum, 77
Thin-airfoil theory, 101
Thixotropic, 122
Three-body abrasion, 137
Thrust bearings, 155, 157
Thrust blocks, 62
Tilting-pad thrust bearings, 157
Time-dependent fluids, 122
Torque, 55
Transient pressure, 53
Transients, 56
Transitional regime, 43
Transition at minimum fluidization, 115
Transitions, 75
Transonic flow, 88
Traversing methods, 204
Tribology, 135
Turbine flow meter, 203
Turbobell, 195
Turbulence, 153
Turbulent boundary layer, 78
Turbulent flow pressure drops, 129, 131
Twin fluid/air blast, 193 *see also* air blast
Two-body abrasion, 137
Two-phase flows
  analysis of, 106
  gas-liquid, 109
  gas-solid/liquid-solid, 114

U

Ultrasonic, 193
Ultrasonic atomizers, 195
Ultrasonic flow meters, 204
Uniform flow, 66
Unrecoverable head loss, 202
U-tube manometer, 3

V

Valves, 50
  air, 58
  air/vacuum, 58
  check, 57
  combination, 58
  control, 54
  release, 58
  selection, 54
Vane types, 184
Vapor deposition, 140
Vee-notch weirs, 69
Velocities, limiting, 53
Velocity
  acoustic, 57
  correction factor, 199
  head pump, 180
  potential, 23
  similar profiles, 77
Vena contracta, 198
Venturi, 199, 201
Vibratory atomizers, 195
Viewpoint
  Eulerian, 11
  Lagrangian, 11
Viscoelastic fluids, 122, 131
Viscometry, 206
Viscosity, 144
  absolute, 19
  apparent, 121
  bulk, 19
  dynamic, 19, 122
  kinematic, 24
  measurements, 206
  second, 19
  zero shear rate, 124
Void fractions, 109, 111
Volume controls, fixed (or deformable), 11
Volume fraction, 106
Volute, 180
Vortex shedding, 203
Vortex stretching, 24
Vorticity, 23 to 24

W

Wake, 75
Water hammer, 53, 56
Wave angle, 100
Wear, 136, 210, 214
Wear reduction, 138

Wedge angle, 100
Weight, specific, 2
Weirs, 69
Weissenburg number, 131
Weld hardfacing processes, 140
Wetted area, 80
Wick lubrication, 172
Wind tunnel, 85
  supersonic, 97

Z

Zero shear rate viscosity, 124